Advanced Wind Energy Systems:

Grid Integration, Markets, and Sustainable Infrastructure

Jackson J. Justo
https://orcid.org/0000-0003-0638-0713
University of Dar es Salaam, Dar es Salaam, Tanzania & ITMO University, St. Petersburg, Russia

Galina Demidova
Hangzhou Dianzi University, China & ITMO University, St. Petersburg, Russia

Francis A. Mwasilu
https://orcid.org/0000-0001-7540-4198
University of Dar es Salaam, Tanzania

Dmitry V. Lukichev
ITMO University, Russia

Ramesh C. Bansal
University of Sharjah, Sharjah, UAE & University of Pretoria, Pretoria, South Africa

IGI Global
Scientific Publishing
Publishing Tomorrow's Research Today

Vice President of Editorial Melissa Wagner
Director of Acquisitions Mikaela Felty
Director of Book Development Jocelynn Hessler
Production Manager Mike Brehm
Cover Design Jose Rosado

Published in the United States of America by
IGI Global Scientific Publishing
701 East Chocolate Avenue
Hershey, PA, 17033, USA
Tel: 717-533-8845 | Fax: 717-533-7115
Website: https://www.igi-global.com E-mail: cust@igi-global.com

Library of Congress Cataloging-in-Publication Data

(CIP Data Pending)
ISBN13: 9798260001066
EISBN13: 9798260001080

British Cataloguing in Publication Data
A Cataloguing in Publication record for this book is available from the British Library.

Table of Contents

Detailed Table of Contents

The fundamental principles governing wind energy conversion, including the power equation and the Betz limit, are explained to establish a theoretical foundation. The chapter further delineates the key components of modern horizontal- and vertical-axis wind turbines, such as rotor blades, nacelles, and drivetrains, and classifies systems by scale and location. A significant focus is placed on the operations and control of wind turbines and wind farms, covering methods like pitch and yaw control, Maximum Power Point Tracking (MPPT), and wake management. Finally, the chapter explores the transformative role of Artificial Intelligence (AI) in condition monitoring, contrasting traditional methods with advanced AI techniques like machine learning and deep learning to highlight their benefits in predictive maintenance, fault diagnosis, and overall system reliability. Through this structured examination, the authors aim to present wind energy as a cornerstone of the global transition to sustainable and renewable power sources.

This chapter examine the technological, environmental, socioeconomic, and governance dimensions of wind energy and its role in the global shift toward low-carbon systems. As nations pursue Paris Agreement and Net-Zero goals, wind power has become a central renewable source due to its low cost, technical maturity, and significant emissions reduction potential. The chapter reviews advancements in turbine design, materials, offshore systems, and digital tools such as forecasting, digital twins, and predictive maintenance. It also examines how wind energy supports sustainable development by creating jobs, enhancing local economies, promoting social inclusion, and contributing to circular-economy practices. Governance structures including market design, permitting, benefit-sharing, environmental

information regarding the requirements of grid codes, as well as the latest monitoring and control strategies that ensure stability, reliability, and the effective operation of wind power system integration. Additionally, it gives some practical worked examples and illustrations to emphasize the necessity for updated policies. The execution of these regulatory frameworks is essential for enabling the effective incorporation of renewable energy into the grid.

This chapter explores the synergy between Artificial Intelligence (AI) and Digital Twin (DT) technologies for advancing smart and autonomous wind-energy systems. It examines how AI-driven analytics and learning models enhance forecasting, control, and predictive maintenance, while DTs provide real-time virtual replicas for simulation and optimization. Their integration enables self-learning, adaptive, and resilient wind farms capable of real-time decision-making and performance optimization. The chapter also addresses implementation challenges, including data reliability, scalability, and cybersecurity, highlighting future research directions for sustainable, intelligent energy infrastructures.

Rapid global expansion of wind energy both onshore and offshore has transformed wind farms into highly digitalized, data-driven infrastructures. While digitalization enhances efficiency, predictive maintenance, and market integration, it also amplifies exposure to cyber threats. This chapter provides a structured overview of cybersecurity landscape specific to wind power systems, highlighting vulnerabilities at the levels of turbine control, supervisory and communication networks, cloud-based analytics, and supply-chain dependencies. It situates these risks within the evolving regulatory and standards framework like IEC 62443, EU NIS2 Directive, and General Data Protection Regulation (GDPR), and explores how compliance requirements intersect with operational realities. It also reviews recent advances in anomaly detection, fault prediction, and privacy-preserving analytics. Finally, it proposes a practical defense-in-depth roadmap integrating technical safeguards, organizational measures, and policy considerations to strengthen the resilience and trustworthiness of wind power systems.

Wind energy is a natural source of energy which is renewable in nature and eco-friendly in nature if used properly. However, in the recent centuries, humans used non-renewable energies for their development and support urbanization, which was a harmful decision in the long run. In the recent decade, efforts on using wind energy have been put to slowly and steadily replace the non-renewable energy up to certain extent in order to save environment from further harm. In this chapter, a number of major Wind Farm Projects will be discussed and challenges or problems faced in their construction. Then, gaps in the current research in regard to wind farm projects are will also be covered in the chapter. Additionally, data from numerous sources will be collected to perform graphical analyses to find out the benefits of constructing Wind Farm projects and how it will be helpful in India's development in the future.

This chapter explores artificial intelligence's transformative impact on wind energy, addressing intermittency challenges to enhance reliability, market integration, and trading efficiency. As wind power expands to meet decarbonization goals, supplying over one-third of global electricity by 2050, its variability strains grids and markets, incurring balancing costs and penalties. The analysis dissects reliability issues, highlighting AI's role in predictive maintenance via machine learning on SCADA data for failure anticipation, deep learning for real-time fault detection using convolutional neural networks, and adaptive controls to minimize downtime and extend asset life. These advancements shift operations from reactive to proactive, leveraging vast datasets to predict and prevent disruptions. It then examines electricity market dynamics, where forecast errors lead to imbalances

A significant logistical challenge inherent to wind energy is that the most abundant and consistent wind resources are frequently situated in remote, inaccessible areas, such as deep offshore locations or isolated plains, far from the urban centers that

Preface

INTRODUCTION

The global energy landscape is undergoing a profound transformation driven by climate imperatives, technological innovation, and the urgent need for sustainable infrastructure. Wind energy has emerged as a cornerstone of this transition, offering scalable, cost-effective, and environmentally responsible electricity generation. Once considered a supplementary renewable source, wind power is now central to national energy strategies, decarbonization roadmaps, and infrastructure modernization efforts worldwide.

However, the expansion of wind energy presents multifaceted challenges that extend far beyond turbine installation. Grid integration, market participation, regulatory compliance, digitalization, cybersecurity, and infrastructure resilience have become equally critical components of successful deployment. As wind penetration increases in modern power systems, its variability and decentralization demand advanced control architectures, innovative transmission technologies, intelligent forecasting, and robust regulatory frameworks.

This volume, *Advanced Wind Energy Systems: Grid Integration, Markets, and Sustainable Infrastructure*, was conceived to address these evolving complexities. Our aim is to provide a comprehensive and interdisciplinary examination of advanced wind energy systems, bridging engineering, digital intelligence, market mechanisms, and policy considerations. Bringing together leading researchers and practitioners from across the globe, this reference work reflects the increasingly interconnected nature of modern wind power systems and their role within intelligent, sustainable energy ecosystems.

CHAPTER OVERVIEW

Chapter 1: Introduction to Wind Energy Systems establishes the theoretical and technical foundation of wind energy conversion. It revisits the governing aerodynamic principles, including the power equation and Betz limit, and details the architecture of horizontal- and vertical-axis wind turbines. The chapter also explores operational strategies such as pitch and yaw control, Maximum Power Point Tracking (MPPT), and wake management. Importantly, it introduces Artificial Intelligence (AI) applications in condition monitoring and predictive maintenance, positioning digital intelligence as a transformative force in modern wind farm operations.

Chapter 2: The Role of Wind Energy in a Sustainable Energy Transition Agenda situates wind power within the broader global decarbonization framework. It examines technological advancements in turbine design, offshore systems, and digital tools such as digital twins and forecasting platforms. Beyond technology, the chapter addresses socioeconomic benefits, governance structures, permitting mechanisms, environmental safeguards, and circular-economy practices, highlighting how wind energy contributes to sustainable development goals and long-term climate commitments.

Chapter 3: Advanced Power Converters in Wind Energy Systems delves into the power electronic interfaces that enable reliable grid integration. It traces the evolution from fixed-speed turbines to fully rated converter-based systems and reviews converter topologies such as two-level and three-level Neutral-Point Clamped (NPC) Voltage Source Converters and modular multilevel converters (MMCs). Control strategies for both machine-side and grid-side converters are examined, including vector control, fault ride-through capability, reactive power support, and power quality enhancement.

Chapter 4: Navigating Grid Code Requirements with Advanced Monitoring and Control addresses the regulatory and technical frameworks governing renewable integration. As conventional generation is progressively displaced, compliance with evolving grid codes becomes paramount. This chapter analyzes monitoring technologies, advanced control strategies, and updated regulatory policies required to maintain grid stability, reliability, and resilience in high-renewable systems.

Chapter 5: AI–Digital Twin Synergy for Smart and Autonomous Wind Energy Systems explores the integration of AI-driven analytics with Digital Twin technologies. By creating real-time virtual replicas of wind assets, digital twins enable predictive optimization, adaptive control, and system-level simulations. When coupled with machine learning models, these tools foster autonomous and resilient wind farms capable of self-learning and real-time decision-making.

Chapter 6: Cybersecurity and Data Privacy in Wind Power Systems examines the growing cyber-physical vulnerabilities associated with digitalized wind infrastructures. It reviews threats across turbine control systems, SCADA networks, and cloud-based analytics platforms, while contextualizing these risks within regulatory frameworks such as the IEC 62443 standards, the EU NIS2 Directive, and the General Data Protection Regulation (GDPR). The chapter proposes a defense-in-depth strategy combining technical safeguards, organizational governance, and compliance mechanisms.

Chapter 7: The Development of Wind Farm Projects in India presents a national case study highlighting infrastructure expansion, policy evolution, and implementation challenges. Through analysis of major wind farm projects, the chapter evaluates economic benefits, construction barriers, research gaps, and the role of wind power in advancing India's renewable energy targets and sustainable infrastructure development.

Chapter 8: Reliability, Market Dynamics, and Energy Trading with AI in Wind Energy explores how artificial intelligence enhances reliability and market participation. The chapter investigates predictive maintenance using machine learning on SCADA data, deep learning for fault detection, adaptive operational controls, and AI-based forecasting to reduce imbalance penalties. It further examines electricity market mechanisms, energy trading strategies, and how intelligent systems optimize participation in competitive power markets.

Chapter 9: HVDC Transmission for Wind Energy Systems in Microgrids and Virtual Power Plants addresses the transmission challenges associated with remote wind resources. High-Voltage Direct Current (HVDC) systems are presented as a superior alternative to conventional AC transmission for long-distance, high-capacity power transfer. The chapter discusses HVDC applications in offshore wind farms, microgrids, and virtual power plants, emphasizing efficiency, stability, and scalability in modern grid architectures.

CONCLUSION

Wind energy is no longer a peripheral renewable option—it is a structural pillar of the global energy transition. Yet its success depends not only on aerodynamic efficiency or turbine scale, but on intelligent integration within increasingly complex power systems. This volume demonstrates that the future of wind energy lies at the intersection of advanced power electronics, digital intelligence, cybersecurity resilience, regulatory alignment, and market adaptability.

Throughout this book, a recurring theme emerges: integration. Integration of renewable resources into legacy grids. Integration of AI and digital twins into physical infrastructure. Integration of cybersecurity frameworks into operational planning. Integration of wind energy into competitive electricity markets. These interconnected dimensions define the next generation of sustainable energy systems.

We hope this reference work serves researchers, engineers, policymakers, and industry professionals as both a technical guide and a strategic resource. By synthesizing foundational principles with emerging innovations, this volume contributes to the collective effort to design resilient, intelligent, and sustainable wind energy infrastructures capable of meeting global climate and development goals.

As editors, we extend our sincere gratitude to the contributing authors for their scholarly rigor and global perspectives. It is our aspiration that this book will inspire continued innovation, interdisciplinary collaboration, and informed policy development in advancing wind energy systems for a sustainable future.

Chapter 1
Introduction to Wind Energy Systems

Samson Mlay
http://orcid.org/0009-0005-2010-7627
University of Dar es Salaam, Tanzania

Daniil Pushkar
http://orcid.org/0009-0005-0439-7580
ITMO University, Russia

Stampinus Stephano
http://orcid.org/0009-0001-8384-9362
University of Dar es Salaam, Tanzania

Aviti T. Mushi
University of Dar es Salaam, Tanzania

Nikolai Poliakov
ITMO University, Russia

ABSTRACT

The fundamental principles governing wind energy conversion, including the power equation and the Betz limit, are explained to establish a theoretical foundation. The chapter further delineates the key components of modern horizontal- and vertical-axis wind turbines, such as rotor blades, nacelles, and drivetrains, and classifies systems by scale and location. A significant focus is placed on the operations and control of wind turbines and wind farms, covering methods like pitch and yaw control, Maximum Power Point Tracking (MPPT), and wake management. Finally, the chapter explores the transformative role of Artificial Intelligence (AI) in condition monitoring, contrasting traditional methods with advanced AI techniques like machine

DOI: 10.4018/979-8-2600-0106-6.ch001

learning and deep learning to highlight their benefits in predictive maintenance, fault diagnosis, and overall system reliability. Through this structured examination, the authors aim to present wind energy as a cornerstone of the global transition to sustainable and renewable power sources.

1. INTRODUCTION TO WIND ENERGY SYSTEMS

Wind energy systems are a cornerstone of renewable energy, converting wind's kinetic energy into electricity to address climate change and reduce fossil fuel dependence. As a rapidly growing sector, wind power is pivotal in the global shift toward sustainability (Wind & Council, 2024). This section explores its historical evolution, fundamental principles, components, types, benefits, challenges, and global status.

Wind energy use began millennia ago with windmills for grinding grain and pumping water in ancient Persia and medieval Europe (Gipe, 2016). The late 19th century marked the advent of electricity-generating turbines, with James Blyth's 1887 turbine powering his home — widely regarded as the first wind turbine to generate electricity for practical use (Price, 2005). The 1970s oil crisis spurred modern development, leading to utility-scale turbines and the first wind farm in 1980 at Crotched Mountain, New Hampshire (Manwell et al., 2009). Today, global installed wind capacity exceeds 1,021 GW as of 2023, with China, the U.S., and Brazil leading deployment. The sector has grown at a compound annual rate of approximately 12% over the past decade (IRENA, 2019).

2. HISTORICAL DEVELOPMENT OF WIND ENERGY

The harnessing of wind as an energy source is a testament to human ingenuity, spanning millennia from rudimentary applications to sophisticated modern systems. Wind energy has evolved from ancient mechanical devices to a cornerstone of renewable electricity, driven by technological advancements, economic needs, and environmental imperatives (Singh et al., 2024). This section traces the historical trajectory of wind energy, highlighting key milestones, technological innovations, policy influences, and the global expansion that has positioned wind as a dominant renewable energy source by 2023.

Ancient Beginnings: Wind in Early Civilizations

The earliest recorded use of wind energy dates back to around 5000 BC, when ancient Egyptians employed wind to propel sailboats along the Nile River, facilitating trade and transportation (Loewe, 1966). This marked humanity's first practical engagement with wind as a motive force. By the 7th century AD, the Persian region of Sistan saw the development of the earliest known windmills. These vertical-axis structures, constructed from wood and reeds, featured sails that captured wind to drive mechanical processes like grinding grain and pumping water for irrigation. Unlike later European designs, these windmills rotated around a vertical shaft, a configuration suited to the region's consistent wind (Righter, 1996). Their simplicity and effectiveness laid the foundational principles for wind-driven mechanical systems.

These early windmills spread across the Middle East and Central Asia, with evidence of their use in China by the 13th century for similar purposes. The technology was labor-intensive but reliable, demonstrating wind's potential as a renewable resource long before the concept of sustainability emerged. The designs were adapted to local needs, with variations in blade materials and configurations, showcasing early engineering adaptability.

Medieval Europe: The Rise of the Windmill

By the Middle Ages, windmills became a hallmark of European technological progress. Their introduction to Europe is often attributed to the 12th century, possibly through trade routes or Crusader influences from the Middle East. In the Netherlands, windmills became iconic for land drainage, a critical task in reclaiming low-lying areas from the sea (Koetsier, 2025). By the 14th century, Dutch windmills, often featuring four large cloth-covered blades, powered pumps to drain polders, enabling agricultural expansion. In England, windmills were primarily used for milling grain, with "post" windmills structures pivoting on a central post to face the wind becoming widespread.

These European windmills were horizontal-axis designs, a significant departure from Persian vertical-axis models. The horizontal-axis windmill, with its ability to yaw (rotate to face the wind), offered greater efficiency and adaptability. By the 16th century, thousands of windmills dotted the European landscape, particularly in the Netherlands, where over 8,000 were operational at their peak. Innovations like the fantail, a small rotor that automatically adjusted the mill's orientation, enhanced efficiency and reduced manual labor (Jenkins et al., 2021). Windmills also powered early industrial tasks, such as sawmilling and oil pressing, contributing to economic growth in pre-industrial Europe.

The Dawn of Electricity: Early Wind Turbines

The transition from mechanical to electrical energy marked a pivotal shift in wind energy's history. In 1887, Scottish engineer James Blyth constructed the world's first electricity-generating wind turbine in Marykirk, Scotland. His 10-meter-tall, cloth-sailed turbine powered his home, a modest but groundbreaking achievement (Price, 2005). Concurrently, in 1888, Charles Brush in Cleveland, Ohio, built a larger turbine with a 17-meter rotor and 144 wooden blades, generating up to 12 kW to charge batteries for his mansion (Righter, 1996). These early turbines were inefficient by modern standards but demonstrated wind's potential for electricity production.

The early 20th century saw limited progress due to the dominance of coal and emerging hydropower. However, small-scale wind turbines gained traction in rural areas, particularly in the U.S., where companies like Jacobs Wind Electric produced systems for farms to power lights and radios. By the 1930s, thousands of these turbines, typically under 3 kW, were in use across the American Midwest, showcasing wind's utility in off-grid settings (van Est, 2022).

The Modern Era: The 1970s Oil Crisis and Beyond

The 1970s oil crises catalyzed a renaissance in wind energy. Soaring oil prices and energy security concerns prompted governments, particularly in the U.S., Denmark, and Germany, to invest in renewable energy research. The U.S. established the National Renewable Energy Laboratory (NREL) in 1977, focusing on wind turbine design and aerodynamics. Denmark, with its strong engineering tradition, pioneered modern turbine development through companies like Vestas, which introduced reliable, grid-connected turbines (Wiser et al., 2015).

A landmark event was the installation of the world's first wind farm in 1980 at Crotched Mountain, New Hampshire. This farm, comprising 20 turbines of 30 kW each, marked the shift to utility-scale wind power (Hau, 2006). Though small by today's standards, it demonstrated the feasibility of aggregating turbines for grid electricity. Denmark followed suit, establishing wind farms that leveraged government subsidies and feed-in tariffs, fostering a robust wind industry.

Technological advancements in the 1980s and 1990s were critical. Early turbines used fixed-speed designs, but the introduction of variable-speed turbines with power electronics improved efficiency and grid compatibility. Blade materials shifted from wood to lightweight composites like fiberglass and carbon fiber, enabling larger, more durable rotors. The development of pitch control systems allowed blades to adjust angles dynamically, optimizing power output and protecting turbines in high winds. By the late 1990s, turbine capacities reached 1 MW, a significant leap from the kilowatt-scale systems of the 1970s (Berger, 2024, 2025).

Global Expansion and the 21st Century

The 21st century has seen exponential growth in wind energy, driven by climate goals, cost reductions, and technological maturity. Global installed capacity grew from 17.4 GW in 2000 to 1,021 GW by 2023, with an average annual growth rate of 12%. China has led this expansion, accounting for over 40% of global capacity, driven by aggressive renewable energy policies (Musial et al., 2023). The U.S., with 147 GW by 2023, ranks second, with Texas, Iowa, and Oklahoma as leading states. Brazil, India, and European nations like Germany and Spain also contribute significantly.

Offshore wind has emerged as a transformative frontier. The first offshore wind farm, Vindeby in Denmark (1991), featured 11 turbines of 450 kW each (Prime et al., 2024). By 2023, offshore capacity reached 64 GW globally, with Europe leading and the U.S. planning 30 GW by 2030 (Berger, 2024, 2025). Offshore turbines, often exceeding 10 MW, benefit from stronger, more consistent winds but face higher costs and engineering challenges, such as deep-water foundations.

Policy has been a key driver. In the U.S., the Production Tax Credit (PTC) since 1992 has incentivized wind development, while Europe's feed-in tariffs and renewable targets have spurred growth (Musial et al., 2023). Cost declines have been dramatic, with the levelized cost of energy (LCOE) for onshore wind dropping to $32/MWh by 2023, competitive with fossil fuels (van Est, 2022). Innovations like larger rotors (up to 230 meters in diameter) and direct-drive generators have further reduced costs and improved reliability.

3. REGIONAL CONTRIBUTIONS AND INNOVATIONS

In the U.S., wind energy became the largest renewable electricity source by 2023, surpassing hydropower. States like Texas leverage vast land and wind resources, while Iowa generates over 50% of its electricity from wind (Zupok et al., 2025). Offshore projects, like Vineyard Wind off Massachusetts, mark the U.S.'s entry into this sector. In Europe, Denmark's wind industry supplies over 50% of national electricity, while Germany's Energiewende policy integrates wind into a decarbonized grid (Canan, 2023).

China's dominance stems from state-driven investment and manufacturing scale. Companies like Goldwind produce turbines at low cost, while offshore projects in the South China Sea push technological boundaries (Agency, 2022). Brazil's wind growth, particularly in the Northeast, capitalizes on consistent trade winds, with capacity doubling from 2015 to 2023 (DOE, 2015).

4. FUNDAMENTAL PRINCIPLES OF WIND ENERGY

Wind energy harnesses the kinetic energy of atmospheric air masses to generate electricity and constitutes a fundamental component of modern renewable energy systems. Its foundational principles are derived from fluid dynamics, thermodynamics, control theory, and meteorology, which collectively guide the design, modelling, and operation of wind turbines for the efficient conversion of wind into electrical power (Barbre & Li, 2025). This section explores the origins of wind energy, the governing physical equations, theoretical efficiency limits such as the Betz limit, operational considerations for turbines, and the environmental and geographical factors influencing wind energy capture.

4.1 Origins of Wind Energy

Wind energy is ultimately a derivative of solar energy, generated through aerodynamic and electromechanical processes that convert atmospheric motion into usable power. Approximately 2% of the solar radiation incident on Earth equivalent to roughly 3.5 million terawatt-hours per year is transformed into atmospheric kinetic energy (Jenkins et al., 2021). This transformation is driven by the uneven heating of the Earth's surface, which creates pressure gradients resulting from thermal contrasts between the equator and poles, land and sea, and day and night cycles. These gradients initiate air movement, while the Coriolis effect arising from Earth's rotation and surface features such as mountains and valleys further modulate wind patterns on both global and local scales (Ackermann, 2012).

As an indirect form of solar energy, wind is renewable, widely distributed, and produces no greenhouse gas emissions during operation. However, its intermittent nature, driven by temporal (e.g., diurnal and seasonal) and spatial variability, presents challenges for energy system integration. Consequently, optimizing wind power generation requires advanced engineering techniques, including site selection, turbine design, control systems, and energy storage strategies to enhance reliability and grid compatibility.

4.2 The Power Equation

The power available in the wind is given by the fundamental wind power equation as in equation (1):

$$P = \frac{1}{2}\rho A v^3 \qquad (1)$$

where P is the power in watts, ρ is air density (kg/m^3) (typically around 1.225 kg/m^3 at sea level), A is the swept area of the rotor (m^2) (i.e., the area covered by the rotating blades, $A = \pi r^2$), and v is the wind speed (m/s). This cubic relationship implies that doubling wind speed increases power output by a factor of eight explaining why turbines are sited in high-wind regions and mounted on tall towers (Hau, 2006). For instance, a 100-meter rotor operating at 10 m/s produces around 4.8 MW, compared to only 600 kW at 5 m/s [12].

4.3 The Betz Limit

In 1919, Albert Betz established that no wind turbine can capture more than 59.3% of the kinetic energy in wind. This theoretical maximum efficiency, known as the Betz Limit, defines the upper bound of aerodynamic efficiency for any wind turbine operating under ideal conditions. It is mathematically expressed as in equation (2):

$$C_p = \frac{P_{turbine}}{P_{wind}} \leq \frac{16}{27} \approx 0.593 \tag{2}$$

Here, C_p is the power coefficient, representing the fraction of available wind power that is extracted by the turbine. While the Betz limit sets a theoretical maximum of 0.593, real-world wind turbines typically achieve C_p values in the range of 0.45 to 0.50 (i.e., 45% to 50%). This deviation is due to unavoidable aerodynamic, mechanical, and electrical losses. Key factors limiting efficiency include blade tip vortices, drag forces, and drivetrain losses, all of which reduce the amount of wind energy that can be effectively converted into usable mechanical and electrical power (Burton et al., 2021).

5. COMPONENTS OF WIND TURBINES

Modern turbines integrate mechanical, electrical, and control systems for efficient energy conversion (Okedu & Muyeen, 2022). Each component plays a critical role in ensuring reliability, performance, and grid compatibility. This section details the primary components rotor blades, nacelle, tower, generator and drivetrain, and foundation and balance of system highlighting their design, function, and technological advancements.

Rotor Blades

Rotor blades are the most visible and aerodynamically critical components, typically three in number for horizontal-axis wind turbines (HAWTs) to balance efficiency and stability. Constructed from lightweight composites like fiberglass, carbon fiber, or hybrid materials, blades are designed to maximize lift and minimize drag, similar to airplane wings (Feng & Shen, 2017). The aerodynamic profile, often based on airfoil shapes, captures wind's kinetic energy, causing the rotor to spin. Modern blades for large offshore turbines can exceed 100 meters in length, with some, like those for 15 MW turbines, reaching 115 meters, yielding a swept area over 40,000 m^2. Advanced manufacturing techniques, such as vacuum infusion, ensure strength while reducing weight. Innovations like bend-twist coupling allow blades to passively adjust shape in high winds, enhancing durability and efficiency(Ackermann, 2012).

Nacelle

The nacelle, mounted atop the tower, serves as the turbine's control and power generation hub. It houses critical components, including the generator, gearbox (in geared designs), control systems, and yaw mechanism. The yaw system, typically driven by electric or hydraulic motors, rotates the nacelle to align the rotor with wind direction, optimizing energy capture. The nacelle also contains sensors for monitoring wind speed, temperature, and vibration, integrated with Supervisory Control and Data Acquisition (SCADA) systems for real-time operation (Musial et al., 2023). Modern nacelles for multi-megawatt turbines are compact yet robust, often weighing 50–100 tons, and are designed for easy maintenance access in harsh environments, particularly offshore.

Tower

The tower elevates the rotor and nacelle to access stronger, less turbulent winds at higher altitudes. Typically made of tubular steel or, increasingly, concrete for larger turbines, towers range from 80–120 meters for onshore installations to over 150 meters for offshore models. Taller towers exploit the wind shear effect, where wind speed increases with height, boosting energy yield due to the cubic relationship with velocity ($P \propto v^3$). Towers must withstand dynamic loads from wind, rotor thrust, and vibrations, requiring robust engineering (Elkodama et al., 2023). Offshore towers often use lattice or hybrid designs to reduce material costs while maintaining stability in marine conditions.

Drivetrain & Generator

The drivetrain transfers rotational energy from the rotor to the generator, which converts it into electricity. In geared turbines, a gearbox increases the rotor's low rotational speed (10–20 RPM) to the high speeds (1,000–3,000 RPM) required by conventional generators and hence becomes prone to maintenance issues (Anagnostopoulos et al., 2023). Direct-drive turbines eliminate the gearbox, using low-speed, permanent-magnet generators to reduce maintenance and improve reliability, especially offshore (Errami et al., 2013). Generators produce alternating current (AC), synchronized with the grid via power electronics. Typical capacities range from 2–5 MW for onshore turbines to 10–15 MW offshore, with efficiency exceeding 95%.

6. TYPES OF WIND ENERGY SYSTEMS

Wind energy systems are diverse, designed to suit varying environmental, economic, and application-specific needs. They are classified by axis orientation, scale, and location, each offering unique advantages and challenges. This section explores horizontal-axis wind turbines and vertical-axis wind turbines, utility-scale and distributed/small-scale systems, and land-based and offshore installations, providing a comprehensive overview of their design, applications, and current trends.

6.1 Axis Orientation

Horizontal-Axis Wind Turbines

Horizontal-axis wind turbines (HAWTs) dominate the global wind energy landscape due to their high efficiency and scalability (Xiao et al., 2021). In HAWTs, the rotor blades are mounted on a hub, rotating parallel to the ground around a horizontal shaft. This design leverages aerodynamic lift, similar to an airplane wing, to drive the rotor, which connects to a generator via a drivetrain. HAWTs typically feature three blades to balance efficiency, structural stability, and manufacturing costs, though two-bladed designs exist for specific applications.

The efficiency of HAWTs stems from their ability to capture wind across a large swept area, with power output proportional to the square of the rotor diameter. Modern HAWTs, particularly offshore models, can have rotors exceeding 230 meters, generating up to 15 MW. Their primary drawback is the need for a yaw mechanism a motorized system that rotates the nacelle to face the wind, ensuring optimal alignment. Misalignment reduces power output significantly, as power decreases with the cube of the cosine of the yaw error angle ($P \propto \cos^3\theta$). Yaw systems, while effective, add complexity and maintenance costs.

HAWTs are versatile, used in both onshore and offshore wind farms. Their widespread adoption is driven by decades of engineering refinement, including advanced blade materials (e.g., carbon fiber composites) and control systems like pitch regulation, which adjusts blade angles to optimize performance and protect against high winds. Despite their dominance, HAWTs are less suitable for turbulent or low-wind urban environments due to their size and yaw requirements.

Vertical-Axis Wind Turbines

Vertical-axis wind turbines (VAWTs), in contrast, feature blades that rotate around a vertical shaft, making them omnidirectional they do not require yaw mechanisms to face the wind. VAWTs come in two main designs: Darrieus (curved, eggbeater-like blades) and Savonius (drag-based, scoop-like blades). Their ability to capture wind from any direction makes them ideal for urban or turbulent wind conditions, such as rooftops or small-scale installations (Kang et al., 2015).

However, VAWTs are generally less efficient than HAWTs, with power coefficients typically below 0.3 compared to HAWTs' 0.45–0.5. This is due to their complex aerodynamics, where blades experience varying wind forces during rotation, leading to lower energy capture. VAWTs also face challenges with structural stability, as the vertical shaft bears significant torque, requiring robust bearings. Despite these drawbacks, VAWTs are quieter and more compact, making them suitable for distributed applications like powering homes, schools, or remote facilities.

Recent innovations aim to improve VAWT efficiency, such as helical blade designs to reduce vibration and hybrid Darrieus-Savonius systems for better low-wind performance. While VAWTs represent less than 5% of global wind capacity, their niche applications in urban and small-scale settings are growing, particularly in regions with space constraints or aesthetic concerns.

6.2 Scale

Utility-Scale Wind Turbines

Utility-scale wind turbines, typically rated at 1 MW or higher, are the backbone of large wind farms supplying grid electricity. In the U.S., the average onshore turbine capacity reached 3.2 MW by 2023, with offshore models exceeding 10 MW. These turbines, almost exclusively HAWTs, are designed for maximum energy output and grid integration, often grouped in wind farms spanning hundreds of megawatts. For example, a single 3.2 MW turbine can power approximately 1,000 homes annually, assuming typical wind conditions (Zhao et al., 2024).

Utility-scale turbines benefit from economies of scale, with larger rotors and taller towers accessing stronger winds. Their design prioritizes reliability, with advanced control systems like Maximum Power Point Tracking (MPPT) and pitch control ensuring optimal performance across varying wind speeds. However, their size rotors up to 160 meters for onshore and 230 meters offshore poses logistical challenges, including transportation and installation. Maintenance costs are significant, but innovations like direct-drive generators reduce downtime by eliminating gearboxes.

Wind farms, often comprising dozens to hundreds of utility-scale turbines, require careful layout to minimize wake effects, where downstream turbines lose efficiency due to turbulence from upwind units. Advanced modeling and AI-driven optimization are increasingly used to design layouts that maximize energy yield. Utility-scale systems dominate global capacity, accounting for over 95% of the 1,021 GW installed by 2023, driven by their cost-effectiveness and ability to meet large-scale energy demands.

Distributed/Small-Scale Wind Turbines

Distributed or small-scale wind turbines, rated under 100 kW, serve localized energy needs, such as homes, farms, or remote communities. These systems, which may be HAWTs or VAWTs, are ideal for off-grid or hybrid applications, often paired with solar panels or batteries (Lundberg & Lee, 2017). A typical 10 kW turbine can power a large home or small business, producing 10,000–20,000 kWh annually in good wind conditions.

Small-scale turbines are simpler and cheaper to install than utility-scale systems, with towers often under 40 meters. They are suited to rural or remote areas where grid access is limited, such as Alaskan villages or agricultural operations. VAWTs are increasingly popular in this segment due to their compact size and ability to handle turbulent winds. However, their lower efficiency and higher cost per kW compared to utility-scale systems limit widespread adoption. Incentives, like net metering or tax credits, have spurred growth in distributed wind, particularly in the U.S., where over 1 GW of small-scale capacity was installed by 2023.

6.3 Location

Land-Based Wind Turbines

Land-based (onshore) wind turbines are the most common and cost-effective type, accounting for the majority of global capacity. They are typically HAWTs installed in open, windy areas like plains, hills, or coastal regions. Onshore turbines benefit from lower installation and maintenance costs compared to offshore systems, with levelized costs of energy (LCOE) $26–50/MWh (*Land-Based Wind Market Report : 2023 Edition*, 2023). Their foundations, usually concrete pads or piles are simpler to construct, and grid connectivity is straightforward in developed regions.

The U.S. leads in onshore wind, with 147 GW installed by 2023, concentrated in states like Texas and Iowa. Onshore sites leverage wind speeds of 6.5–9 m/s, with taller towers (80–120 meters) accessing stronger winds. Challenges include land-use conflicts, visual and noise concerns, and wildlife impacts, though mitigation strategies like careful siting and radar-based shutdowns reduce these issues.

Offshore Wind Turbines

Offshore wind turbines harness stronger, more consistent winds over bodies of water, offering higher energy yields. They are typically HAWTs mounted on fixed foundations (monopiles or jackets) in shallow waters or floating platforms in deeper seas (Natarajan, 2014). Offshore turbines are larger, with capacities of 10–15 MW and rotors up to 230 meters, capitalizing on wind speeds of 8–10 m/s. However, their high costs LCOE often exceeds $100/MWh and complex installation limit deployment.

Global offshore capacity reached 64 GW by 2023, led by Europe (e.g., UK and Denmark). The U.S. lags with 42 MW but aims for 30 GW by 2030, driven by projects like Vineyard Wind. Floating platforms, enabling access to deep-water sites, are a growing trend, with Norway and Japan leading innovation. Offshore wind's potential is vast, with global resources estimated at 70,000 TWh/year, but challenges include corrosion, marine logistics, and grid integration.

7. OPERATIONS AND CONTROL OF WIND ENERGY SYSTEMS

Wind energy systems are engineered to convert the kinetic energy of wind into reliable, grid-compatible electricity while ensuring operational safety and efficiency. The operations of wind turbines and wind farms involve a complex interplay of mechanical, electrical, and control systems, all designed to maximize energy capture across variable wind conditions. Control systems are pivotal, enabling turbines to adapt to fluctuating wind speeds, protect structural integrity, and optimize performance. This section provides an in-depth exploration of the basic operations of wind turbines, control methods including generator speed, blade pitch, and yaw control, advanced control techniques like Maximum Power Point Tracking (MPPT), and the operational strategies for wind farms, including wake management.

Basic Operations of Wind Turbines

The operation of a wind turbine is governed by the interaction between wind and the turbine's aerodynamic and electrical systems. Wind turbines function within four distinct operational regions, defined by wind speed, which dictate power output and control strategies:

1. **Below Cut-In Speed**: At low wind speeds, typically below 3–5m/s, the turbine remains idle as the wind lacks sufficient energy to overcome frictional losses and initiate rotation. The rotor may be locked or allowed to idle without generating power.
2. **Partial Load Region**: As wind speed increases above the cut-in threshold, the turbine begins generating power. The power output scales with the cube of wind speed. In this region, the turbine operates below its rated capacity, and control systems aim to maximize energy capture by optimizing rotor speed and blade angle.
3. **Rated Power Region**: Once the wind speed reaches the turbine's rated speed typically 10–15 m/s, the turbine achieves its maximum designed power. Control systems, such as blade pitch adjustment, maintain constant power by regulating rotor speed to prevent overloading the generator or structural components.
4. **Above Cut-Out Speed**: At very high wind speeds, typically above 25 m/s, the turbine shuts down to avoid structural damage from excessive aerodynamic loads (Tchakoua et al., 2014). The blades are feathered (pitched to minimize lift), and the rotor is stopped, ensuring safety during extreme weather.

Daily operations involve continuous monitoring through Supervisory Control and Data Acquisition (SCADA) systems, which collect data on wind speed, rotor speed, power output, and component health. SCADA enables real-time diagnostics, fault detection, and remote control, minimizing downtime. Maintenance, both preventive and corrective, is critical, with technicians inspecting blades, gearboxes, and electrical systems regularly. Grid integration ensures that the variable power output aligns with grid requirements, often using power electronics to stabilize voltage and frequency. In 2023, advanced SCADA systems reduced downtime by up to 15%, enhancing operational efficiency.

Control Methods

Control systems are the backbone of wind turbine performance, ensuring optimal energy capture, structural safety, and grid compatibility. The primary control methods include generator speed control, blade pitch control, and yaw control, each addressing specific operational challenges.

Generator Speed Control

Generator speed control regulates the rotational speed of the turbine's rotor to optimize power output and protect the drivetrain. In variable-speed turbines, which dominate modern designs, power electronics such as frequency converters adjust the rotor speed to maintain an optimal tip-speed ratio $\lambda = \omega R / \nu$ where λ is angular velocity, R is rotor radius, and ν is wind speed. This ensures the turbine operates at peak C_p, typically 0.45–0.50, in the partial load region.

Variable-speed turbines use doubly-fed induction generators (DFIGs) or permanent-magnet synchronous generators (PMSGs) coupled with full or partial power converters. These converters decouple the generator from the grid frequency, allowing independent speed control and enabling grid compliance with standards like IEEE 1547. For example, a 3 MW turbine might adjust rotor speed from 10 to 20 RPM to track optimal λ, increasing energy capture by 10–20% compared to fixed-speed designs. Fixed-speed turbines, now less common, rely on simpler squirrel-cage induction generators but are less efficient due to their inability to adapt to wind variations.

Blade Pitch Control

Blade pitch control is a critical mechanism in modern wind turbines, responsible for regulating rotor speed and power output by adjusting the angle of the blades relative to the wind. By altering the angle of attack, pitch control systems can optimize aerodynamic lift in low wind conditions or reduce it during high wind speeds to prevent mechanical overload and maintain safe operation.

There are two primary approaches to pitch regulation:

- Active Pitch Control: This method uses hydraulic or electric actuators to rotate each blade around its longitudinal axis. In the partial load region (below rated wind speed), blades are pitched to a low angle (typically near 0°) to maximize lift and energy capture. When wind speeds exceed the rated threshold, the blades are pitched toward a feathered position (approaching 90°) to reduce lift and maintain constant power output. Active pitch control is the dominant strategy in modern variable-speed turbines due to its precision and adaptability. Typical actuator response times range from 1° to 5° per second (Tchakoua et al., 2014).

- Stall Control: Commonly found in older or fixed-speed turbines, stall control relies on blade geometry to induce aerodynamic stall at high wind speeds, passively limiting lift. While simpler and requiring fewer moving parts, stall-controlled turbines are generally less efficient and less responsive to changing wind conditions, making them less common in contemporary designs.

Beyond power regulation, pitch control significantly contributes to turbine safety and durability. During extreme wind events, blades can be feathered to minimize aerodynamic loads, reducing the risk of structural damage. Additionally, individual pitch control (IPC) a more advanced strategy enables independent adjustment of each blade. IPC compensates for asymmetric loads caused by wind shear, turbulence, or yaw misalignment, and has been shown to extend blade lifespan by up to 10% through load mitigation.

Yaw Control

Yaw control aligns the nacelle with the wind direction to maximize energy capture. Misalignment reduces power output, as the effective wind speed is $v\cos\theta$, leading to a power loss proportional to $\cos^3\theta$. For instance, a >10° misalignment reduces power by approximately 15% (Yang et al., 2019). Yaw systems use wind vanes or anemometers to detect wind direction and electric or hydraulic motors to rotate the nacelle, typically at 0.5–1° per second.

Active yaw control is standard in HAWTs, with response times of 10–30 seconds to track wind changes. Some modern turbines employ predictive yaw, using LiDAR to measure incoming wind direction, improving alignment accuracy. Yaw systems must balance responsiveness with energy consumption, as frequent adjustments increase wear on bearings and motors.

Advanced Control Techniques

Advanced control strategies enhance turbine performance beyond basic methods:

- **Maximum Power Point Tracking (MPPT)**: MPPT algorithms dynamically adjust rotor speed and blade pitch to maintain the optimal tip-speed ratio across varying wind speeds [68]. By tracking the power curve's peak, MPPT can increase annual energy production by 5–10%. For example, a hill-climbing MPPT algorithm iteratively adjusts speed based on power output measurements, converging on the optimal operating condition.
- **Stall Control for Fixed-Speed Turbines**: While less common, stall-regulated turbines use fixed blade angles designed to induce stall at high speeds, limit-

ing power output. This passive approach reduces mechanical complexity but sacrifices efficiency, making it suitable for smaller or older systems.

- **Model Predictive Control (MPC)**: MPC uses real-time wind and turbine data to forecast and optimize performance over a short time horizon (e.g., 10 seconds). By solving optimization problems, MPC minimizes structural loads and maximizes power, particularly in turbulent conditions. Studies show MPC can reduce fatigue loads by 15%, extending turbine lifespan.

Wind Farm Operations

Wind farms, comprising multiple turbines, require coordinated operations to optimize energy production and grid integration. Centralized control centers use SCADA to monitor turbine performance, weather conditions, and grid demands, enabling predictive maintenance and fault detection [70]. For instance, SCADA data can identify a failing gearbox bearing, scheduling repairs before catastrophic failure, saving up to $100,000 per incident.

Wake Control

Wake effects, where upwind turbines create turbulence that reduces downwind turbine efficiency, are a significant challenge. Wake losses can reduce farm output by 10–20%. Wake control strategies include:

- **Yaw-Based Wake Steering**: Adjusting the yaw angle of upwind turbines to deflect wakes away from downstream units. This can increase farm output by 3–8%, as demonstrated in field tests [69].
- **Pitch-Based Wake Management**: Modifying blade pitch to reduce wake intensity, though less common due to energy trade-offs.
- **Layout Optimization**: Using computational fluid dynamics (CFD) and AI, farm layouts are designed to minimize wake interactions, with turbines spaced 5–10 rotor diameters apart.

Advanced wind farm control uses AI-driven models to predict wind patterns and optimize turbine settings in real-time, improving overall efficiency. For example, reinforcement learning algorithms have increased farm output by 3–7% in recent studies.

Challenges and Future Directions

Modern wind energy systems face several operational challenges that must be addressed to ensure reliable and cost-effective integration into the electrical grid. One of the primary challenges is wind variability, which necessitates robust real-time control strategies to maintain grid stability. This includes the need for inertia emulation and reactive power support to compensate for the reduced system inertia associated with inverter-based renewable sources (Gu et al., 2021). Additionally, operations and maintenance (O&M) costs represent a significant portion approximately 20–25% of the total lifetime expenses of a wind turbine, underscoring the importance of predictive maintenance and system optimization.

Looking ahead, artificial intelligence (AI) and digital twin technologies are emerging as transformative tools in wind energy. These technologies enable real-time fault forecasting, performance diagnostics, and layout optimization, with the potential to reduce the levelized cost of energy (LCOE) by 5–10% by 2030 (Gu et al., 2021). AI-driven predictive control algorithms can dynamically adjust turbine behaviour based on wind forecasts and operational data, while LiDAR (Light Detection and Ranging) systems provide upstream wind measurements, allowing for anticipatory pitch and yaw adjustments. Furthermore, the development of hybrid systems that integrate wind energy with battery storage or other renewable sources can help smooth power output and enhance dispatchability. Collectively, these advancements are expected to significantly improve the reliability, efficiency, and cost-competitiveness of wind power, further solidifying its role as a cornerstone of global energy systems by the end of the decade.

8. CONDITION MONITORING USING ARTIFICIAL INTELLIGENCE

Condition monitoring (CM) is a critical aspect of wind energy systems, ensuring the reliability, longevity, and optimal performance of wind turbines. By continuously assessing the health of turbine components such as blades, gearboxes, generators, and bearings CM helps prevent unexpected failures, reduces maintenance costs, and maximizes energy output. Traditional CM methods have laid the foundation for turbine maintenance, but they often fall short in predictive capabilities. Artificial Intelligence (AI) has revolutionized CM by enabling proactive, data-driven fault prediction and diagnosis, transforming reactive maintenance into predictive strategies (Lundberg & Lee, 2017).

This section delves into the role of AI in CM, contrasting traditional and AI-based methods, exploring key AI techniques such as machine learning (ML), deep learning (DL), and explainable AI (XAI), and discussing benefits, challenges, and future trends, including integration with the Internet of Things (IoT).

8.1 The Role of Condition Monitoring in Wind Turbines

Wind turbines operate in harsh environments, exposed to variable wind loads, extreme weather, and mechanical stresses, leading to failures in critical components. Common faults include blade cracks, gear wear, bearing spalling, generator winding failures, and pitch system malfunctions. Unplanned downtime for a single multi-megawatt turbine can cost operators upwards of $100,000 per day in lost revenue and emergency repair costs (Lundberg & Lee, 2017).

CM involves collecting and analyzing data from a network of sensors to detect anomalies early. Sensors monitor variables such as: Vibration (accelerometers on gearbox, generator, main bearing), temperature (oil, bearing, generator winding), acoustic emissions (high-frequency stress waves from crack propagation), oil quality (particle counters, spectrometry for wear metals), and SCADA parameters (power output, rotor speed, wind speed, pitch angle, yaw position)

AI enhances CM by processing vast, multivariate datasets from these sensors, identifying subtle patterns and correlations invisible to traditional rule-based methods. As of 2025, AI-driven CM systems have demonstrably reduced unexpected turbine failures by up to 30%, according to industry deployment reports (Lundberg & Lee, 2017). AI's predictive power stems from its ability to learn complex, non-linear relationships from historical operational data, forecast impending faults with high confidence, and optimize maintenance schedules a capability that is indispensable as wind farms grow in scale and complexity, making manual inspections increasingly impractical.

8.2 Traditional vs. AI-Based Methods

Traditional Condition Monitoring Methods

Traditional CM relies on physical measurements and predefined thresholds. Key methods include:

Vibration Analysis: Uses accelerometers to detect imbalances, misalignments, or bearing defects. Fast Fourier Transform (FFT) decomposes time-domain signals into frequency spectra to identify fault signatures (e.g., characteristic bearing defect frequencies). However, it requires expert interpretation and is highly sensitive to operational noise and varying load conditions (Djalel et al., 2014)

Oil Monitoring: Analyzes lubricant for wear particles, water contamination, or viscosity changes using spectroscopy or ferrography. Effective for gearbox health but typically limited to periodic, offline sampling, missing real-time degradation trends (Djalel et al., 2014).

Acoustic Emission Monitoring: Detects high-frequency stress waves generated by crack initiation or friction. Useful for early-stage fault detection but challenged by high ambient noise levels in operational wind farms (Djalel et al., 2014).

Thermography: Uses infrared cameras to identify hotspots from electrical faults, bearing overheating, or brake drag. Non-invasive but highly weather-dependent and provides limited predictive capability (Djalel et al., 2014).

Strain Gauge Monitoring: Measures structural loads and fatigue on blades or towers. Accurate but expensive to deploy at scale and primarily used for R&D or critical structural validation (Djalel et al., 2014).

These methods are cost-effective for basic fault detection but suffer from significant limitations: they are often reactive, detecting faults only after they have manifested; they require specialized human expertise for interpretation; they struggle with complex, non-linear interactions between multiple sensor inputs; and they frequently generate false positives or negatives, especially under variable wind conditions (Djalel et al., 2014). For instance, a fixed vibration threshold might overlook a subtle, early-stage bearing fault developing under low-wind, low-load conditions.

AI-Based Methods

AI shifts CM to a proactive, predictive paradigm by leveraging ML and DL algorithms to learn from data. AI models process high-dimensional, multivariate data streams from SCADA, vibration, acoustic, and other sensors, automatically learning the "normal" operating envelope of a turbine and flagging deviations as potential anomalies.

Unlike traditional methods, AI:

Handles big data volumes and high-velocity sensor streams.

Adapts to turbine-specific behaviors and site conditions, improving accuracy over time.

Excels in scenarios where traditional methods fail for example, predicting a gearbox failure from subtle, evolving vibration shifts over weeks, or forecasting blade icing by correlating temperature, humidity, and power curve deviations.

A 2024 comparative study demonstrated that an AI-based fault diagnosis system achieved 95% accuracy in separating multiple concurrent fault types, compared to 70–80% accuracy achieved by traditional vibration analysis alone (Musial et al., 2023). Those methods are as summarized in Table 1 below.

Table 1. Comparison of traditional and AI-based methods for wind turbines monitoring and fault diagnosis

Aspect	Traditional Methods	AI-Based Methods
Approach	Rule-based, threshold-driven	Data-driven, learning-based
Prediction Capability	Reactive (post-fault detection)	Proactive (fault prediction)
Data Handling	Limited to specific, isolated signals	Multivariate, big data integration
Accuracy	70–85% (highly dependent on expertise)	90–98% (with sufficient quality data)
Cost	Lower initial cost, higher labour cost	Higher initial (model training), lower long-term O&M
Limitations	Noise sensitivity, false alarms	Data quality dependency, "black-box" interpretability
Examples	Vibration FFT, oil spectroscopy	ML classification, DL CNNs, anomaly detection models

This table illustrates AI's clear superiority in scalability, adaptability, and predictive accuracy for modern wind farm operations (Musial et al., 2023).

8.3 AI Techniques in Condition Monitoring

AI techniques for CM span supervised/unsupervised machine learning, deep learning, and explainable AI (XAI), each tailored to specific diagnostic and prognostic tasks.

Machine Learning Techniques

ML algorithms learn patterns from labeled (supervised) or unlabeled (unsupervised) data to detect, classify, and predict faults.

Supervised Learning for Classification: Algorithms like Support Vector Machines (SVM), Random Forests, or Gradient Boosting Machines (e.g., XGBoost) are trained on labeled datasets to classify operational states (e.g., "Normal," "Bearing Fault," "Gear Tooth Crack"). A 2023 study using ensemble ML on SCADA data achieved 98% accuracy in multi-fault classification for a 3 MW onshore turbine fleet (Subedi et al., 2025).

Regression for Anomaly Detection: Models like Support Vector Regression (SVR) or Gaussian Process Regression (GPR) learn the normal power curve ($P=f(v,\theta)$, where v is wind speed and θ represents other parameters like air density or pitch angle). Deviations from the predicted curve (measured by metrics like Mean Squared Error, MSE= indicate potential faults such as blade erosion or pitch misalignment. These methods routinely achieve >90% anomaly detection rates (Subedi et al., 2025).

Unsupervised Learning: Techniques like k-Means clustering or Isolation Forests identify outliers without needing pre-labeled fault data, making them ideal for detecting novel or rare failure modes. The k-Means algorithm minimizes within-cluster variance: $J = \sum_{i=1}^{k} \sum_{x \in C_i} \|x - \mu_i\|^2$.

Hybrid approaches, such as combining genetic algorithms for optimal feature selection with SVM classifiers, have been shown to improve diagnostic accuracy in noisy, real-world wind farm data environments (Subedi et al., 2025).

Deep Learning Techniques

DL, a subset of ML, uses multi-layered neural networks to automatically learn hierarchical features from raw, high-dimensional data particularly effective for time-series and image-like sensor data.

Convolutional Neural Networks (CNNs) for Vibration Data: CNNs excel at processing 1D time-series vibration signals or 2D spectrogram images. By applying convolutional filters, pooling, and fully connected layers, CNNs can automatically extract fault-related features. A 2024 study using 1D-CNNs on raw accelerometer data from wind turbine gearboxes achieved 95% accuracy in detecting and classifying bearing faults [77].

Recurrent Neural Networks (RNNs) and LSTMs for Prognosis: These networks are designed for sequential data. Long Short-Term Memory (LSTM) networks, with their gating mechanisms, are particularly effective for predicting Remaining Useful Life (RUL) by learning long-term dependencies in SCADA or sensor time-series data [77].

Autoencoders for Anomaly Detection: These unsupervised DL models learn to compress and reconstruct normal operational data. High reconstruction errors on new data points indicate anomalies. Variational Autoencoders (VAEs) add a probabilistic layer, making them more robust to noise (Mbasso et al., 2025).

DL's "end-to-end" learning capability eliminates the need for manual feature engineering, a major advantage. However, it demands large, high-quality datasets and significant computational resources for training.

Explainable AI (XAI) Techniques

A major challenge with complex DL models is their "black-box" nature. XAI techniques provide interpretable insights into model decisions, which is crucial for building operator trust and facilitating actionable maintenance decisions in safety-critical wind energy systems (Aledhari et al., 2020).

SHAP (SHapley Additive exPlanations): Based on cooperative game theory, SHAP values quantify the contribution of each input feature to the model's output.

Applied to a fault diagnosis model, SHAP can reveal, for example, that a spike in a specific vibration frequency band was the primary driver for a "Bearing Fault" prediction, helping engineers pinpoint the issue (Aledhari et al., 2020).

LIME (Local Interpretable Model-agnostic Explanations): Creates simple, interpretable models (like linear regressions) that approximate the complex model's behavior locally around a specific prediction, providing intuitive explanations.

Gaussian Process Regression (GPR): While technically an ML method, GPR is inherently interpretable as it provides not just a prediction but also a measure of uncertainty ($f(x) \sim GP(m(x), k(x,x'))$). This is invaluable for risk-based maintenance scheduling (Aledhari et al., 2020). Those monitoring techniques for wind turbines are as summarized in Table 2.

Table 2. AI Techniques for wind turbine condition monitoring

AI Technique	Primary Application in CM	Key Mathematical Feature	Performance Example
ML (SVM/ RF)	Fault classification	Hyperplane optimization / Ensemble trees	98% multi-fault accuracy [76]
ML (SVR/ GPR)	Power curve anomaly detection	MSE minimization / Kernel covariance	>90% anomaly detection rate [76]
DL (1D-CNN)	Vibration-based fault diagnosis	Convolutional filters	95% bearing fault detection [77]
DL (LSTM)	Remaining Useful Life (RUL)	Gated recurrent units	85% accuracy in RUL prediction
XAI (SHAP)	Feature importance / Diagnosis	Shapley values (game theory)	Identifies root cause of blade fault [78]
XAI (GPR)	Anomaly detection with uncertainty	Gaussian kernel / Uncertainty bounds	Quantifies confidence in predictions [78]

This table summarizes the most effective AI techniques, their applications, and representative performance metrics in wind turbine CM.

Benefits of AI in Condition Monitoring

The adoption of AI in CM delivers substantial, quantifiable benefits:

Reduced Downtime and Costs: Predictive maintenance enabled by AI can cut unplanned outages by 20–30%, translating to millions of dollars in annual savings for a large wind farm operator. By optimizing maintenance schedules, AI also extends the operational life of critical components like gearboxes and bearings (Aledhari et al., 2020).

Improved Accuracy and Efficiency: AI systems consistently achieve fault detection and classification accuracies of 90–98%, far surpassing the 70–85% typical of traditional methods. Real-time AI analytics can process terabytes of sensor data from entire wind farms within seconds, enabling rapid decision-making (*Land-Based Wind Market Report : 2023 Edition*, 2023).

Scalability: AI solutions are inherently scalable, capable of monitoring hundreds or thousands of turbines simultaneously by integrating data from diverse sensors and SCADA systems, providing a holistic view of fleet health.

Environmental Impact: By enhancing turbine reliability and operational efficiency, AI helps maximize energy yield and reduce the Levelized Cost of Energy (LCOE), indirectly lowering CO_2 emissions per MWh generated. A 2024 market analysis projects that the global AI-driven CM market for wind energy will exceed $5.4 billion by 2028, driven overwhelmingly by these economic and operational advantages (*Land-Based Wind Market Report : 2023 Edition*, 2023).

Challenges in AI Implementation

Despite its transformative potential, the implementation of AI in CM faces several persistent challenges:

Data Quality and Availability: Wind turbine data is often noisy, incomplete, or imbalanced (with far more "normal" operation data than fault data). Poor data quality can lead to model overfitting or underfitting, reducing real-world performance. Robust preprocessing (imputation, normalization, filtering) is essential but time-consuming (Velandia-Cardenas et al., 2021).

Model Interpretability and Trust: Deep learning models, while powerful, are often opaque "black boxes." While XAI methods like SHAP and LIME help, they are not yet universally adopted or fully trusted by field engineers, slowing deployment (Velandia-Cardenas et al., 2021).

Computational Resources and Edge Deployment: Training sophisticated DL models requires significant GPU resources. For real-time, on-turbine inference, "edge AI" solutions are needed, but deploying and maintaining these on remote or offshore turbines presents logistical and cost challenges (Velandia-Cardenas et al., 2021).

Integration with Legacy Systems: Many wind farms operate with older SCADA and control systems that were not designed for AI integration, requiring costly middleware or system upgrades.

Regulatory and Ethical Considerations: As AI takes on more decision-making roles, ensuring these decisions are fair, transparent, and compliant with safety regulations becomes paramount.

Addressing these challenges requires a multi-faceted approach, including the development of hybrid physics-informed ML models, investment in high-quality, labeled datasets, and the establishment of industry-wide standards for AI in CM (Velandia-Cardenas et al., 2021).

9. FUTURE TRENDS: INTEGRATION WITH IOT AND BEYOND

The future of AI in wind turbine CM is inextricably linked with the Internet of Things (IoT) and other emerging technologies, creating truly intelligent, self-monitoring energy systems.

IoT and Edge AI Integration: The proliferation of low-cost, wireless IoT sensors enables denser, real-time data collection from every critical component. When combined with edge AI where data is processed locally on the turbine or substation this allows for instantaneous fault detection and response, drastically reducing latency and bandwidth requirements (Ghosh et al., 2025).

Federated Learning: This privacy-preserving technique allows AI models to be trained across multiple wind farms without sharing raw, proprietary operational data. Each farm trains a local model, and only model updates (not data) are shared and aggregated, preserving data security and competitive advantage (Yin et al., 2022).

Hybrid AI-Physics Models: The next generation of CM will combine data-driven AI with first-principles physics-based models (e.g., finite element analysis for structural loads, thermodynamic models for generator heating). This hybrid approach improves model generalizability and reliability, especially for rare or unseen fault conditions.

Advanced XAI and Causal Inference: Future XAI will move beyond feature importance to provide causal explanations (e.g., "This bearing failed because of misalignment caused by a faulty yaw motor"), enabling more targeted and effective maintenance.

Sustainability and Circular Economy Focus: AI will increasingly be used to optimize not just for cost and uptime, but for sustainability for example, predicting the optimal time to refurbish a component versus replacing it, or optimizing operations to minimize lifetime carbon footprint.

As global wind capacity is projected to reach 2,000 GW by 2030, the integration of AI and IoT will be pivotal in ensuring this massive infrastructure operates reliably, efficiently, and sustainably (Ghosh et al., 2025; Li et al., 2020)

In conclusion, AI is fundamentally transforming wind turbine condition monitoring from a reactive, manual process into an intelligent, predictive, and automated system. By harnessing the power of data, AI promises not only enhanced reliability and reduced costs but also a more sustainable and resilient future for wind energy.

REFERENCES

Ackermann, T. (2012). Wind Power in Power Systems. In T. Ackermann (Ed.), *Wind Power in Power Systems* (2nd ed.). wiley. DOI: 10.1002/9781119941842

Agency, I. E. (2022). International Energy Agency (IEA) World Energy Outlook 2022. Https://Www.Iea.Org/Reports/World-Energy-Outlook-2022/Executive-Summary

Aledhari, M., Razzak, R., Parizi, R. M., & Saeed, F. (2020). Federated Learning: A Survey on Enabling Technologies, Protocols, and Applications. *IEEE Access : Practical Innovations, Open Solutions*, *8*, 140699–140725. DOI: 10.1109/ACCESS.2020.3013541 PMID: 32999795

Anagnostopoulos, S. J., Bauer, J., Clare, M. C. A., & Piggott, M. D. (2023). Accelerated wind farm yaw and layout optimisation with multi-fidelity deep transfer learning wake models. *Renewable Energy*, *218*, 119293. DOI: 10.1016/j.renene.2023.119293

Barbre, Z., & Li, G. (2025). Enhanced Wind Energy Forecasting Using an Extended Long Short-Term Memory Model. *Algorithms 2025, Vol. 18, Page 206, 18*(4), 206. DOI: 10.3390/a18040206

Berger, R. (2024). *Lazard' S Levelized Cost of Energy Analysis—Version 9.0. 24*, 1–3.

Berger, R. (2025). *Lazards LCOE 2025. June.*

Burton, T., Jenkins, N., Sharpe, D., Bossanyi, E., & Graham, M. (2021). *Wind energy handbook.* https://www.wiley.com/en-us/Wind+Energy+Handbook%2C+3rd+Edition-p-9781119451167

Canan, A. (2023). Offshore wind energy policy paths: A comparative analysis of Denmark and Germany. *Economics and Policy of Energy and the Environment*, *1*(1), 35–59. DOI: 10.3280/EFE2023-001003

Djalel, D., Abdallah, G., & Hocine, L. (2014). STUDY OF THE LIGHTNING IMPACT ON THE WIND-TURBINE. *Energy Research Journal*, *5*(1), 17–25. DOI: 10.3844/erjsp.2014.17.25

DOE. U. (2015). *Wind Vision. A New Era for Wind Power in the United States (Executive Summary, Full Report, and Appendices); U.S. Department of Energy (DOE), NREL (National Renewable Energy Laboratory).* DOI: 10.2172/1215051

Elkodama, A., Ismaiel, A., Abdellatif, A., Shaaban, S., Yoshida, S., & Rushdi, M. A. (2023). Control Methods for Horizontal Axis Wind Turbines (HAWT): State-of-the-Art Review. *Energies*, *16*(17), 6394. Advance online publication. DOI: 10.3390/en16176394

Errami, Y., Maaroufi, M., & Ouassaid, M. (2013). Maximum Power Point Tracking of a Wind Power system based on the PMSG using Sliding Mode Direct Torque Control. *Proceedings of 2013 International Renewable and Sustainable Energy Conference, IRSEC 2013*, 218–223. DOI: 10.1109/IRSEC.2013.6529706

Feng, J., & Shen, W. Z. (2017). Design optimization of offshore wind farms with multiple types of wind turbines. *Applied Energy*, *205*, 1283–1297. DOI: 10.1016/j.apenergy.2017.08.107

Ghosh, D., Adhikary, A., Nayek, A. K., & Hazra, S. (2025). Renewable energy: Opportunities, applications, and future scope of artificial intelligence and internet of things. *Next Generation Artificial Intelligence Driven Smart and Renewable Energy*, 1–12. https://doi.org/DOI: 10.1201/9781003628279-1/RENEWABLE-ENERGY-DIPANWITA-GHOSH-ARPAN-ADHIKARY-ASIT-KUMAR-NAYEK-SUNANDA-HAZRA

Gipe, P. (2016). *Wind energy for the rest of us : a comprehensive guide to wind power and how to use it : introducing electricity rebels and how they are changing the face of wind energy*. 560.

Gu, H., Liu, W., Gao, Q., & Zhang, Y. (2021). A review on wind turbines gearbox fault diagnosis methods. *Journal of Vibroengineering*, *23*(1), 26–43. DOI: 10.21595/jve.2020.20178

Hau, E. (2006). Wind turbines: Fundamentals, technologies, application, economics. *Wind Turbines: Fundamentals, Technologies, Application, Economics*, 1–783. DOI: 10.1007/3-540-29284-5

IRENA. (2019). Future of wind: Deployment, investment, technology, grid integration and socio-economic aspects. In *International Renewable Energy Agency*.

Jenkins, N., Burton, T., Bossanyi, E., Sharpe, D., & Graham, M. (2021). *Wind Energy Handbook* (3rd ed.). Wind Energy Handbook., DOI: 10.1002/9781119451143

Kang, J., Schwartz, R., Flickinger, J., & Beriwal, S. (2015). Machine Learning Approaches for Predicting Radiation Therapy Outcomes: A Clinician's Perspective. *International Journal of Radiation Oncology*Biology*Physics, 93*(5), 1127–1135. DOI: 10.1016/j.ijrobp.2015.07.2286

Koetsier, T. (2025). 6 Advances in Historical Studies Windmills in The Netherlands. *Advances in Historical Studies*, *2025*(1), 16–35. DOI: 10.4236/ahs.2025.141002

Land-Based Wind Market Report : 2023 Edition. (2023).

Li, T., Sahu, A. K., Talwalkar, A., & Smith, V. (2020). Federated Learning: Challenges, Methods, and Future Directions. *IEEE Signal Processing Magazine*, *37*(3), 50–60. DOI: 10.1109/MSP.2020.2975749

Loewe, M. (1966). Science and civilisation in China. Vol. 4: Physics and physical technology. Part II: Mechanical engineering. By J. Needham. pp. lv, 759; 158 plates. Cambridge University Press, 1965. £9 10 s. *Journal of the Royal Asiatic Society of Great Britain & Ireland*, *98*(2), 160–163. DOI: 10.1017/S0035869X0012502X

Lundberg, S. M., & Lee, S. I. (2017). *A unified approach to interpreting model predictions* (Vol. 2017). Advances in Neural Information Processing Systems. Decem.

Manwell, J. F., McGowan, J. G., & Rogers, A. L. (2009). *Wind energy explained : theory, design and application*. 689. https://www.wiley.com/en-us/Wind+Energy+Explained%3A+Theory%2C+Design+and+Application%2C+2nd+Edition-p-9781119994367

Mbasso, W. F., Harrison, A., Dagal, I., Jangir, P., Khishe, M., Kotb, H., Shaikh, M. S., Smerat, A., Fendzi Donfack, E., & Kumar, R. (2025). Digital twins in renewable energy systems: A comprehensive review of concepts, applications, and future directions. *Energy Strategy Reviews*, *61*, 101814. Advance online publication. DOI: 10.1016/j.esr.2025.101814

Musial, W., Spitsen, P., Duffy, P., Beiter, P., Shields, M., Mulas Hernando, D., & Sathish, S. (2023). *Offshore Wind Market Report: 2023 Edition (No. NREL/TP-5000-87232). National Renewable Energy Laboratory*. NREL.

Natarajan, A. (2014). An overview of the state of the art technologies for multi-MW scale offshore wind turbines and beyond. *Wiley Interdisciplinary Reviews. Energy and Environment*, *3*(2), 111–121. DOI: 10.1002/wene.80

Okedu, K. E., & Muyeen, S. M. (2022). Comparative Performance of DFIG and PMSG Wind Turbines during Transient State in Weak and Strong Grid Conditions Considering Series Dynamic Braking Resistor. *Energies*, *15*(23), 9228. Advance online publication. DOI: 10.3390/en15239228

Price, T. J. (2005). James Blyth - Britain's first modern wind power pioneer. *Wind Engineering*, *29*(3), 191–200. DOI: 10.1260/030952405774354921

Prime, J., Abdulkadir Ahmed, I., Akande, D., Elhassan, N., Escamilla, G., Jamdade, A., Melnikov, Y., & Whiteman, A. (2024). Renewable Energy Statistics 2024, International Renewable Energy Agency. *International Renewable Energy Agency*, 1–1099. https://www.irena.org/Publications/2024/Jul/Renewable-energy-statistics-2024

Righter, R. W. (1996). *Wind energy in America : a history*. 361. https://wind-works.org/wind-energy-in-america-a-history-by-robert-righter/

Singh, M. B., Jain, P., & Singh, P. (2024). Wind Energy. *Wind Energy Storage and Conversion*, 1–16. DOI: 10.1002/9781394204564.ch1

Subedi, S., Gui, Y., & Xue, Y. (2025). Applications of Data-Driven Dynamic Modeling of Power Converters in Power Systems: An Overview. *IEEE Transactions on Industry Applications*, *61*(2), 2434–2456. DOI: 10.1109/TIA.2025.3529797

Tchakoua, P., Wamkeue, R., Ouhrouche, M., Slaoui-Hasnaoui, F., Tameghe, T. A., & Ekemb, G. (2014). Wind turbine condition monitoring: State-of-the-art review, new trends, and future challenges. *Energies*, *7*(4), 2595–2630. DOI: 10.3390/en7042595

van Est, R. (2022). The Success of Danish Wind Energy Innovation Policy: Combining Visionary Politics and Pragmatic Policymaking. *Successful Public Policy in the Nordic Countries: Cases, Lessons. Challenges*, •••, 45–66. DOI: 10.1093/oso/9780192856296.003.0003

Velandia-Cardenas, C., Vidal, Y., & Pozo, F. (2021). Wind Turbine Fault Detection Using Highly Imbalanced Real SCADA Data. *Energies 2021, Vol. 14, Page 1728*, *14*(6), 1728. DOI: 10.3390/en14061728

Wind, G., & Council, E. (2024). *GWEC | GLOBAL WIND REPORT 2024*.

Wiser, R., Lantz, E., Mai, T., Zayas, J., DeMeo, E., Eugeni, E., Lin-Powers, J., & Tusing, R. (2015). Wind Vision: A New Era for Wind Power in the United States. *The Electricity Journal*, *28*(9), 120–132. DOI: 10.1016/j.tej.2015.09.016

Xiao, C., Liu, Z., Zhang, T., & Zhang, X. (2021). Deep learning method for fault detection of wind turbine converter. *Applied Sciences (Basel, Switzerland)*, *11*(3), 1–22. DOI: 10.3390/app11031280

Yang, Q., Liu, Y., Chen, T., & Tong, Y. (2019). Federated machine learning: Concept and applications. *ACM Transactions on Intelligent Systems and Technology*, *10*(2), 1–19. Advance online publication. DOI: 10.1145/3298981

Yin, X., Zhu, Y., & Hu, J. (2022). A Comprehensive Survey of Privacy-preserving Federated Learning: A Taxonomy, Review, and Future Directions. *ACM Computing Surveys*, *54*(6), 1–36. Advance online publication. DOI: 10.1145/3460427

Zhao, Y., Song, Z., Li, D., Qian, R., & Lin, S. (2024). Wind Turbine Gearbox Fault Diagnosis Based on Multi-Sensor Signals Fusion. *Protection and Control of Modern Power Systems*, *9*(4), 96–109. DOI: 10.23919/PCMP.2023.000241

Zupok, S., Chomać-Pierzecka, E., Dmowski, A., Dyrka, S., & Hordyj, A. (2025). A Review of Key Factors Shaping the Development of the U.S. Wind Energy Market in the Context of Contemporary Challenges. *Energies 2025, Vol. 18, Page 4224, 18*(16), 4224. DOI: 10.3390/en18164224

Chapter 2
The Role of Wind Energy in a Sustainable Energy Transition Agenda

Kübra Nur Şahin
http://orcid.org/0000-0001-9786-6270
Abdullah Gul University, Turkey

ABSTRACT

This chapter examine the technological, environmental, socioeconomic, and governance dimensions of wind energy and its role in the global shift toward low-carbon systems. As nations pursue Paris Agreement and Net-Zero goals, wind power has become a central renewable source due to its low cost, technical maturity, and significant emissions reduction potential. The chapter reviews advancements in turbine design, materials, offshore systems, and digital tools such as forecasting, digital twins, and predictive maintenance. It also examines how wind energy supports sustainable development by creating jobs, enhancing local economies, promoting social inclusion, and contributing to circular-economy practices. Governance structures including market design, permitting, benefit-sharing, environmental safeguards, and grid planning are discussed as key factors shaping deployment success. The chapter concludes by identifying future research needs, highlighting the importance of hybrid renewable systems, digital intelligence, and adaptive policies.

1. INTRODUCTION

With the rapid advancement of modern society, accelerating industrialization, and global population growth, the demand for energy has increased dramatically over the past century. The continuous expansion of consumption has placed unprecedented

DOI: 10.4018/979-8-2600-0106-6.ch002

pressure on global energy systems, which still depend heavily on fossil fuels (Ang et al., 2022). According to the International Energy Agency's Global Energy Review (2025), total energy-related carbon dioxide (CO_2) emissions rose by approximately 0.8% in 2024, the highest annual increase ever recorded (Global Energy Review 2025 – Analysis - IEA, n.d.). Fueled by persistent fossil fuel reliance, this trend poses a critical threat to ecosystems and to global economic and social stability. The rising emissions trajectory highlights the widening imbalance between escalating energy demand and the global objective of achieving net-zero emissions. As conventional energy production deepens environmental degradation and climate vulnerability, transitioning toward sustainable and renewable energy sources has become a defining necessity for ensuring long-term planetary and human well-being.

Despite global efforts such as the Net Zero targets, the United Nations 2030 Agenda, and the Paris Agreement, fossil fuel–based energy sources continue to dominate the global energy production mix (Global Energy Review 2025 – Analysis - IEA, n.d.). The transition toward renewable energy sources, particularly solar, wind, and geothermal energy has therefore become an urgent necessity for reducing greenhouse gas emissions and mitigating climate change. Increasing the proportion of renewables in the global energy mix not only decreases environmental and health impacts but also supports economic growth, fosters social development, and provides a long-term pathway to achieving energy sustainability (Kardooni et al., 2016).

With the pressure of increased concerns about climate change, wind energy has gained attention as a promising alternative. Based on the installatio7n specifications, wind energy production capacity can differ. With the technological improvement in wind turbines and control mechanisms, the adaptation of wind power has been significantly increased all around the world. The global wind generation capacity reach to 113 GW by 2024 (Prime et al., 2025) China, the USA, Germany, India, Brazil, and the UK have the highest wind power capacity as can be seen in Figure 1 (Janet L. Sawin, n.d.). With the growing attention to wind energy all around the world the wind energy becomes one of the biggest contributors to national grids.

Wind energy provides societies with a clean, sustainable, and abundant source of power can contribute to meet growing demand without harming the environment. It contributes significantly to reducing greenhouse gases and mitigating the impacts of climate change since it is not related with direct carbon emission. Besides its benefits on the environment, wind energy has substantial advantages on the economy and society. The development of wind power plants creates employment opportunities in construction, maintenance, and manufacturing sectors and generating revenue streams for local communities and regional authorities. Its increasing cost competitiveness has positioned wind energy as one of the most affordable options for energy generation. Furthermore, by decreasing dependence on imported fossil fuels, wind energy enhances national energy security and promotes greater grid independence, key elements in achieving a resilient and sustainable energy future.

Figure 1. Wind power capacity by country, 2023 and 2024 (GW)

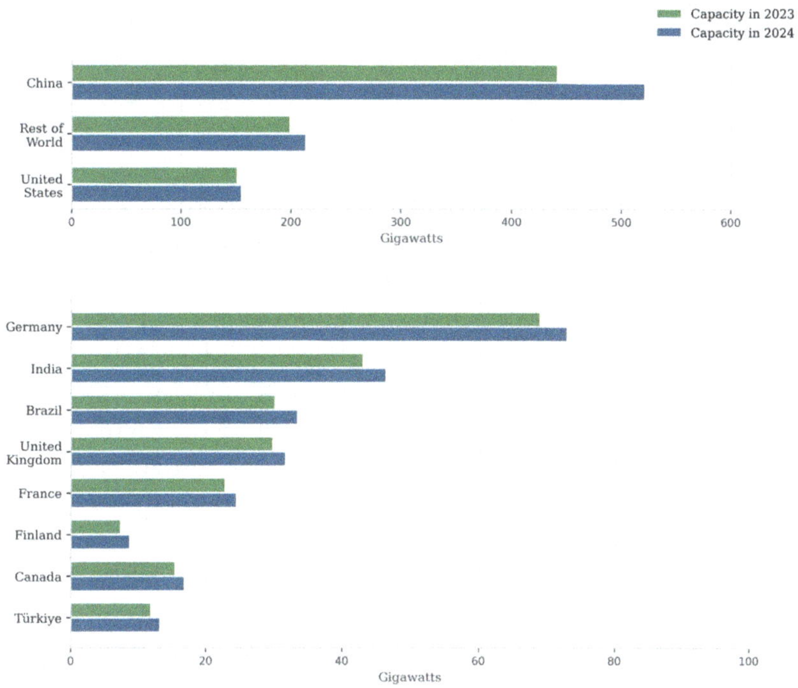

Data source: REN21, Renewables 2025 Global Status Report (2025), based on GWEC, Global Wind Report 2025.

However, this benefit comes with its own challenges. Wind power, due to its nature, is intermittent and highly variable. This variability makes the integration of wind energy into smart grids challenging since this variability can jeopardize grid reliability and stability. Thus, integration of wind energy into smart grids requires advanced forecasting and control strategies to overcome these issues. Since the generation from wind power plants continues at nighttime during the low-energy demand period, smart storage solutions are needed. In addition to that, as wind turbines reach their end of useful life, recycling the waste generated from their blades creates an environmental concern since they are made of composite materials.

Apart from the technological environmental concerns, there are several broader challenges that hinder the large-scale deployment of wind energy. Due to the uneven distribution of wind resources, the potential of this abundant source differs both at the local and national level. In addition to this variability, since wind projects require high initial investment, financing is challenging for developing countries. Moreover, the grid infrastructure of many areas may have limited transmission capacity, especially in offshore regions where the installation and maintenance costs are even higher. Another crucial aspect is societal acceptance, since public opposition may arise due to visual effects, noise, and land-use disputes. The impacts of wind power plants on wildlife, marine ecosystems, and other ecological disruptions adversely affect the adaptation of this technology. Thus, addressing these discrepancies via well-coordinated policy frameworks and national and international cooperation is vital to assure harnessing the potential of wind energy in order to contribute to the global sustainability transition.

In addition to these structural difficulties, digital innovation continues to impact the development of wind energy. Digital-twin environments, sensor-based condition monitoring, AI-assisted forecasting, and data-driven predictive maintenance techniques are revolutionizing the operation of wind power plants. These technologies strengthen the circular-economy concepts highlighted in sustainable development agendas, increase efficiency, and prolong component lifetimes. Additionally, digital intelligence enhances the reliability of renewable energy sources, which have an intermittent generation structure, allowing for improved grid integration and supporting sophisticated market processes such as demand-response coordination and flexibility services.

Building on this foundation, the following sections explore the technological advancements that have influenced the evolution of wind power, assess its contribution to sustainability objectives, and examine the policy mechanisms and future pathways needed to support a resilient and low carbon energy future.

2. TECHNOLOGICAL ADVANCEMENTS IN WIND ENERGY

The utilization of wind energy dates back centuries, initially through mechanical windmills, and has progressively evolved into modern wind-turbine systems capable of generating large-scale electricity. As a cleaner energy resource, it plays a vital role in reducing the negative effects of greenhouse gases, while the share of fossil fuel–based energy production in the energy mix remains significant. The global expansion of wind energy is fundamentally shaped by broader sustainability imperatives driving contemporary energy transitions. As expressed in the IPCC AR6 Synthesis Report (2023), wind energy is one potential contributor to net emission reduction by 2030, since it is expected to reduce global emissions by at least half of the 2019 level (Lee & Romero, n.d.). Besides, political conflicts and economic instabilities, together with the rising cost of fossil fuels, have underscored the urgent need for action for energy security and to avoid external dependency (Khan et al.,

2024). Both onshore and offshore wind have experienced significant cost reductions since 2010, becoming increasingly competitive with fossil-fuel electricity (Khan et al., 2024). Due to its cost effectiveness and efficiency in energy production, wind energy significantly supports decarbonization, continuous energy supply, and a reliable and resilient power system. Technological developments in this sustainable setting demonstrate how wind energy serves as a platform for engineering, environmental, and socio-technical innovation in addition to being a clean power source.

With rapid advancements in technology in recent years, wind energy production technologies have evolved substantially. As a result, the generation capacity of wind turbines has significantly increased, and today their power output is commonly expressed in megawatts (MW). With improved efficiency and adaptability, modern wind turbines can be smoothly integrated into contemporary power systems. Due to the high potential of wind energy, aerodynamic optimization has gained significant attention from both scientists and industry. Blade design is critical for maximizing aerodynamic efficiency and energy production. There are mainly two types of wind turbine designs: horizontal-axis and vertical-axis turbines. Depending on the application purpose and the location, both technologies offer renewable energy solutions, each with advantages that make them more appropriate under different operating conditions. These types of technological advancements in wind turbines significantly increase annual energy production and contribute to lowering the levelized cost of electricity.

Furthermore, innovations in materials research have enhanced the fatigue resistance and structural performance of wind turbines. Longer blades with less mass are made possible by carbon fiber–reinforced polymers (CFRP), glass fiber–reinforced polymers (GFRP), and hybrid composites, which reduce the overall mass of the blade and improve resistance to fatigue stresses and gravitational effects. Structural health monitoring (SHM) systems that use fiber Bragg grating sensors, acoustic emission techniques, and vibration-based diagnostics enable real-time damage assessment and support lifetime extension (Brøndsted et al., 2018).

Power electronics and control systems have also advanced. To maximize energy extraction in unstable environments while preserving structural safety, high-bandwidth pitch and torque control algorithms and computational fluid dynamics and artificial intelligence techniques are used in modern turbines. The application of digital technologies such as simulation engines and coordinated set point optimization contributes to understanding and analyzing the effects of wake flow on turbine energy production performance. Power electronic converter improvements, such as fault ride through and reactive power correction, allow wind power plants to integrate into the grid structure and provide ancillary services (Luo et al., 2014).

Offshore wind technologies have advanced considerably. Improved foundation systems such as monopile XL and jacket structures and advanced installation equipment that decrease construction costs are advantageous for fixed-bottom installations. A game changing invention that makes deployment possible in deep oceans is floating offshore wind technology, which has become a major innovation enabling deployment in previously inaccessible locations. Spar-buoy, semi-submersible, and tension-leg platforms provide cost-competitive generation for offshore wind turbines (Song et al., 2024).

Digitalization is one of the biggest milestones for the integration of wind energy into modern energy systems. Machine-learning-based forecasting models have improved short-term power generation forecasts, thereby reducing instability issues in energy production planning. With digitalization, integrated sensors enable the acquisition of real-time data, which is critical for developing aeroelastic simulations to build digital twins and replicate turbine behavior. Through the analysis of real-time turbine data, predictive maintenance and fault-diagnosis models can be developed. Condition monitoring systems, such as thermographic imaging and vibration diagnostics, provide early fault detection and extend the useful lifetime of components (Civera & Surace, 2022). Thus, integrating digitalization into wind energy systems helps reduce operational costs and increase system reliability.

Technological advancements hold the key to making wind energy a competitive, high-performance, and reliable renewable-energy alternative. To strengthen the contribution of wind power to the global energy transition, from traditional fossil-based systems to cleaner and more sustainable solutions requires sustained progresses in turbine engineering, advanced control mechanisms, offshore deployment, and data driven analytics. These technical developments provide the engineering foundation upon which socio-economic, environmental, and policy evaluations can be built, as discussed in the subsequent sections.

3. THE ROLE OF WIND ENERGY IN SUSTAINABLE DEVELOPMENT

Wind energy has become a vital tool for promoting sustainable development in social, economic, and environmental areas. With the Paris Agreement and Net-zero targets, countries have placed significant attention on their national policy frameworks to achieve long-term sustainability goals. Wind energy not only provides low-carbon electricity production, but its contributions extend far beyond that. It drives new job opportunities, industrial diversification, technological innovation, economic progress, environmental protection, community empowerment, and energy system resilience.

The onshore and offshore wind industry has created job opportunities in manufacturing, engineering, supply-chain logistics, civil works, environmental monitoring, operations and maintenance, and digital services for approximately 1.5 million people globally in 2023. The employment rate in the wind industry was highest in Asia (with a share of 59%) and Europe (22%) (IRENA & ILO, 2025). The geographical diversity in job opportunities fosters economic and social development in rural areas, coastal regions, and port cities that host manufacturing facilities or offshore wind turbines. Through wind energy investments, ports such as Esbjerg in Denmark, Bremerhaven in Germany, and Hull in the United Kingdom have been transforming into significant industrial hubs supporting shipping logistics, turbine assembly, blade manufacturing, and specialized maintenance operations. In Asia, South Korea's Ulsan and Taiwan's Taichung Port have also developed into major centers within the offshore wind supply chain. In developing nations such as Vietnam, Brazil, Morocco, and South Africa, wind energy deployment has been used to foster local manufacturing clusters and attract foreign investment, enabling the domestic production of wind turbines, blades, and nacelle components.

Furthermore, wind energy promotes innovation at various areas. Aerospace engineering, composite manufacturing, robotics, maritime engineering, and advanced analytics are all affected by the developments in aerodynamics, materials science, digital twin platforms, and power electronics. By establishing national test centers, offshore demonstration sites, and simulation laboratories that promote industry-academia partnerships, research institutions and universities contribute significantly to these innovation ecosystems. Denmark, Germany, and the United States are among the nations that have demonstrated long-term innovation-driven competitiveness, resulting in the production of high-value intellectual property and advanced products through early investments in wind research. This innovation capacity is becoming increasingly essential as the global clean energy sector moves toward smart grid integration and hybrid energy systems, which are required for modern energy infrastructures.

The societal influence of wind energy cannot be underestimated since there are multiple interconnected factors which impact how communities perceive and respond to renewable energy development. Community acceptance is a critical concern, as perceptions of fairness, trust, and procedural transparency of wind power plant project create a significant impact on how communities evaluate the projects (Devine-Wright & Wiersma, 2020; Horbaty et al., 2012). Wind energy provides significant economic value to host communities through land lease payments, job opportunities i n various sectors, and community benefit funds, these communities generally want to ensure that financial gains are retained locally and aligned with regional development priorities. Since the host communities are the ones who is adversely affected by the projects. Apart from financial considerations, participatory governance and local ownership models that are widely used in Denmark and Germany and progressively adopted in Scotland, Canada, and New Zealand show that when communities are the part of decision-making processes or influence share in project revenues, the resistance to project decrease and social legitimacy improves (Aitken, 2010; Glasson, 2021; Toke, 2005). Furthermore, distributed wind systems increase local resilience in underserved or rural areas by supplying reliable and reasonably priced electricity and reducing dependency on imported fossil fuels. Wind systems contribute to continuous energy supply and energy reliability, affordable energy prices, and decreasing fossil-fuel dependency. Thus, community resilience especially in rural and underdeveloped areas can be strengthened. When all these aspects are combined, the societal impact of wind energy reaches far beyond local acceptance and becomes one of the key determinants of the success of wind projects.

As stated in the IRENA 2025 report, initiatives within wind energy projects that promote gender equality and the engagement and representation of vulnerable communities increase the social value and fairness of the energy transition, and wind energy also supports broader social justice and inclusion goals (Alfaro-Pelico, 2025). Although gender inequalities exist in the industry, wind projects create technical, engineering, and managerial roles for women. Based on the IRENA 2020 report, 21% of the global wind workforce is represented by women, which is higher than in the fossil-fuel sector (Renewable Energy Agency, 2020). Yet, there is still a need for greater investment to support gender equality. Thus, programs such as "Women in Wind" and the "Women in Wind Global Leadership Program" have been launched to increase the representation of women in the wind energy sector.

One of the biggest drawbacks of the society is the end-of-life management of the wind turbines as they are made of composite materials which are challenging to recycle. Majority of the wind turbines reach their end of lifetime around 20 to 25 years (Tapoglou, 2024). The waste created by the wind turbines is expected to reach 43 million by 2050 (Global Wind Report, 2025). The materials used in the production of turbine blades contribute to acidification potential, global warming potential, eutrophication, and human toxicity potential (Atilgan Turkmen & Germirli Babuna, 2024). To overcome these risks several recycling techniques have been proposed. With mechanical recycling techniques such as shredding and grinding the composite materials used in the turbines can be recycle similar to filler in building materials, polymer goods ad cement manufacturing (Jani et al., 2022). In addition to that there are thermal technologies such as pyrolysis and fluidized bed process that break down the resin matrix at high temperatures (Ren et al., 2026). In that way, the glass and carbon fibers can be recovered while some mechanical properties are preserved. moreover, chemical recycling techniques for example solvolysis uses pressured solvent for recovering premium fibers (Atilgan Turkmen & Germirli Babuna, 2024; Hu et al., 2024). These recycling strategies provide viable solutions for reducing waste, recycling valuable materials, and reducing negative environmental impact of wind turbines.

Beyond end-of-life considerations, wind energy offers substantial environmental advantages. As illustrated in Figure 2, the life-cycle carbon dioxide emissions of wind power are noticeably low, typically ranging between 12–14 g CO_2/kWh, which represents a cleaner energy alternative when compared to conventional fossil fuel based electricity generation and even solar electricity generation. In contrast, especially fossil fuel based power plants emit several hundred grams of CO_2 per kilowatt-hour. This major difference underlines the critical role of wind energy in reducing greenhouse gas emissions and mitigating climate change impacts. With comprehensive and effective recycling and circular-economy strategies for turbine components, wind energy emerges as one of the most environmentally sustainable energy sources for electricity.

Besides the benefits of wind energy, one of its most significant advantages is its contribution to reducing greenhouse gas emissions. Global wind plants are estimated to reduce up to 32,864 Mt of CO_2 by 2100 and decrease global surface temperature by 0.64°C, which supports the goals of the Paris Agreement (Elkelawy et al., n.d.). With emission reductions, the levels of air pollutants also decrease, which is closely related to lower risks of cardiovascular and respiratory diseases. The environmental benefits of wind energy go beyond enhanced decarbonization, contributing to improved air quality, public health, and long-term ecological sustainability.

Figure 2. Estimated carbon footprints of energy sources

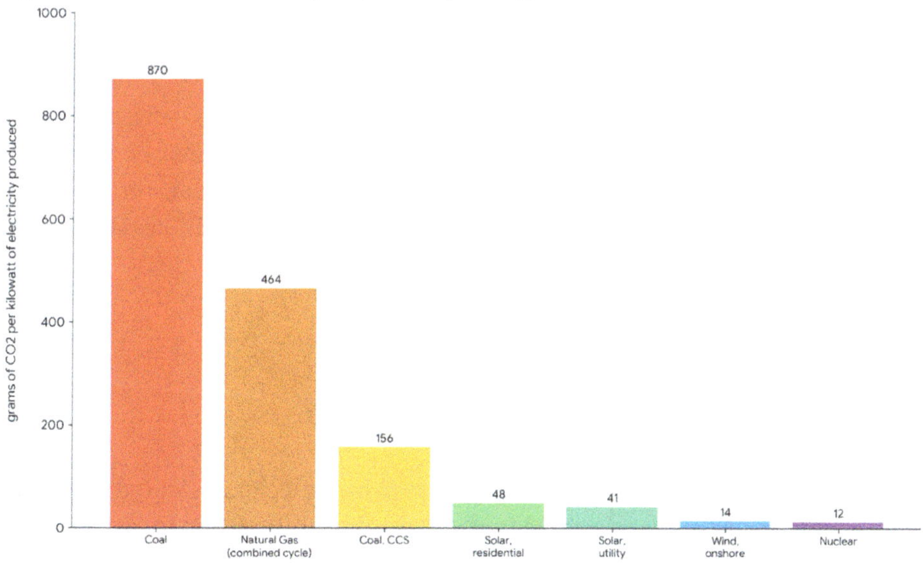

(Source: Rhodes, 2017).

The ecological impact of wind energy has raised serious concerns and requires modern mitigation strategies as the deployment of wind turbines expands. During high risk periods for wildlife especially birds and bats smart detection and curtailment strategies, radar-assisted shutdown systems, and camera-based detection technologies are used to limit turbine operations. In the installation phase, siting regulations protect ecologically sensitive areas to avoid harm to wildlife. These strategies have been shown to effectively reduce fatalities, with only a small reduction in annual energy production (Smallwood & Bell, 2020; Voigt et al., 2024). Thus, monitoring protocols and adaptive management frameworks are becoming an inevitable part of modern wind projects to reduce the ecological risks associated with them.

Apart from environmental and social benefits, with integration of wind energy as renewable energy source into energy system strengthens reliance, reliability flexibility. With storage technologies and integration of combination of other types of renewable energy sources such as solar power enables wind energy to enhance stability of energy supply. Integration of wind energy penetration into grid systems reduce balancing constraint, enhance system performance and improve resilience when digital technologies including AI-based forecasting, demand response system, smart inverters and virtual power plants implemented. With these technologies, wind energy can be integrated into complex and dynamic energy systems smoothly and support national and international agendas. for sustainable development.

Overall, economic, societal, environmental and systemic contributions emphasize the essential role of wind energy to multidimensional driver for sustainable development. Wind energy support industrial expansion, employment, community engagement, circular economy ecological protection, social justice and. reliance of modern energy systems. these diversified and profound advantages indicate that wind energy is a transforming instrument, capable of shaping economies, empowering communities, restoring ecosystems, and improving energy systems for the future. Harnessing this potential, however, involves solid governance and a sustained policy commitment, which are discussed in the next section.

4. GOVERNANCE, MARKET DESIGN, AND POLICY FRAMEWORKS IN WIND ENERGY

Large spread of wind energy is not only Driven by the technological advancement and social acceptance. It is also shaped governance structure, regulatory policy frameworks and market design that affect investment decisions of the wind projects. For development of wind projects long-term cooperation among governments, industry players, grid service providers, financial services providers, and communities at large are required. Policies and regulations are essential determinants of the pace, equity, and resilience of the wind energy transition (Wojtaszek et al., 2025a). When policy violations, complicated bureaucratic processes, or unstable market structures exist in an environment, delayed projects, investor reluctance, and higher capital costs are highly likely to be experienced.

In the past, wind energy governance relied on policy design to stimulate investment, decrease associated risks, and address market failures. In the early phases of wind deployment especially in Europe feed-in tariffs were introduced to offer favorable prices for electricity generated from renewable sources and to encourage renewable energy deployment. This tariff structure provided revenue certainty, and manufacturers were encouraged to expand energy generation from renewables while minimizing their costs. Yet, as the number of manufacturers increased, the transition moved toward more competitive market mechanisms, which shaped the modern global wind market.

Policies that accelerate authorization processes, incorporate environmental protections, and provide smoother grid access are essential for wind development in addition to price-based support mechanisms. As wind energy penetration increases, well-structured grid integration regulations become essential. Many countries have implemented priority dispatch rules for renewable energy sources. Wind turbines are increasingly capable of providing ancillary services that support grid reliability and stability. Grid code requirements have been developed to facilitate and promote these services. Permitting procedures additionally must find a balance between deployment pace and environmental protection (Gao et al., 2025; Wretling et al., n.d.). Delays in wind projects raise costs and slow down progress in climate action. Streamlined permitting frameworks supported by ecological exclusion criteria, pre-identified renewable energy zones, and digital mapping tools have proven to be effective approaches.

Benefits and responsibilities have to be shared fairly among national authorities and local communities. The distribution of these benefits and responsibilities is determined by governance frameworks. One of the critical components of these frameworks is spatial planning, which determines the location of wind turbines, regulates land-use conflicts, and integrates public opinion into project planning. For example, the Danish Planning Act requires local authorities to designate special zones for wind farms. The German Federal Spatial Planning Act identifies priority and restricted zones to balance environmental protection with energy production objectives (Westholm, 2025; Wojtaszek et al., 2025b). Similarly, to prevent potential conflicts among fishers, shipping, and marine wildlife conservation in offshore wind development, the United Kingdom uses a marine spatial planning system. These types of planning actions are essential components of governance frameworks to ensure that the wind deployment process aligns with environmental, societal, and financial goals.

Power market reform and long-term grid planning are vital components of wind governance. Wind turbines especially offshore installations often require upgrades to transmission infrastructure. The EnLAG network expansion plan in Germany, the ultra–high-voltage transmission strategy in China, and the offshore transmission owner regime in the United Kingdom are examples of proactive grid planning frameworks (Hiorns et al., 2010; Huang et al., 2009; Reitz et al., n.d.). These planning frameworks have enabled the integration of large volumes of wind-generated energy into national energy systems. On the other hand, insufficient transmission capacity and inconsistent planning may result in interruptions in energy supply, inefficiencies, and financial losses. As intermittent energy production from renewable energy sources increases, electricity markets should incentivize adaptability, storage solutions, and demand response mechanisms as the penetration of renewables in energy systems rises. In order to integrate wind energy smoothly into national energy systems, market capacity, flexibility, and ancillary services play key roles.

For wind governance, financial policy and investment risk management cannot be understated. Large-scale wind projects need substantial capital investment. A stable regulatory environment, predictable project timelines, and transparent contracting protocols are prerequisites for accessing affordable finance for wind projects. Emerging markets increasingly rely on blended-finance models to mitigate the risks of early-stage projects and attract private capital. In such circumstances, solid financial governance becomes essential to wind deployment success, allowing nations to expand renewable energy sources more rapidly while maintaining investor trust.

In conclusion, wind energy is at the forefront of the global sustainability transition, as it offers a unique combination of socioeconomic value, ecological benefits, and technological advancement. The combination of engineering innovation with responsible governance, transparency, and long-term policy stability will be fundamental to its continued success. Wind energy may go beyond its function as a renewable power source and become a key component of resilient, reliable, and low-carbon energy systems by integrating digital intelligence and enabling sustainable and equitable development. Moreover, wind energy has the potential to significantly influence the development of a more trustworthy and sustainable global energy future through comprehensive institutional frameworks and continuous international cooperation.

5. BARRIERS TO FASTER WIND ENERGY DEPLOYMENT

With technological advancements and the innovations, cost reduction and establishment of energy agencies, wind energy has reached a maturity stage over the last decade. It has become a cost competitive alternative energy source to fossil fuels. Therefore, wind energy has emerged as one of the leading renewable energy technologies in terms of its total installed capacity after hydropower (Renewable Energy Agency, 2019). Based on the report of the IRENA, projections indicate an accelerated growth in global installed wind capacity driven by ongoing logistic improvements, cost reductions and enhanced infrastructure and support of the policy frameworks and financial incentives.

Figure 3. Global offshore wind cumulative installed capacity and projections, 2000–2050 (GW)

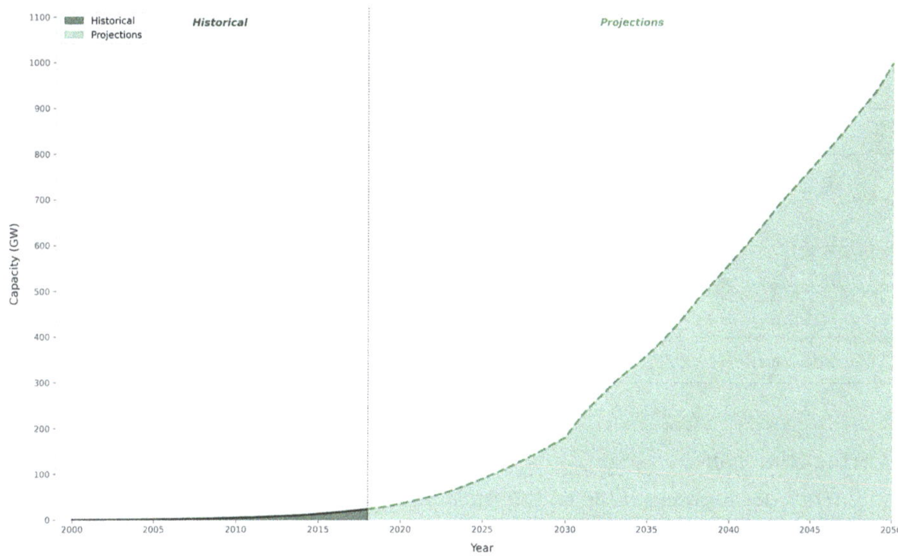

(Adapted from IRENA, 2019)

As illustrated in Figure 3, the projection of cumulative installed offshore wind capacity is expected to reach 1000 GW by 2050 with increasing turbine capacities and deployment of large-scale offshore wind projects. In addition, Figure 4 shows that the cumulative installed onshore wind capacity presents sustained growth, and it is expected to reach approximately 5044 GW by 2050. The estimated values highlight the roles of offshore and onshore wind technologies in supporting the global energy agenda and achieving decarbonize future.

Figure 4. Global onshore wind cumulative installed capacity: historical growth (2000–2018) and projected expansion (2019–2050) in gigawatts

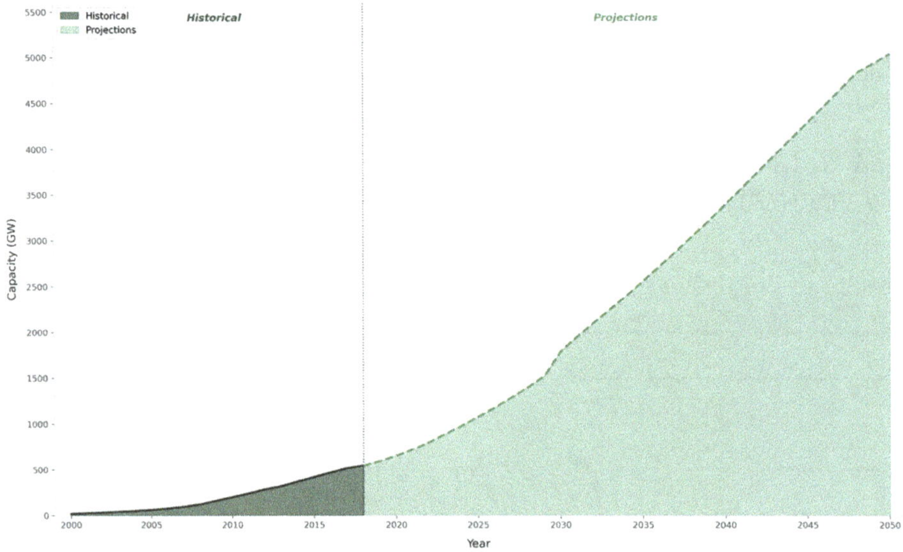

(Adapted from IRENA, 2019)

On the other hand, wind projects often confront with serious barriers accelerated wind energy deployment. One of the most significant barriers to accelerated wind energy deployment is the inadequacy of grid connection and integration and lack of supporting distribution infrastructure. Transmission challenges particularly for offshore wind projects, insufficient grid capacity, limited connections, and ineffective grid reinforcement planning makes the integration of variable renewable energy sources challenging. Overall, grid and transmission bottlenecks represent systemic barriers that diminish the economic advantages of wind integrated energy systems.

Regulatory and policy challenges further diminished the deployment for the wind projects due to complex and outdated regulatory frameworks and scarce financial policy support. Inconsistent assessment requirements, and complicated governance structures often cause project delays or cancellations which create uncertainty for investors. These regulatory inefficiencies contribute to slower deployment and reduce the ability of wind energy to comply with climate goals. The frameworks should support deployment of the projects in order to increase energy accessibility and reduce energy cost. It is important to understand that the wind power can be catalyst of sustainable economic and social development with brooder polices and frameworks.

Financial barriers also have a critical role for large scale wind projects. Even though the levelized cost of wind power has been decrease over the last decade the wind project still requires high initial investment. Since the energy market is highly volatile, gathering finance is generally difficult.

In addition to that, social acceptance and spatial planning is one of the obstacles for deployment of wind projects. Land use conflict, environmental concerns and the visual impacts of wind turbines may create local discomfort. Even though general public opinion support renewable energy integration, localized resistance may emerge in the absence of effective benefit sharing mechanisms.

Overall, infrastructural, social, political inconsistences between climate goals and implementation strategies hinder wind energy deployment. Stable, long-term policy frameworks are therefore required for enabling wind energy to fulfill its central role in the global energy transition. Overcoming these barriers through proactive grid investment, regulatory reform, financial risk mitigation, and social engagement is indispensable to utilize the complete economic and environmental benefits of wind energy.

6. FUTURE PATHWAYS AND RESEARCH OUTLOOK

The dynamic intersection of technological innovation, evolving digital technologies, and emerging societal and environmental needs will determine the future of wind energy. Ongoing advances in turbine engineering—such as larger rotors, increased hub heights, mobile offshore platforms, and advanced recyclable composite materials—will enhance the technological and geographical capabilities of wind energy while reducing costs and adverse environmental effects. As the share of renewable energy increases in the energy production mix, ongoing advancements in digital intelligence, including digital twins, AI-enhanced forecasting, advanced supervisory control, and data-driven predictive maintenance, will make it possible for wind farms to operate more efficiently and integrate more smoothly into energy systems.

Environmental sustainability and societal legitimacy increasingly shape future energy research agendas. Developing ecological monitoring technologies and circular-economy strategies for component recycling will support global wind capacity expansion. Ensuring fair benefit-sharing, community engagement, and inclusive workforce participation will strengthen public trust and the continued adoption of wind energy as countries move toward a decarbonized future. These aspects influence wind project design, spatial planning, and governance frameworks, making them central to the long-term deployment of wind energy.

In the future, wind energy will be closely linked to new infrastructures and technologies. Reliability will increase, and system-balancing issues will be reduced through hybrid renewable energy systems that combine solar, wind, storage, and flexible demand. Local power flows will be transformed by offshore energy hubs and multi-terminal HVDC networks. The integration of wind energy into energy systems will create new opportunities for decarbonizing transportation, industry, and heating. These changes will require interdisciplinary research that integrates engineering, economics, environmental science, and social policy.

With the ambitious global decarbonization targets, wind energy holds a key role in the global energy transition. Comprehending the potential of wind energy depends on following innovations supported by robust policy frameworks, good governance, investment strategies, and community involvement in decision-making processes. Wind energy may serve as a driving force for constructing a resilient, affordable, and decarbonized future.

Publisher Note

Correction Notice: A correction to this manuscript has been published:

The author requested that the figures be replaced because of ethical and copyright concerns. One image was removed while three others were replaced.

Correction Date: [March 20, 2026]

REFERENCES

Aitken, M. (2010). Wind power and community benefits: Challenges and opportunities. *Energy Policy, 38*(10), 6066–6075. DOI: 10.1016/j.enpol.2010.05.062

Alfaro-Pelico, R. (2025). *A just energy transition for communities large-scale wind and solar projects in sub-Saharan Africa The report benefited from the review and inputs of IRENA colleagues Babucarr Bittaye, Mirjam Reiner, Ntsebo Sephelane.* www.irena.org

Ang, T. Z., Salem, M., Kamarol, M., Das, H. S., Nazari, M. A., & Prabaharan, N. (2022). A comprehensive study of renewable energy sources: Classifications, challenges and suggestions. *Energy Strategy Reviews, 43*, 100939. DOI: 10.1016/j.esr.2022.100939

Atilgan Turkmen, B., & Germirli Babuna, F. (2024). Life Cycle Environmental Impacts of Wind Turbines: A Path to Sustainability with Challenges. *Sustainability 2024, Vol. 16, Page 5365, 16*(13), 5365. DOI: 10.3390/su16135365

Brøndsted, P., Lilholt, H., & Lystrup, A. (2018). Composite materials for wind power turbine blades. *Renewable Energy: Four Volume Set, 3–4*, 182–216. DOI: 10.4324/9781315793245-94/COMPOSITE-MATERIALS-WIND-POWER-TURBINE-BLADES-POVL-BR

Civera, M., & Surace, C. (2022). Non-Destructive Techniques for the Condition and Structural Health Monitoring of Wind Turbines: A Literature Review of the Last 20 Years. *Sensors 2022, Vol. 22, Page 1627, 22*(4), 1627. DOI: 10.3390/s22041627

Devine-Wright, P., & Wiersma, B. (2020). Understanding community acceptance of a potential offshore wind energy project in different locations: An island-based analysis of 'place-technology fit.'. *Energy Policy, 137*, 111086. DOI: 10.1016/j.enpol.2019.111086

Elkelawy, M., Wael, N., Alm-Eldin Bastawissi, H., & Seleem, H. (n.d.). Cutting-Edge Innovations in Wind Power: Enhancing Efficiency, Sustainability, and Grid Integration. [PESJ]. *Pharos Engineering Science Journal, 02*, 2025. DOI: 10.21608/pesj.2025.369017.1020

Gao, X., Zhou, S., & Luo, T. (2025). Navigating wind energy siting in the U.S.: The role of permitting and siting regulations and public engagement. *Energy Policy, 205*, 114692. DOI: 10.1016/j.enpol.2025.114692

Glasson, J. (2021). Community Benefits and UK Offshore Wind Farms: Evolving Convergence in a Divergent Practice. Https://Doi.Org/10.1142/S1464333221500010, *22*(1–2). DOI: 10.1142/S1464333221500010

Global Energy Review 2025 – Analysis - IEA. (n.d.). Retrieved October 27, 2025, from https://www.iea.org/reports/global-energy-review-2025

Global Wind Report. (2025). https://www.gwec.net/events/2025-global-wind-report-launch

Hiorns, A., Smith, R., & Wright, D. (2010). An integrated approach to offshore electricity transmission. *IET Conference Publications, 2010*(570 CP). DOI: 10.1049/cp.2010.0977

Horbaty, R., Huber, S., & Ellis, G. (2012). Large-scale wind deployment, social acceptance. *Wiley Interdisciplinary Reviews. Energy and Environment, 1*(2), 194–205. DOI: 10.1002/wene.9

Hu, Y., Zhang, Y., Li, Y., Wang, Y., Li, G., & Liu, X. (2024). Wind turbine blade recycling: A review of the recovery and high-value utilization of decommissioned wind turbine blades. *Resources, Conservation and Recycling, 210*, 107813. DOI: 10.1016/j.resconrec.2024.107813

Huang, D., Shuu, Y., Ruan, J., & Hu, Y. (2009). Ultra high voltage transmission in China: Developments, current status and future prospects. *Proceedings of the IEEE, 97*(3), 555–583. DOI: 10.1109/JPROC.2009.2013613

IRENA (2019). *Future of Wind: Deployment, Investment, Technology, Grid Integration and Socio-Economic Aspects*. International Renewable Energy Agency, Abu Dhabi. Available at: https://www.irena.org/publications/2019/oct/future-of-wind

IRENA & ILO. (2025). *Renewable energy and jobs: Annual review 2024*. www.irena.org

Janet, L. Sawin. (n.d.). *REN21. 2025. Renewables 2025 Global Status Report Collection*. Retrieved October 28, 2025, from https://www.ren21.net/gsr-2025/technologies/wind-power/

Jani, H. K., Singh Kachhwaha, S., Nagababu, G., & Das, A. (2022). A brief review on recycling and reuse of wind turbine blade materials. *Materials Today: Proceedings*, *62*(P13), 7124–7130. DOI: 10.1016/j.matpr.2022.02.049

Kardooni, R., Yusoff, S. B., & Kari, F. B. (2016). Renewable energy technology acceptance in Peninsular Malaysia. *Energy Policy*, *88*, 1–10. DOI: 10.1016/j.enpol.2015.10.005

Khan, K., Khurshid, A., Cifuentes-Faura, J., & Xianjun, D. (2024). Does renewable energy development enhance energy security? *Utilities Policy*, *87*, 101725. DOI: 10.1016/j.jup.2024.101725

Lee, H., & Romero, J. (n.d.). *Climate Change 2023 Synthesis Report IPCC, 2023: Sections. In: Climate Change 2023: Synthesis Report. Contribution of Working Groups I, II and III to the Sixth Assessment Report of the Intergovernmental Panel on Climate Change [Core Writing Team*. 35–115. DOI: 10.59327/IPCC/AR6-9789291691647

Luo, N., Vidal, Y., & Acho, L. (2014). *Wind Turbine Control and Monitoring*. DOI: 10.1007/978-3-319-08413-8

Prime, J., Ahmed, I. A., Akande, D., Elhassan, N., Melnikov, Y., & Whiteman, A. (2025). *Renewable capacity statistics 2025 statistiques de capacité renouvelable 2025 estadísticas de capacidad renovable 2025 About IRENA*. https://www.irena.org/-/media/Files/IRENA/Agency/Publication/2025/Mar/IRENA_DAT_RE_Capacity_Statistics_2025.pdf

Reitz, S., Goshen, L., & Ohlhorst, D. (n.d.). *Trade-offs in German wind energy expansion: building bridges between different interests, values and priorities*. DOI: 10.1186/s13705-022-00365-1

REN21. (2025). *Renewables 2025 Global Status Report Collection*. Available at:https://www.ren21.net/gsr-2025/technologies/wind-power/

Ren, K., Liu, S., Kong, S., Ma, S., & Feng, R. (2026). Thermal recycling of wind turbine blade composites: Optimizing pyrolysis and oxidation for fiber recovery. *Composites. Part A, Applied Science and Manufacturing*, *202*, 109460. DOI: 10.1016/j.compositesa.2025.109460

Renewable Energy Agency. I. (2019). *Renewable Energy Statistics 2019*. www.irena.org

Renewable Energy Agency. I. (2020). *Wind energy: A gender perspective*. www .irena.org

Renewables 2025 Global Status Report. (2025). https://www.ren21.net/gsr-2025/figures/figure_W_2/

Rhodes, J.D. (2017). *Nuclear and Wind Power Estimated to Have Lowest Levelized CO Emissions*. Energy Institute, The University of Texas at Austin. Available at: https://energy.utexas.edu/news/nuclear-and-wind-power-estimated-have-lowest-levelized-co2-emissions

Smallwood, K. S., & Bell, D. A. (2020). Effects of Wind Turbine Curtailment on Bird and Bat Fatalities. *The Journal of Wildlife Management*, *84*(4), 685–696. DOI: 10.1002/jwmg.21844

Song, D., Fan, T., Li, Q., & Joo, Y. H. (2024). Advances in Offshore Wind. *Journal of Marine Science and Engineering 2024, Vol. 12, Page 359, 12*(2), 359. DOI: 10.3390/jmse12020359

Tapoglou, E. (2024). *Wind energy circularity challenges*. European Commission.

Toke, D. (2005). Will the government catch the wind? *The Political Quarterly*, *76*(1), 48–56. DOI: 10.1111/j.1467-923X.2005.00655.x

Voigt, C. C., Bernard, E., Huang, J. C. C., Frick, W. F., Kerbiriou, C., MacEwan, K., Mathews, F., Rodríguez-Durán, A., Scholz, C., Webala, P. W., Welbergen, J., & Whitby, M. (2024). Toward solving the global green–green dilemma between wind energy production and bat conservation. *Bioscience*, *74*(4), 240–252. DOI: 10.1093/biosci/biae023 PMID: 38720909

Westholm, A. (2025). *Ocean Development & International Law The Role of Planning in Offshore Wind Power Deployment*. DOI: 10.1080/00908320.2025.2558538

Wojtaszek, H., Borowski, P. F., Handschke, M., Miciuła, I., Stecyk, A., Bielawa, A., Ozdyk, S., Kowalczyk, A., & Czepło, F. (2025a). Wind Energy in Transition: Development, Socio-Economic Impacts, and Policy Challenges in Europe. *Energies 2025, Vol. 18, Page 2811, 18*(11), 2811. DOI: 10.3390/en18112811

Wojtaszek, H., Borowski, P. F., Handschke, M., Miciuła, I., Stecyk, A., Bielawa, A., Ozdyk, S., Kowalczyk, A., & Czepło, F. (2025b). Wind Energy in Transition: Development, Socio-Economic Impacts, and Policy Challenges in Europe. *Energies 2025, Vol. 18, Page 2811, 18*(11), 2811. DOI: 10.3390/en18112811

Wretling, V., Balfors, B., & Mörtberg, U. (n.d.). *Balancing wind power deployment and sustainability objectives in Swedish planning and permitting*. DOI: 10.1186/s13705-022-00376-y

Chapter 3
Advanced Power Converters in Wind Energy Systems:
Topologies, Control Strategies, and Grid Integration

Sutanu Chatterjee
http://orcid.org/0000-0002-7888-5764
ITMO University, Russia & Harbin Engineering University, China

Yue Zhao
http://orcid.org/0009-0003-6800-3849
Harbin Engineering University, China

Qingyi Zhou
http://orcid.org/0009-0007-6453-0173
Harbin Engineering University, China

Dmitry Lukichev
ITMO University, Russia

Aleksandr Mamatov
http://orcid.org/0000-0002-1887-3059
ITMO University, Russia

Galina Demidova
ITMO University, Russia

ABSTRACT

The integration of wind energy into electrical grids is essential for a sustainable energy future and relies on advanced power electronic converters. These converters provide the interface between variable-speed wind turbines and modern grid requirements. This chapter examines power converter topologies and control strategies for wind energy systems, outlining the transition from early fixed-speed turbines to fully rated converter-based designs used in today's multi-megawatt installations. It reviews key converter architectures, including two-level and three-level Neutral-Point Clamped (NPC) Voltage Source Converters (VSCs), and highlights the emerging

DOI: 10.4018/979-8-2600-0106-6.ch003

use of modular multilevel converters (MMCs) in high-power applications. The chapter also discusses control strategies for machine-side and grid-side converters, covering maximum power point tracking (MPPT), vector control, fault ride-through capability, reactive power support, and power quality considerations, emphasizing the role of power converters in reliable grid integration.

1. INTRODUCTION

With advancement of power converters, modern wind turbines have changed from simple fixed-speed units to advanced multi-megawatt plants that feed power into transmission grids. Power converters allow turbines to run at variable speeds. They also control active and reactive power. In addition, they help grid stability with voltage support, fault ride-through, and services like frequency control and synthetic inertia.

This chapter examines these key technologies. It covers system architectures and converter types, then control methods and grid rules. The text compares partial-scale converters in doubly fed induction generators (DFIGs) systems with full-scale converters in permanent magnet synchronous generators (PMSG) setups. It notes their differences in cost, efficiency, and performance. Next, it reviews multilevel topologies that cut losses and improve power quality. Finally, it shows how control strategies make wind farms active grid supporters that meet current standards.

2. THE ROLE OF POWER ELECTRONICS IN MODERN WIND TURBINES

Modern wind turbines rely on power electronic converters to decouple the mechanical dynamics of the rotor from the electrical requirements of the grid, enabling variable-speed operation, better energy capture, and flexible provision of ancillary services (Muyeen, 2012; Polinder et al., 2006). Compared with early fixed-speed, directly grid-connected induction generators, converter systems can actively regulate power, voltage, and frequency, which is crucial as wind penetration increases in low-inertia grids (IRENA, 2019; Polinder et al., 2006).

3. SYSTEM ARCHITECTURES FOR VARIABLE-SPEED WIND TURBINES

Partial-Scale Converters (PSCs)

Variable-speed wind turbines typically adopt either partial-scale back-to-back converters in DFIG-based systems or full-scale converters interfacing synchronous or induction generators, each with distinct cost, efficiency, and control characteristics (Desalegn et al., 2022; Okedu & Muyeen, 2022; Polinder et al., 2006). The choice of architecture strongly influences drivetrain design, maintenance strategy, and ability to meet stringent grid codes (IRENA, 2019; Okedu & Muyeen, 2022; Polinder et al., 2006).

PSCs such as DFIG-based systems use slip-ring induction machines with the stator directly connected to the grid and the rotor connected via a partial-rated back-to-back converter, typically 20–30% of machine rating (Desalegn et al., 2022; Polinder et al., 2006). Advantages include reduced converter cost and losses, mature technology, and good efficiency near rated power, while drawbacks include slip-ring maintenance and more complex LVRT protection because the stator is directly exposed to grid faults (Desalegn et al., 2022; Hansen & Michalke, 2007; Polinder et al., 2006).

DFIG-based wind turbines face particular challenges during three-phase grid faults due to direct stator-grid coupling. Advanced FRT techniques combining crowbar protection, series dynamic braking resistors, and coordinated converter control enable DFIG systems to remain connected during severe voltage sags while limiting rotor overcurrents and supporting grid voltage recovery through reactive current injection (Makolo et al., 2018).

Full-Scale Converter Systems (FSCs)

FSCs systems with PMSG, wound-rotor synchronous generators (WRSG), or squirrel-cage induction generators (SCIG) connect the machine entirely through back-to-back voltage source converters (VSCs), often with medium-speed or direct-drive drivetrains (Desalegn et al., 2022; Muyeen, 2012; Okedu & Muyeen, 2022). These systems decouple the generator from the grid, offer a wider speed range, enhance FRT and reactive power support, and avoid slip-rings, at the expense of higher converter cost, larger filters, and thermal management challenges at multi-MW scale (Desalegn et al., 2022; Okedu & Muyeen, 2022; Polinder et al., 2006).

The comparative analysis of PSCs and FSCs systems are as in Table 2.1 (Desalegn et al., 2022; Okedu & Muyeen, 2022; Polinder et al., 2006). DFIG-based architectures remain competitive onshore where lower initial investment and proven designs are prioritized, especially in markets with established LVRT solutions and mature grid code compliance (Desalegn et al., 2022; Hansen & Michalke, 2007; Okedu & Muyeen, 2022; Polinder et al., 2006)

Table 1. Comparative analysis of partial and full-scale converter systems

Architecture	Generators	Converter Rating	Cost	Efficiency	Performance	Key Advantages	Key Drawbacks
PSCs	DFIG	20-30% machine rating	Lower initial investment	92-95% near rated power	Good onshore, mature LVRT	Reduced losses, proven designs	Slip-ring maintenance, fault exposure
FSCs	PMSG, SCIG, WRSG	100% machine rating	Higher upfront (20-30% more)	94-97%, wider speed range	Superior offshore, FRT/reactive support	No slip-rings, grid decoupling	Larger filters, thermal challenges

4. ADVANCED POWER CONVERTER TOPOLOGIES

Two-Level Voltage Source Converters (2L-VSC)

2L-VSCs are still widely deployed in wind turbines due to their simplicity, robust control, and compatibility with standard IGBT or SiC devices at medium voltage and power levels (Muyeen, 2012; Polinder et al., 2006). As turbine ratings increase and grid codes require better harmonic performance and dynamic support, multilevel and modular topologies are increasingly adopted to reduce switching losses, improve waveform quality, and enable higher voltage operation (Muyeen, 2012; Polinder et al., 2006; Teodorescu et al., 2011).

2L-VSCs use pulse-width modulation (PWM) to synthesize AC waveforms from a DC link, which requires sizable filters to meet harmonic limits at high power (Muyeen, 2012; Polinder et al., 2006). Their straightforward structure simplifies protection and fault management but can become less efficient or impractical at very high voltages due to device stress and switching losses (Polinder et al., 2006; Teodorescu et al., 2011).

A typical 2L-VSC as shown in Figure 3.1 consists of a DC power source with a DC-link capacitor forming the DC bus voltage V_{DC}, a three-phase six-switch bridge, it should be noted that the upper and lower switches of the same phase leg must never conduct simultaneously; otherwise, a DC bus short circuit (shoot-through) will occur. Therefore, complementary switching with an appropriate dead time, t_d is mandatory.

V_{DC}, which is split by two series DC-link capacitors C1 creating a midpoint N (each capacitor ideally at $V_{DC}/2$); the inverter has three legs with power switches T1~T6 (driven by gate signals g1~g6) and anti-parallel freewheeling diodes D1~D6, generating switching nodes A, B, and C that supply phase currents i_A, i_B, and i_C to a Y-connected three-phase load whose neutral is n; the load phase-to-neutral voltages are v_{An}, v_{Bn}, and v_{Cn}, referenced to the star point n, while O denotes the negative DC bus/return reference.

The 2L-VSC synthesizes the desired AC voltage or current fundamental component within each switching period Ts, while shifting harmonic components toward the switching frequency range for filtering. In practice, sinusoidal PWM (SPWM) or space vector PWM (SVPWM) is commonly adopted to map continuous voltage references into discrete switching states (Chattejee et al., 2023; Hazra et al., 2016).

Figure 1. Typical two-level voltage source converter topologies.

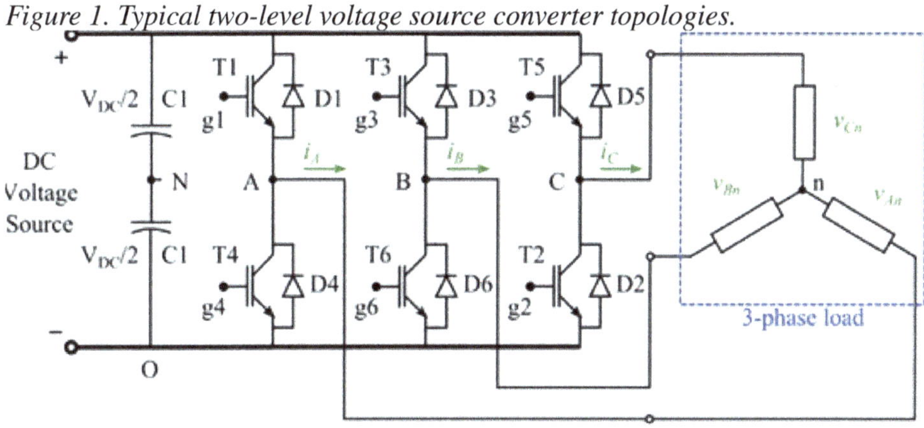

For modeling purposes, the switching function of the upper device in each phase leg is defined as

$$S_x \in \{0, 1\}$$

0 means that the upper is turned off and the lower is turned on; 1 means that the upper is turned on and the lower is turned off.

With three phases, the converter has eight possible switching states Figure 3.2, among which six generate non-zero-line voltages (active vectors), while two states (0,0,0) and (1,1,1) produce zero-line voltage (zero vectors). This vector set forms the basis of SVPWM. Compared with SPWM, SVPWM offers improved DC bus voltage utilization and well-defined harmonic characteristics, which are advantageous for grid-connected wind power converters under voltage variations and fault ride-through conditions (Hazra et al., 2016).

Figure 2. Reference vector generation in different sectors of space-vector hexagon.

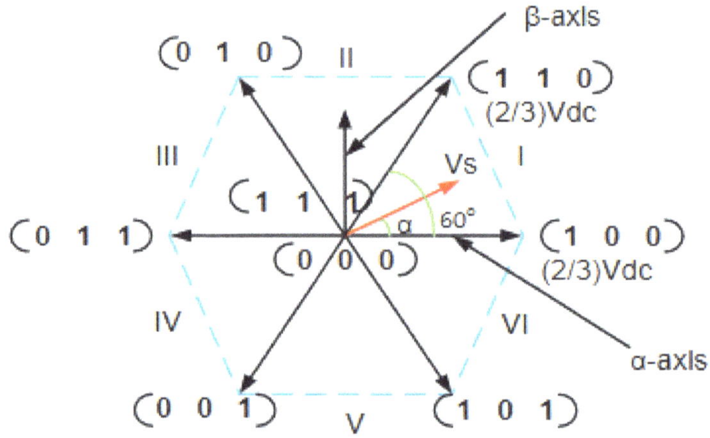

Despite its structural simplicity and ease of modulation, the 2L-VSC exhibits pronounced switching-frequency harmonics and high dv/dt. Grid-side filters are therefore required, introducing resonance and additional damping requirements in control design. Furthermore, high dv/dt and common-mode voltage transitions can cause EMI, insulation stress, and bearing currents, especially in high-voltage or long-cable applications (Hazra et al., 2016). Non-ideal effects such as dead time and device freewheeling further degrade low-frequency waveform quality. As a result, dead-time compensation and careful gate-drive and switching-sequence design are essential in high-performance wind power converter systems.

Multilevel Converters: NPC, Flying Capacitor, Cascaded H-Bridge

Multilevel converters such as diode NPC, flying-capacitor, and cascaded H-bridge (CHB) structures generate stepped voltage waveforms that significantly reduce dv/dt and total harmonic distortion, easing filter design (Blaabjerg et al., 2024; Polinder et al., 2006; Teodorescu et al., 2011).

The primary objective of multilevel converters is to generate multiple discrete voltage levels on the AC side, thereby improving waveform quality. Compared with two-level converters, multilevel topologies reduce device voltage stress, lower dv/dt, and improve harmonic performance, which is particularly beneficial for medium- and high-voltage, high-power applications.

Neutral-Point-Clamped (NPC) Converter

The three-level NPC converter is the most widely used multilevel topology (Rodríguez et al., 2007). Its DC side consists of two series-connected capacitors forming a neutral point, while each phase leg employs four active switches and clamping devices to generate three voltage levels such that;

$$v_{aN} \in \left\{ +\frac{V_{dc}}{2}, 0, -\frac{V_{dc}}{2} \right\}$$

Figure 3. Circuit configuration of 3L NPC-VSC

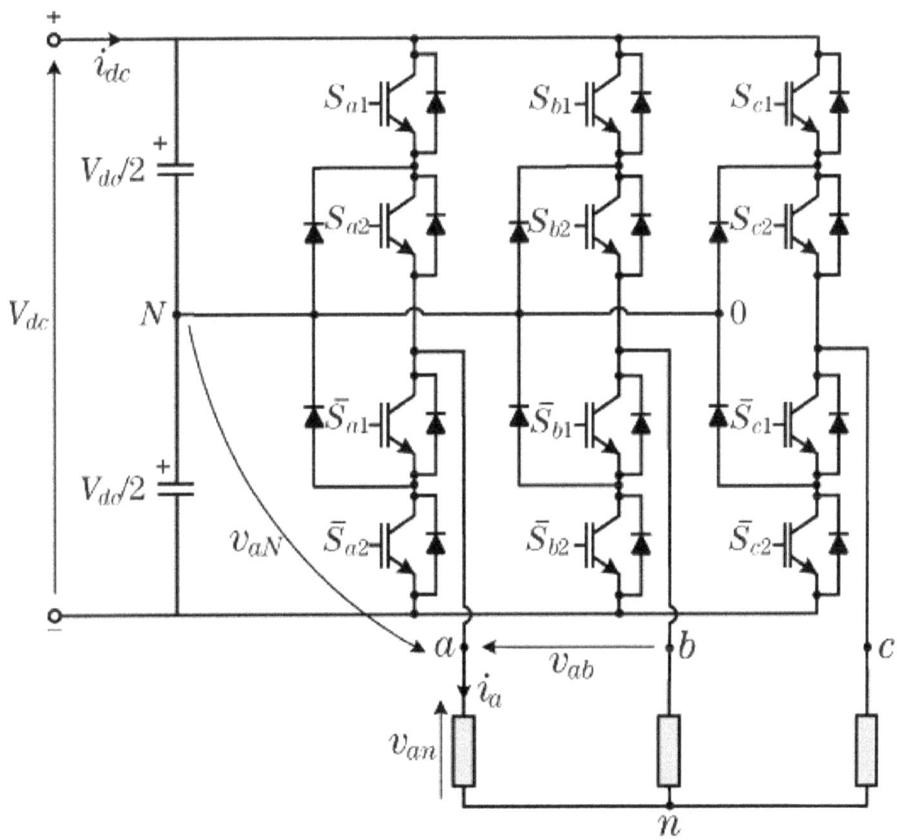

Figure 3 represents a three-phase, 3L-NPC voltage-source inverter fed by a split DC link: the total DC-link voltage is V_{dc}, shared by two series capacitors (each ideally at $V_{dc}/2$), whose midpoint is the neutral point N (often also denoted the converter zero potential "0"); i_{dc} is the DC-link current. Each phase leg (a, b, c) uses four controlled switches per leg, with binary switching functions S_{a1},S_{a2} (upper devices) and $\underline{S}_{a1}$,$\underline{S}_{a2}$ (lower devices), similarly to other legs; the diodes shown provide the NPC clamping paths to N, enabling three output voltage levels per phase; i_a is the phase-a current delivered to the load; v_{aN} is the pole (phase-leg) voltage of node a with respect to the DC midpoint N; v_{an} is the load phase-to-neutral voltage (node a to load neutral n); and v_{ab} is the line-to-line voltage between nodes a and b.

Under balanced conditions, each switch blocks approximately half of the DC bus voltage, significantly reducing device voltage stress compared with two-level converters. The main challenge of NPC converters lies in maintaining DC-link capacitor voltage balance. Neutral-point voltage deviation leads to output distortion and uneven device stress, affecting reliability. Owing to their maturity and suitability for high-power systems, three-level NPC converters are often preferred when 2L topologies reach their voltage or *dv/dt* limits (Rodríguez et al., 2007).

Flying Capacitor (FC) Converter

FC converters generate multiple voltage levels by inserting floating capacitors into each phase leg. Each additional voltage level typically requires an extra layer of flying capacitors to provide the corresponding voltage potential. Multiple switching state redundancies exist for the same output voltage level, offering theoretical fault-tolerance advantages.

Figure 3.4 is a 4-level FC VSC (4L FC-VSC). The DC link voltage is V_{dc}, split into $\pm V_{dc}/2$ by two series capacitors, with the midpoint denoted as N, and id is the DC-link current. Each phase leg (a, b, c) is built from three upper switching functions S_{a1},S_{a2},S_{a3} and their complementary lower devices $\underline{S}_{a1}$, $\underline{S}_{a2}$, $\underline{S}_{a3}$ (similarly for phases b and c). Two FCs per phase leg are inserted to generate intermediate voltage levels and to support capacitor-voltage balancing. By selecting appropriate switching states, the pole voltage v_{aN} can synthesize four discrete voltage levels.

Figure 4. Circuit configuration of 4L FC-VSC

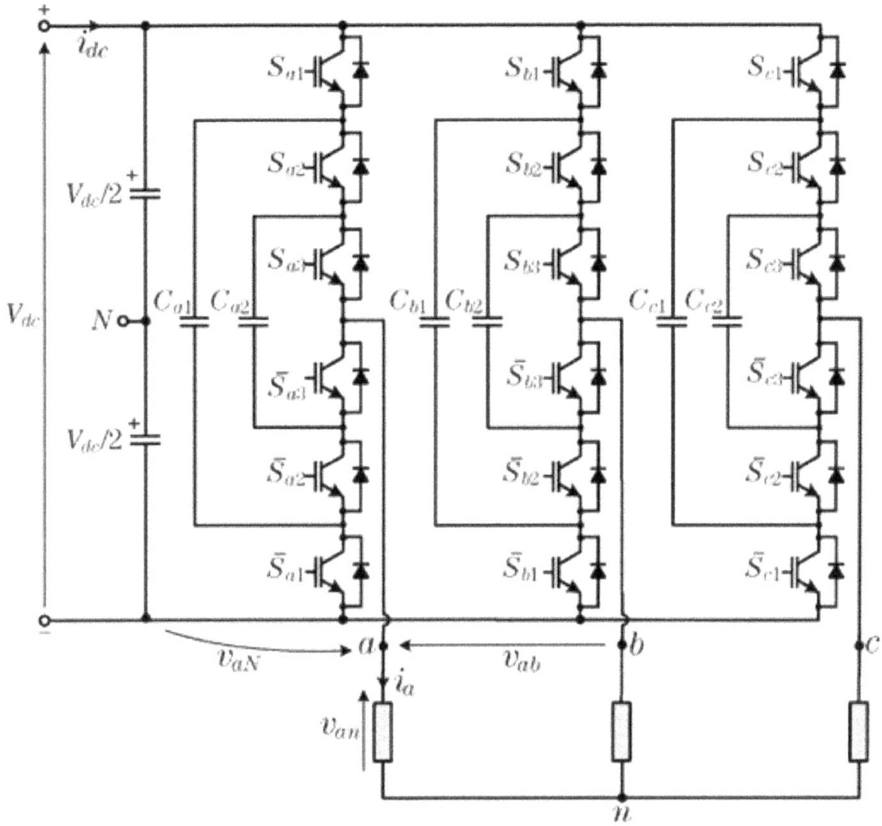

However, this topology requires many capacitors, resulting in increased volume, reduced reliability, and complex pre-charging and voltage-balancing requirements. Stable operation strongly depends on modulation and feedback-based balancing control (Lezana et al., 2007; Rodríguez et al., 2007). Due to these practical challenges, FC converters are less commonly adopted in large-scale industrial systems compared with NPC and CHB topologies.

Cascaded H-Bridge (CHB) Converter

A CHB converter consists of multiple single-phase H-bridge cells connected in series, each supplied by an independent DC source. A single H-bridge can generate multiple voltage levels at its output that varies as $v_{cell} \in \{+V_{de,cell}, 0, -V_{de,cell}\}$

By cascading N H-bridge cells in series, the phase voltage, v_a as in (1) becomes the algebraic sum of the individual cell outputs

$$v_a = \sum_{k=1}^{N} v_{cell,k} \quad (1)$$

When all DC sources have equal voltage ratings, the number of achievable voltage levels reaches 2N+1.

CHB converters feature a highly modular structure, excellent waveform quality, and low device voltage stress, making them well suited for medium- and high-voltage applications. Their main limitation is the requirement for multiple isolated DC sources. When such sources are available, CHB converters offer an attractive balance between performance and scalability; otherwise, system complexity and energy-balancing requirements increase significantly (Lezana et al., 2007; Rodríguez et al., 2007).

Modular Multilevel Converters (MMCs) for High-Power Applications

MMCs, consisting of cascaded submodules in each arm, are well established in HVDC and are emerging in high-power wind collection and transmission. (Muyeen, 2012; Teodorescu et al., 2011). Their high submodule count allows fine-grained control of energy distribution and fault management (Blaabjerg et al., 2024; Muyeen, 2012; Teodorescu et al., 2011).

A typical three-phase MMC consists of three phase legs; each composed of an upper arm and a lower arm. Each arm contains a series connection of multiple submodules (SMs), together with an arm inductor and an equivalent arm resistance, shown in Figure 3.5 and Figure 3.6. If each arm includes N submodules, a single phase can synthesize approximately 2N+1 or even finer voltage levels at the line output, depending on the modulation and redundancy strategy (Debnath et al., 2015). Two key variables characterize MMC operation: the insertion number, and the submodule capacitor energy. The arm voltage (v_{arm}) can be approximated as (2)

$$v_{arm}(t) \approx \sum_{k} v_{c,k}(t) \quad (2)$$

where $v_{c,k}$ is the capacitor voltage of the k-th submodule. The stored energy (E_k) of each submodule capacitor is given as in (3).

$$E_k = \frac{1}{2} C_{SM} v^2_{c,k} \quad (3)$$

The multilevel voltage synthesis of the MMC is therefore realized through large-scale periodic energy exchange between the AC and DC sides via the submodule capacitors, making capacitor voltage balancing and energy regulation central control objectives.

Figure 5. Schematic representation of the MMC.

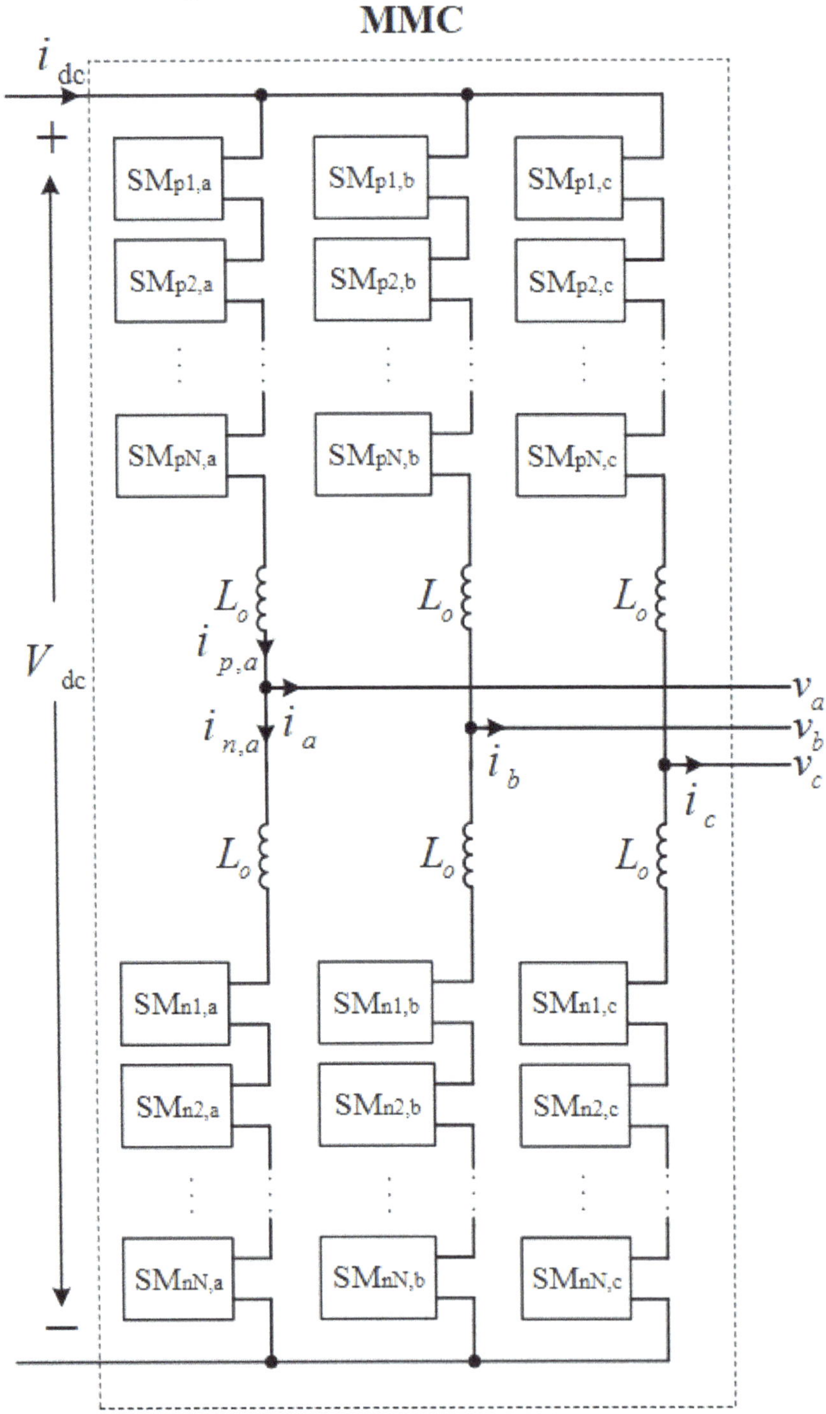

The primary advantages of the MMC are its high modularity and scalable voltage capability. By cascading many submodules, MMCs can generate high-level-count output waveforms at high voltage and power ratings, significantly reducing dv/dt, harmonic distortion, and AC-side filtering requirements. These features make MMCs particularly suitable for applications such as VSC-HVDC and offshore wind power transmission, where high power quality, reactive power support, and redundancy-friendly maintenance are required. However, these benefits come at a substantial engineering cost. The large number of submodule capacitors increases system volume, cost, and lifetime stress (Debnath et al., 2015; Lezana et al., 2007). Control systems must simultaneously address capacitor voltage balancing and circulating current suppression, leading to high sensing, communication, and computational complexity. In addition, DC-side fault handling and protection impose stringent requirements on fast isolation schemes and topology selection at the system level. Table 3.1 presents the summary of those converter topologies as discussed in above section. Moreover, the selection criterions of those converter topologies are summarized in Table 3.2 (Muyeen, 2012; Polinder et al., 2006; Teodorescu et al., 2011).

Figure 6. Various SM topologies: (a) the half-bridge, (b) the full bridge, (c) the clamp-double, (d) the three-level FC, (e) the three-level NPC, and (f) the five-level cross-connected SM.

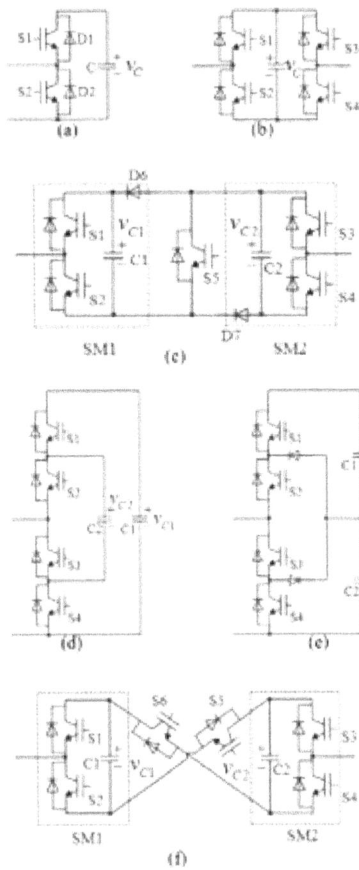

Table 2. Summary of different power converter topologies

Topology	Key Features	Applications	Advantages	Limitations
2L-VSC	PWM AC synthesis from DC link	Medium power turbines	Simple, robust IGBT/SiC control	Large filters, high dv/dt losses
Multilevel NPC	Diode-clamped, stepped waveforms	MV wind turbines	Low THD, reduced dv/dt	Neutral-point imbalance
FC	Modular cells per leg	Medium-voltage drives	Scalable levels, redundancy	Capacitor voltage balancing

continued on following page

Table 2. Continued

Topology	Key Features	Applications	Advantages	Limitations
CHB	Isolated DC sources per cell	Multi-terminal HVDC	Modular, fault-tolerant	Multiple transformers needed
MMC	Cascaded submodules per arm	High-power offshore/HVDC	Scalable, low harmonics	Complex control/ balancing

Table 3. Selection criteria of power converters

Criterion	Description	Preferred Topologies	Considerations
Cost	Initial investment, component scaling	2L-VSC (lowest)	MMCs higher due to submodules
Efficiency	Switching losses, thermal performance	Multilevel NPC/CHB, SiC-enabled	Wide-bandgap devices boost all
Semiconductor Tech	IGBT/SiC/GaN compatibility	All, SiC for high-freq	Higher temp/freq reduces filters
Reliability	Fault tolerance, redundancy	MMC/CHB (modular)	Two-level simplest protection
Grid Code Compliance	Harmonics, FRT, voltage levels	Multilevel/MMC	Low THD meets stringent codes

Beyond conventional converter topologies, unconventional power electronic interfaces offer enhanced power quality through advanced switching strategies and active filtering capabilities (Gidwani et al., 2013). These innovative approaches can significantly reduce harmonic distortion and improve grid compliance, particularly important as wind penetration levels increase and power quality requirements become more stringent.

5. CONTROL STRATEGIES FOR WIND ENERGY CONVERTERS

Machine-Side Converter (MSC) Control:

MSC control seeks to maximize energy capture while protecting mechanical components, typically through MPPT and torque control across a wide speed range(Ackermann, 2012; Okedu & Muyeen, 2022; Polinder et al., 2006). Grid-side converter (GSC) control ensures DC-link stability, grid synchronization, and accurate delivery of active and reactive power according to grid codes and ancillary service requirements (Hansen & Michalke, 2007; IRENA, 2019; Muyeen, 2012).

The MSC is located between the generator and the DC link. Its primary control task is to regulate the generator-side currents in a fast and stable manner under given current or power references and operating constraints, while injecting energy into the DC bus in a controlled way and ensuring the electrical safety of both the power devices and the generator (Atallah & El Tantawy, 2015; Zarei et al., 2020; Zhang et al., 2014).

In full-scale wind power conversion systems, the main control objectives of the MSC include: accurate and fast current tracking with current limitation to satisfy thermal and electromagnetic constraints; voltage feasibility and modulation margin management to ensure that the required voltage commands can be realized under the available DC-link voltage and modulation scheme, thereby avoiding modulation saturation and current instability; and controllable power injection into the DC link, enabling rapid reduction of generator power during DC overvoltage conditions, grid-side power limitation, or fault scenarios (Zarei et al., 2020; Zhang et al., 2014, 2015).

The MSC is most commonly implemented using vector control, also known as field-oriented control (FOC). The key idea of FOC is to project the three-phase stator currents into a synchronously rotating *dq* reference frame aligned with the rotor flux, thereby decoupling torque and flux (or excitation) control(Atallah & El Tantawy, 2015).

Figure 7. Configuration of a direct-drive PMSG wind turbine connected to a power grid.

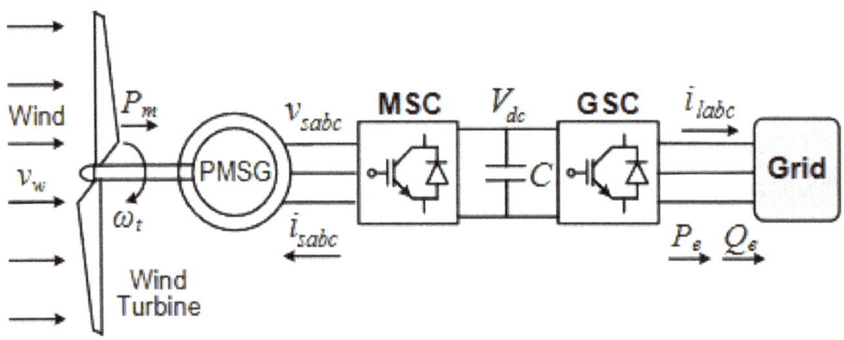

Figure 7 shows a Type-4 wind turbine with a PMSG interfaced to the grid through a full-scale back-to-back converter: wind speed v_w drives the turbine producing mechanical power P_m at rotor speed ωt, which the PMSG converts to three-phase electrical variables (v_{sabc}, i_{sabc}); the MSC regulates generator currents/torque and transfers power into the DC link, whose capacitor C buffers energy and decouples dynamics as reflected by the DC-link voltage V_{dc}; the GSC controls the injected grid current i_{abc} to maintain V_{dc} while independently setting active and reactive power (P_g,Q_g)delivered to the grid.

In the synchronous reference frame, the voltage equations of a PMSG can be expressed as (4) through (5).

$$v_d = R_s i_d + L_d \frac{di_d}{dt} - \omega_e L_q i_q \tag{4}$$

$$v_q = R_s i_q + L_q \frac{di_q}{dt} + \omega_e L_d i_d + \omega_e \psi_f \tag{5}$$

where *Rs* is the stator resistance, L_d and L_q are the d- and q-axis inductances, ω_e is the electrical angular speed, and ψ_f is the equivalent permanent-magnet flux linkage.

The electromagnetic torque T_e of the PMSG is given using (6) such that;

$$T_e = \frac{3}{2} p\left(\psi_f i_q + \left(L_d - L_q\right) i_d i_q\right) \tag{6}$$

where p is the number of pole pairs. For surface-mounted PMSGs, $L_d \approx L_q$, and the torque is therefore mainly determined by the q-axis current (Zarei et al., 2020; Zhang et al., 2014, 2015). As a result, a widely adopted control strategy sets the d-axis current for flux or field-weakening regulation, while the q-axis current directly controls the electromagnetic torque. This yields a clear control structure with good robustness and engineering practicality.

The main advantage of FOC is that, after dq decoupling, the controlled plant becomes approximately linear, which enables straightforward controller design, tuning, and predictable dynamic performance. This framework readily accommodates feedforward decoupling, field-weakening operation, and current or voltage constraints, and has been widely validated in industrial applications. However, its performance relies on accurate electrical angle estimation and machine parameters, and is therefore sensitive to parameter variations, delays, saturation, overmodulation, and dead-time effects. Sensorless FOC further increases implementation difficulty, particularly at low speed and during fast transients (Zhang et al., 2014, 2015).

Maximum Power Point Tracking (MPPT) Algorithms

Machine-side MPPT algorithms, including tip-speed ratio control, optimal torque control, and power signal feedback, regulate generator torque to track the aerodynamic optimum under varying wind speeds (Ackermann, 2012; Okedu & Muyeen, 2022; Polinder et al., 2006). Vector control and direct torque control (DTC) are widely used to decouple flux and torque in DFIGs and PMSGs, allowing fast current regulation, limitation of stator and rotor currents during faults, and smooth transition between operating modes (Ackermann, 2012; Desalegn et al., 2022).

The mechanical power extracted from the wind turbine can be approximately expressed as

$$P_m = \frac{1}{2}\rho A v^3 C_p(\lambda, \beta) \tag{7}$$

where v is the wind speed, $\lambda = \omega R/v$ is the tip-speed ratio (defined as the blade tip speed divided by the wind speed), and β is the blade pitch angle. For fixed-pitch operation, which is typically the case below rated conditions ($\beta \approx 0$), there exists an optimal tip-speed ratio λopt that maximizes the power coefficient C_p. This operating point corresponds to the maximum aerodynamic energy capture at a given wind speed (Errami et al., 2012, 2013). Accordingly, MPPT objective below rated wind speed can be stated succinctly as: adjust the rotor speed ω in response to wind speed variations such that the tip-speed ratio λ remains as close as possible to λ_{opt}, thereby maximizing C_p and the extracted power. Common MPPT strategies, including TSR control, optimal torque (OT) control, and power signal feedback (PSF), all generate reference signals in the form of desired rotor speed or electromagnetic torque (Errami et al., 2012, 2013).

Figure 8. Wind generator power curves at various wind speeds.

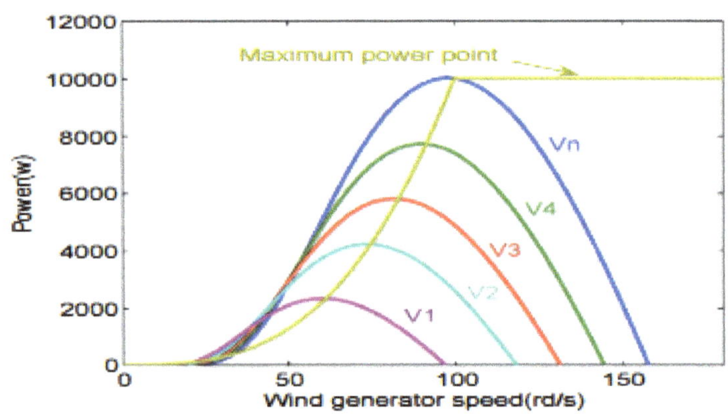

Figure 8 shows wind generator power–speed characteristics for several wind speeds (V1~Vn): for each wind speed, electrical/mechanical power rises with generator speed to a single peak and then drops, and the MPP marks the optimal operating speed that yields the highest extractable power for that particular wind speed (the basis of MPPT by tracking the peak across curves).

Tip-Speed Ratio (TSR) Control

The core idea of (TSR) control is to directly regulate the tip-speed ratio λ to its λ_{opt}. This requires measurement of both the wind speed v and the rotor speed ω, based on which the reference rotor speed is calculated as

$$\omega^* = \frac{\lambda_{opt}}{R} v \tag{8}$$

A speed control loop (typically PI-based) is then used to regulate the T_e so that the actual rotor speed tracks the reference. When wind speed measurements are reliable and model parameters are accurate, TSR control provides precise tracking of the optimal operating point. However, it relies on wind speed sensors, which introduce additional cost and reliability concerns. Measurement errors due to sensor placement, turbulence, and gusts are common, and wind speed signals usually require filtering, which introduces phase delay and degrades dynamic tracking performance.

Optimal Torque Control (OTC)

The key feature of OTC is that it does not require wind speed measurement. Under below-rated and fixed-pitch conditions, the wind turbine exhibits an approximate relationship (9) at the optimal power coefficient.

$$T_e^* = K_{opt}\omega^2 \tag{9}$$

This implies that higher rotor speed naturally allows a larger T_e following a quadratic law, causing the operating point to converge to the optimal power capture curve. OTC requires only rotor speed information, resulting in a simple structure, low cost, and high reliability. It is therefore widely adopted in industrial applications. However, variations in air density, blade aging, and icing can degrade accuracy, and practical implementations must be combined with torque or current limits as well as cut-in and cut-out wind speed logic.

Power Signal Feedback (PSF)

PSF is based on a predefined optimal power–speed characteristic and does not rely on wind speed measurement. Common implementations include lookup-table methods, perturb-and-observe schemes, and hill-climbing algorithms. Like OTC, PSF avoids wind speed sensors and is straightforward to implement in practice when a lookup table is used. However, search-based methods may exhibit oscillatory behavior under turbulent wind conditions and therefore require additional measures such as limiting, dead zones, and low-pass filtering. For lookup-table-based PSF, performance strongly depends on the quality of the calibrated data; inaccurate tables can lead to persistent deviation from the optimal operating point.

Generator Torque and Speed Control (Vector Control, DTC)

If FOC can be interpreted as "regulating the currents first and then generating torque through the controlled currents," DTC follows a different philosophy: it directly monitors torque and flux errors and immediately selects the most suitable inverter switching state to drive these errors toward zero. As a result, DTC exhibits a very fast dynamic response (Atallah & El Tantawy, 2015; Errami et al., 2013).

The basic operation of DTC typically proceeds as follows. The stator flux and electromagnetic torque are estimated online and compared with their respective reference values. The resulting errors are processed by hysteresis comparators or decision logic, and the corresponding voltage vector is then selected from a predefined switching table to determine the inverter switching state for the next control interval (Errami et al., 2013; Shukla & Singh, 2019).

By directly selecting inverter voltage vectors based on instantaneous flux and torque errors, DTC achieves superior torque dynamics. However, conventional DTC implementations suffer from relatively large torque and flux ripple as well as variable switching frequency. In practical applications, improved schemes such as DTC with space vector modulation (DTC-SVM) are commonly adopted to mitigate these issues and to satisfy grid-side harmonic limits and protection requirements (Shukla & Singh, 2019).

Grid-Side Converter Control

GSC control typically adopts a hierarchical structure consisting of grid synchronization via a phase-locked loop (PLL), an inner current control loop, and outer power or DC-link voltage control loops. The PLL establishes a grid-aligned dq reference frame. Under grid-voltage-oriented control, the d-axis current primarily regulates active power to stabilize the DC-link voltage, while the q-axis current controls reactive power to meet power factor requirements and provide voltage support (Davari & Mohamed, 2017; Hu et al., 2011; Vukosavić, 2018).

Figure 9. VSC connected to a weak ac grid

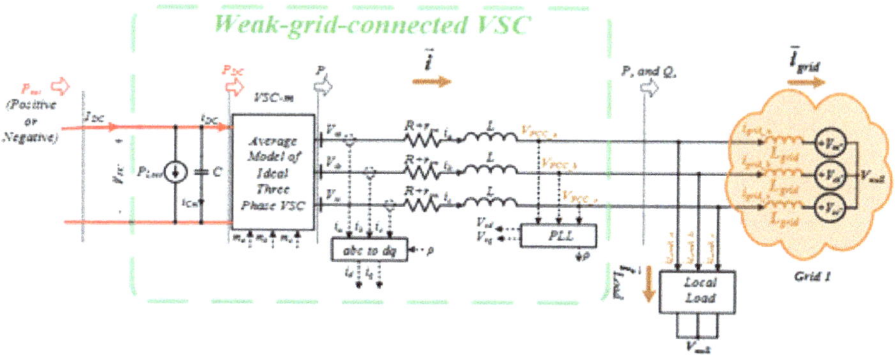

This is a weak-grid-connected three-phase VSC in Figure 4.3 with a DC link capacitor C supplying/absorbing DC power P_{dc}, an averaged VSC model generating controlled AC-side voltages, and an L–R output filter interfacing the converter to the point of common coupling (PCC) where the PCC voltages v_{pcc} and converter current i are measured. A PLL locks to the PCC voltage to provide the electrical angle for the *abc–dq* transformation, enabling current/power control that regulates injected active and reactive power (P, Q). The PCC feeds a local load and connects to a weak grid modeled with significant grid impedance, so the resulting grid current i_{grid} and PCC voltage are strongly influenced by grid strength.

Under weak-grid conditions and during grid faults, appropriate bandwidth tuning, decoupling compensation, and current limiting with defined priority schemes are applied to ensure stable grid connection and compliance with protection and grid-code requirements.

Grid Synchronization

For the GSC to inject active power or exchange reactive power with the grid, the phase angle θ_g and frequency ω_g of the grid voltage must be accurately known. The purpose of the PLL is therefore to align the control reference frame with the grid voltage vector, providing a well-defined directional reference for power control (Blaabjerg et al., 2006; Hu et al., 2011). Accordingly, the PLL outputs the estimated grid frequency is estimated using (10).

$$\theta_g(t),\ \omega_g(t) = \hat{\theta}_g(t) \tag{10}$$

The most widely used scheme in practice is the synchronous reference frame PLL (SRF-PLL). The three-phase grid voltages v_{abc} are first sampled and transformed into the *dq* frame via Clarke and Park transformations, yielding v_d and v_q as described in (11) and (12) respectively.

$$v_d = v_\alpha cos\theta + v_\beta sin\theta \tag{11}$$

$$v_q = -v_\alpha sin\theta + v_\beta cos\theta \tag{12}$$

When the reference frame is perfectly aligned with the grid voltage vector, the entire voltage appears on the d-axis and $v_q \approx 0$. The control objective of the SRF-PLL is therefore to drive v_q to zero.

Figure 10. General structure of the three-phase dq PLL method.

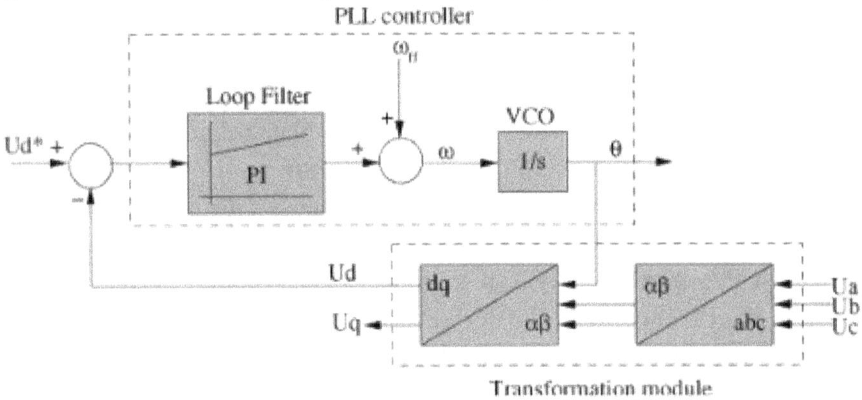

Figure 10 illustrates a synchronous-reference-frame PLL (SRF-PLL): the measured three-phase voltages (Ua,Ub,Uc) are transformed to αβ and then to dq using the estimated angle θ; the q-axis voltage U_q is driven to zero by a PI loop filter, whose output adjusts the estimated ω; integrating ω produces the updated θ, so the loop continuously aligns the d-axis with the grid voltage vector, yielding a synchronized phase θ and frequency ω for control and transformations.

From an engineering perspective, the PLL offers fast phase locking under strong-grid conditions and supports stable operation under weak-grid scenarios. However, it also has inherent limitations. In strong grids, voltage distortion, harmonics, or oscillations may be interpreted by the PLL as phase variations, resulting in control jitter. In weak grids, phase locking tends to be slower and the dynamic response degrades, which can adversely affect overall system performance (Blaabjerg et al., 2006).

DC-Link Voltage Regulation:

The fundamental principle of DC-link voltage regulation is to stabilize the DC-link voltage V_{dc} by controlling active power flow. The energy stored in the DC-link capacitor is given by (13).

$$E_{dc} = \frac{1}{2} C_{dc} V^2{}_{dc} \tag{13}$$

and its energy variation is determined by the power imbalance between the machine-side converter and the grid-side converter:

$$\frac{dE_{dc}}{dt} = P_{MSC} - P_{grid} \tag{14}$$

or

$$C_{dc} V_{dc} \frac{dV_{dc}}{dt} = P_{MSC} - P_{grid} \tag{15}$$

Expressed in terms of voltage as in (14) and (15), this relationship shows that when $P_{MSC} > P_{grid}$, excess energy is stored in the DC-link capacitor and V_{dc} increases; conversely, when $P_{MSC} < P_{grid}$, the capacitor releases energy to supply the power deficit and V_{dc} decreases. Therefore, the GSC regulates V_{dc} by adjusting the grid-side active power P_{grid}, which is achieved by controlling the grid-side active current (Blaabjerg et al., 2006; Davari & Mohamed, 2017; Hu et al., 2011).

In practice, DC-link voltage regulation adopts a cascaded control structure with an outer voltage loop and an inner current loop. The outer loop is a DC-link voltage PI controller that generates the reference for the d-axis current. Under grid-voltage-oriented control, the d-axis current predominantly corresponds to the active power channel. The inner loop then regulates the dq-axis current to track this reference with fast dynamics (Vukosavić, 2018).

Current limiting is an essential part of DC-link voltage regulation. The maximum grid-side current is constrained by the ratings of the power devices, the filter inductance, and grid-code requirements. Accordingly, the voltage-loop PI controller must incorporate anti-windup mechanisms; otherwise, integrator saturation during current limiting can lead to slow recovery and voltage overshoot (Blaabjerg et al., 2006; Davari & Mohamed, 2017).

Active & Reactive Power Control(PQ Control):

PQ control is a standard GSC control framework. With grid synchronization provided by a PLL and grid-voltage-oriented control (aligning the synchronous d-axis with the grid-voltage vector such that *vq≈0*), as shown in Figure 4.5 the power references (P*, Q*) are mapped to current commands (i_d*, i_q*), which are then enforced by a high-bandwidth inner current controller. This enables regulated injection of active power and reactive power into the grid (Vukosavić, 2018).

Figure 11. PQ Control principle structures.

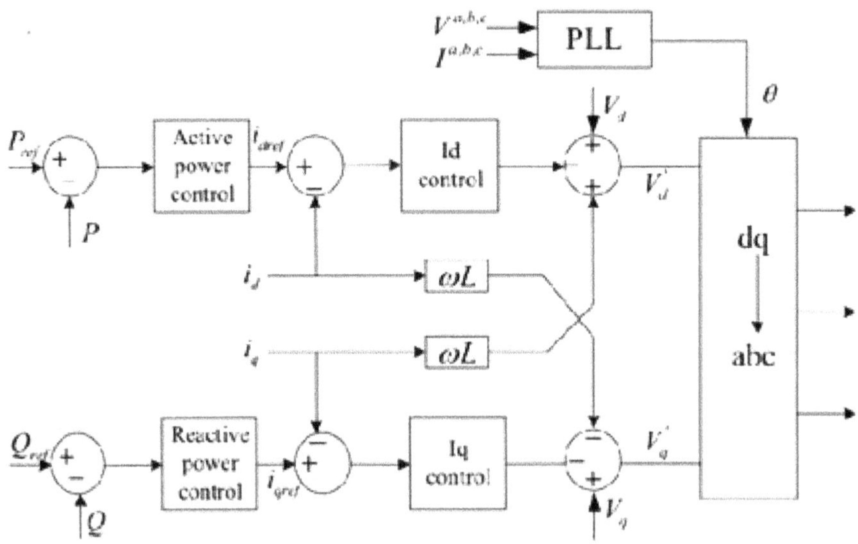

In the synchronous dq frame, instantaneous power can be written as (16)

$$P = \frac{3}{2}\, v_d i_d + v_q i_q \,,\; Q = \frac{3}{2}\, v_q i_d - v_d i_q \,; \tag{16}$$

Under voltage orientation (v_q*≈0*), the system is approximately decoupled then instantaneous powers are given as (17).

$$P \approx \frac{3}{2} v_d i_d \,,\; Q \approx -\frac{3}{2} v_d i_q , \tag{17}$$

the d-axis primarily regulates P, while the q-axis primarily regulates Q (the sign convention depends on the adopted reactive-power and current directions).

Accordingly, a common reference mapping is

$$i_d^* \approx \frac{2}{3}\frac{P}{v_d}, i_q^* \approx -\frac{2}{3}\frac{Q}{v_d} \tag{18}$$

PQ control enables a GSC to regulate active and reactive power in a PLL-synchronized, grid-voltage-oriented dq frame by mapping (P*, Q*) into (i_d*,i_q*) and enforcing them with a fast inner current loop. In practice, it serves as the primary interface for (i) DC-link voltage stabilization via active-power modulation to maintain power balance between machine-side and grid-side stages, and (ii) grid-code compliance through reactive-power support and power-factor control (Blaabjerg et al., 2006).

Its main advantages are a clear and dispatchable (P, Q) control objective, near-decoupled d/q regulation under voltage orientation with mature tuning practices, and straightforward integration of current limiting and anti-windup protection. Key limitations are its reliance on accurate PLL synchronization and the validity of the decoupling assumption in weak, harmonic-rich, or unbalanced grids, and the fact that during voltage sags the converter becomes current-limited, forcing a trade-off between tracking P and Q (Hu et al., 2011; Vukosavić, 2018). Those control strategies are as discussed in Table 4.1.

Table 4. Summary of control strategies for wind energy converters

Strategy	**Application**	**Advantages**	**Limitations**
Vector Control (FOC)	Machine-side torque/ speed	Decoupled control, fast response	Parameter sensitivity
DTC	DFIG/PMSG MPPT	Simpler, ripple-free torque	Variable switching frequency
MPPT (TSR/PSF)	Power optimization	Max energy capture	Wind measurement needs
SMC	Grid-side faults	Robust to uncertainties	Chattering issues
MPC	Both sides	Multi-objective optimization	High computation

6. AI-ASSISTED MODELLING AND SIMULATION AND OF POWER CONVERTERS IN WIND ENERGY SYSTEMS

Power converters in wind turbines—such as the partially rated converter in DFIG-based systems and the MSC and GSC in full-scale Type-4 systems—carry the core functions of energy conversion and control in modern wind power plants. As wind penetration increases and grid operating conditions become more complex, engineering-oriented modeling must strike a balance between dynamic fidelity and computational scalability: models should capture multi-timescale dynamics relevant to control and grid interaction, while remaining suitable for large-scale, multi-scenario simulation studies. In practice, purely physics-based modeling is often constrained by limited access to controller internals, pronounced operating-point variability, and the computational burden of high-fidelity EMT simulations (Subedi et al., 2025). Figure 5.1 formalizes the resulting "physics–data–AI–applications" workflow for AI-assisted converter modeling and simulation. The left branch (physics-based model → EMT simulation → analysis & design) delineates the conventional high-fidelity route, whose scalability can be limited in extensive parametric studies and wind-farm aggregation analyses. The right branch captures AI/ML-enabled augmentations that exploit operational and/or EMT-generated data to (i) construct rapidly callable surrogate models for accelerated simulation, (ii) impose physics-consistent constraints and support online calibration, and (iii) enable online monitoring and health-management functions. Incorporating confidence/uncertainty awareness further supports safe deployment and systematic fallback, thereby improving simulation efficiency, robustness, and scalability for converter-level and system-level studies.

Figure 12. Physics–Data–AI–Application Workflow for AI-Assisted Modeling of Wind Power Converters

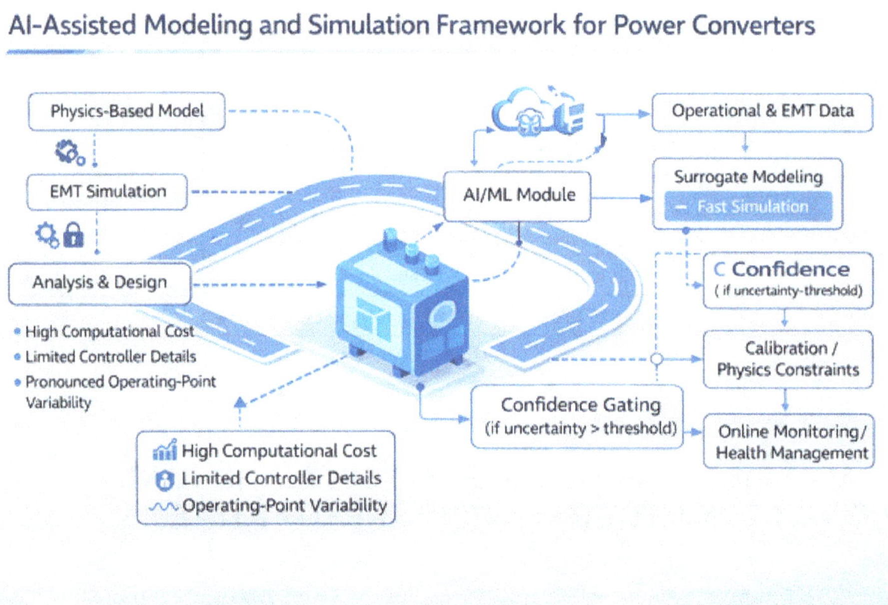

Data-Driven Modelling of Power Converters Using Machine Learning

Data-driven modelling of power converters in wind energy systems complements physics-based models by learning complex nonlinear relationships from operational data, including switching dynamics, thermal behavior, and degradation patterns. Machine learning techniques such as neural networks, ensemble methods, and Gaussian processes can approximate converter losses, temperature profiles, and fault signatures, enabling faster simulations and improved control design in large-scale studies

Figure 13. Multi-Time-Scale Dynamics of Wind Power Converters and Model Abstraction Levels

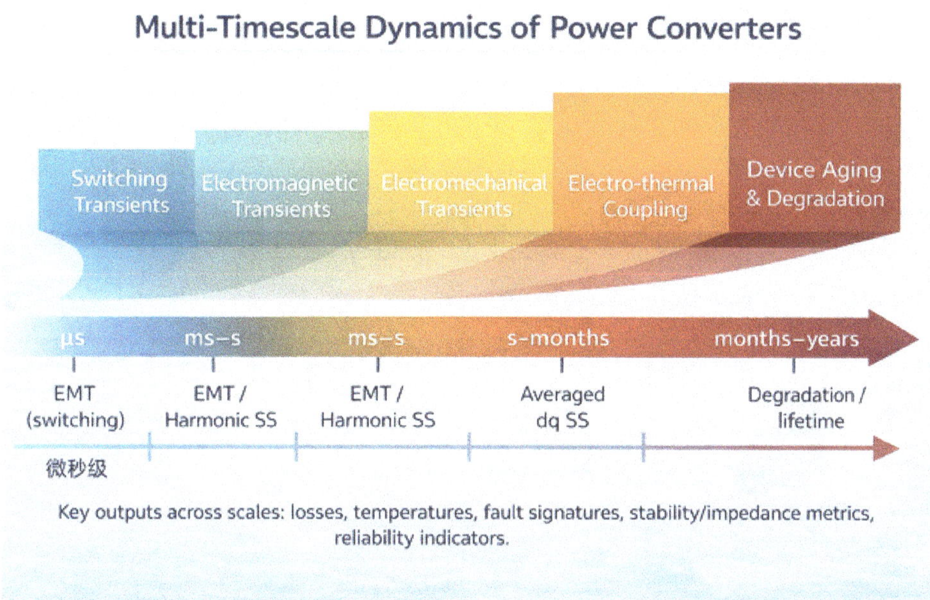

As illustrated in Figure 5.2, converter dynamics in wind energy systems span multiple time scales. Switching and electromagnetic transients (*µs–ms*) dominate waveform-level behavior and directly shape current-control performance and grid interaction, whereas electro-thermal coupling (hours–days) and device aging and degradation (months–years) govern long-term efficiency drift, thermal stress accumulation, and reliability evolution. This multi-time-scale nature implies that a single modeling abstraction is rarely optimal: high-fidelity EMT models are indispensable for unit-level validation and transient compliance studies, while averaged/dq and

RMS-level representations are typically preferred for wind-farm aggregation, grid-strength sensitivity analyses, and extensive controller-parameter sweeps, where EMT-level computation becomes prohibitive (Subedi et al., 2025). Accordingly, data-driven modeling is commonly introduced as a complementary layer to physics-based modeling, providing fast equivalence and rapid prediction with acceptable accuracy for large-scale, multi-scenario engineering analyses.

Under this objective, a practical starting point is to construct a rapidly callable surrogate that replaces repetitive, high-cost simulation with a learned input–output mapping such that;

$$\hat{y} = f_{\theta}(x) \tag{19}$$

$$L(\theta) = \sum_{k=1}^{N} \left(y_k - f_{\theta}(x_k)\right)^2 \tag{20}$$

$$\theta^{\backslash *} = argmin\ L(\theta)\ (over\ \theta) \tag{21}$$

The input feature vector x can include v_{dq}, i_{dq} the DC-link voltage V_{dc}, switching frequency, modulation index, ambient temperature, and operating-state features constructed from SCADA data; the output y may correspond to engineering-critical quantities such as losses, temperature trajectories, and fault signatures.

Figure 14. Surrogate Model Training Workflow for Data-Driven Converter Modeling

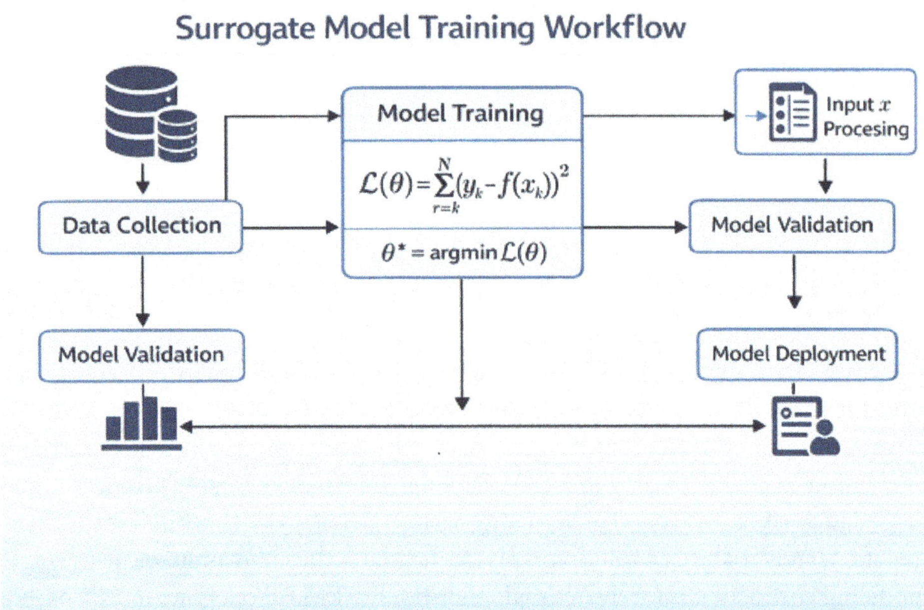

The associated workflow—data acquisition, feature processing, training, validation, and deployment—is summarized in Figure 5.3, highlighting how surrogate models can be deployed for fast screening tasks (e.g., device/filter selection, MSC/GSC gain scheduling exploration, robustness maps) and for accelerating multi-scenario studies with reduced dependence on extensive EMT simulations (Subedi et al., 2025).

However, if the surrogate is trained purely as a black box, generalization may degrade across operating regimes not well represented in the training data—for example, under weak-grid impedance variations, FRT events, or control-mode transitions. To restore physical consistency, it is often necessary to embed physics-informed residual constraints into the learning objective. As conceptually contrasted in Fig. 5.4, the physics-guided surrogate augments the data loss with a physics loss as in (22).

$$L = L_{data} + \lambda L_{phys} = \sum_k \| y_k - \hat{y}_k \|^2 + \lambda \sum_k \| R(x_k, \hat{y}_k) \|^2 \tag{22}$$

Where the residual $R(\cdot)$ can be constructed from dq-frame state-space relations, harmonic state-space models, energy-balance constraints, or impedance-consistency constraints, thereby encouraging the learned model to respect fundamental physical laws and improving robustness under parameter drift (Xu et al., 2025).

Figure 15. Black-Box versus Physics-Guided (Gray-Box) Surrogate Modeling

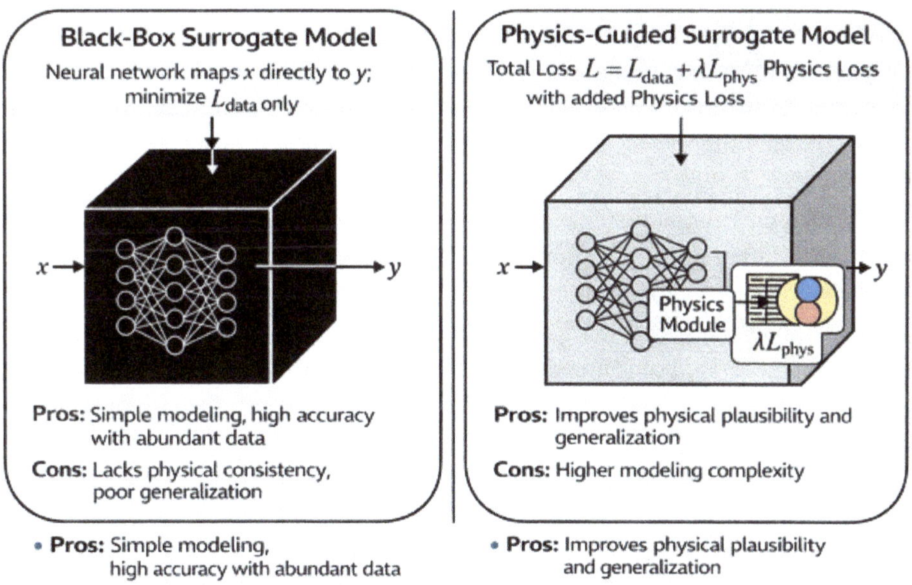

Even with physical constraints, rare or extreme operating conditions—such as FRT events—still require quantifying the "degree of trust" in model outputs to enable safe use and engineering fallback strategies. Uncertainty prediction can be expressed as;

$$p(y \mid x, D) = \int p(y \mid x, \theta) p(\theta \mid D) d\theta \tag{23}$$

In engineering workflows, this uncertainty estimate supports confidence gating: when confidence degrades or the operating condition is out of distribution, the system can revert to conservative physics-based models, tighten safety margins, or trigger additional data collection and online adaptation—consistent with the "physics–data–AI–applications" pipeline summarized in Figure 5.4 (Xu et al., 2025).

Deployment of Lightweight AI Models on Embedded Systems for Low-Latency Converter Control

Deployment of lightweight AI models on embedded platforms allows low-latency inference for tasks like sensor fault detection, parameter estimation, and adaptive tuning of control loops without significant computational overhead. Techniques such as model compression, pruning, and quantization make it feasible to implement these models in digital signal processors or microcontrollers already present in turbine converters.

Embedded AI targets the existing DSP/MCU platforms used in wind-turbine converter control systems, enabling data-driven functionality to move from offline analysis toward online inference. Typical embedded applications include anomaly detection, sensor-health assessment, parameter estimation, and health-index computation, which support local and scalable condition monitoring even under limited connectivity.

To be engineering-actionable, AI execution must satisfy hard real-time constraints jointly with sampling, conventional control computation, and modulation updates; otherwise, accuracy alone is not deployable. As illustrated by the timing budget in Figure 5.5, one control period T_S must accommodate data acquisition/conditioning T_{acq}, AI inference T_{inf}, conventional control computation T_{ctrl}, and PWM/modulation update overhead T_{PWM}, leading to the feasibility constraint described in equation (24).

$$T_S \geq T_{acq} + T_{inf} + T_{ctrl} + T_{PWM} \tag{24}$$

Where, T_s captures sensing, filtering, and basic preprocessing; T_{inf} is the runtime of the embedded model; T_{ctrl} corresponds to the established control stack; and T_{PWM} accounts for modulation/PWM update latency. Figure 5.5 highlights that T_{inf} directly consumes the same deadline-limited budget required by certified control and modulation tasks, which motivates a conservative role for embedded AI in converter control (Subedi et al., 2025).

Figure 16. Real-Time Timing Budget for Embedded AI in Converter Control

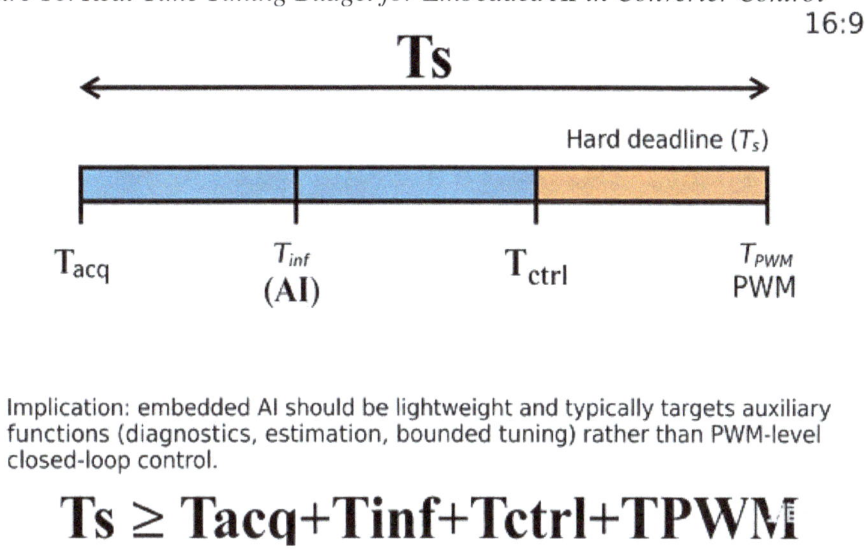

Accordingly, a robust integration strategy is to position AI as a supervisory/assistive layer and restrict its outputs to bounded, auditable interfaces, rather than generating PWM-level commands. In this layered design, the certified inner–outer loops, PLL, and protection logic preserve stability and compliance boundaries, while AI improves perception and decision support without compromising fallback capability (Subedi et al., 2025).

From an implementation standpoint, tight computational windows and limited DSP/MCU resources favor a "feature extraction + lightweight model" pipeline: high-rate electrical measurements are condensed into a tractable feature set, and model size/latency are reduced via pruning, quantization, or distillation. SCADA variables provide low-rate operating context and are naturally aligned with O&M-oriented inference, enabling scalable deployment at the wind-farm level.

Intelligent Control Strategies That Adapt to Grid Disturbances, Wind Variability, and Converter Aging

Intelligent control strategies use AI to adapt to grid disturbances, wind variability, and converter aging, for example by online tuning of PI gains, predictive adjustment of current limits, or selection of optimal operating points that minimize thermal stress –. Reinforcement learning and hybrid model-based/AI controllers are being explored to improve fault ride-through performance and optimize ancillary service provision under uncertain conditions

Figure 17. Hierarchical Architecture of AI Supervisory Control over Conventional MSC/GSC Loops.

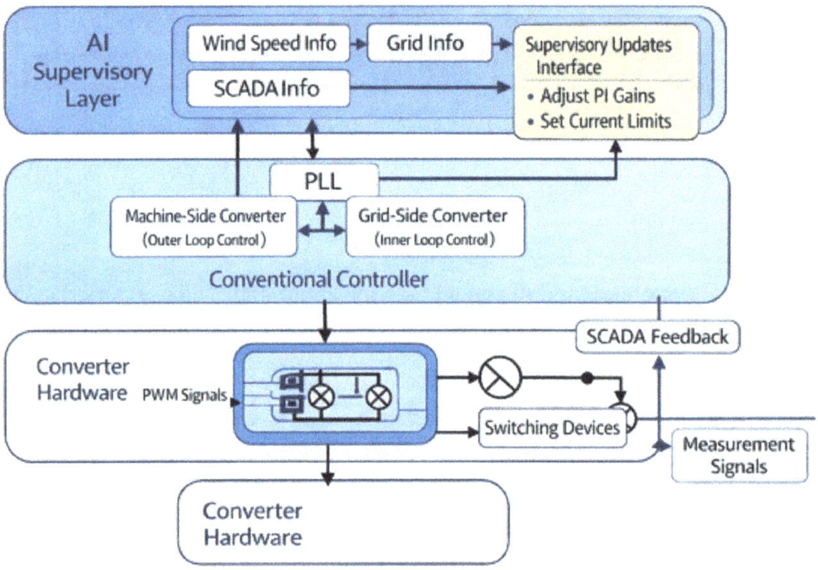

In grid-connected wind-converter control, AI is typically introduced as a supervisory layer rather than being embedded in the PWM-generation path (Subedi et al., 2025). As shown in Figure 5.6, the conventional controller layer—including the MSC/GSC current inner loops, outer loops (voltage/power), PLL-based synchronization, limiters, and protection—defines the certified stability and compliance boundaries and handles fast dynamics and protective actions. On top of this foundation, the AI supervisory layer consumes operating-context inputs together with SCADA/measurement feedback, and issues bounded updates through the Supervisory Updates Interface in Fig. 5.6. In this architecture, AI provides saturable "parameter- and policy-layer" adjustments—rather than generating PWM directly—thereby improving adaptivity under wind fluctuations, grid-strength variations, and slow parameter drift while preserving verifiability and fallback capability (Subedi et al., 2025).

Typical supervisory actions (cf. Fig. 5.6) include PI/PR gain scheduling for the MSC/GSC loops, current limiting and power ramp-rate management, thermal-derating recommendations, and synchronization/bandwidth/mode-switching guidance under weak-grid conditions. During disturbances or fault events, the supervisory layer can temporarily tighten constraints to reduce risk exposure while the conventional protection functions remain authoritative.

To mitigate misjudgment under out-of-distribution conditions, the supervisory layer should incorporate confidence/consistency monitoring and fallback logic (also reflected in Figure 5.6 by restricting AI outputs to an auditable interface): when confidence degrades or the operating point approaches constraint boundaries, the system automatically reverts to baseline or conservative scheduling (Xu et al., 2025). If supervisory decision-making is formalized, a unified objective can be written as;

$$J(\pi) = E\left[\sum_{t=0}^{\infty} \gamma^t r_t\right] \quad (25)$$

and the supervisory policy is selected as$\pi^* = argmax_\pi J(\pi)$. Here, r_t can aggregate tracking error, penalties for constraint violations, and thermal-stress proxy metrics, enabling explicit trade-offs among performance, constraint satisfaction, and lifetime objectives. The supervisory outputs must remain saturable, rate-limited, and auditable to ensure compatibility with existing control and protection structures.

Real-Time Monitoring, Predictive Maintenance, and Performance Optimization Using AI-Enhanced Digital Twins

AI-enhanced digital twins combine real-time SCADA data, physics-based models, and data-driven components to monitor turbine and converter behavior, perform predictive maintenance, and optimize performance at both turbine and farm levels ,. Recent studies demonstrate digital twin frameworks that use machine learning to forecast wind power output and detect anomalies, integrated with cloud or edge infrastructure and high-speed communication for scalable deployment.

Figure 18. Closed-Loop AI-Enhanced Digital Twin for Real-Time Monitoring and O&M Decision Support.

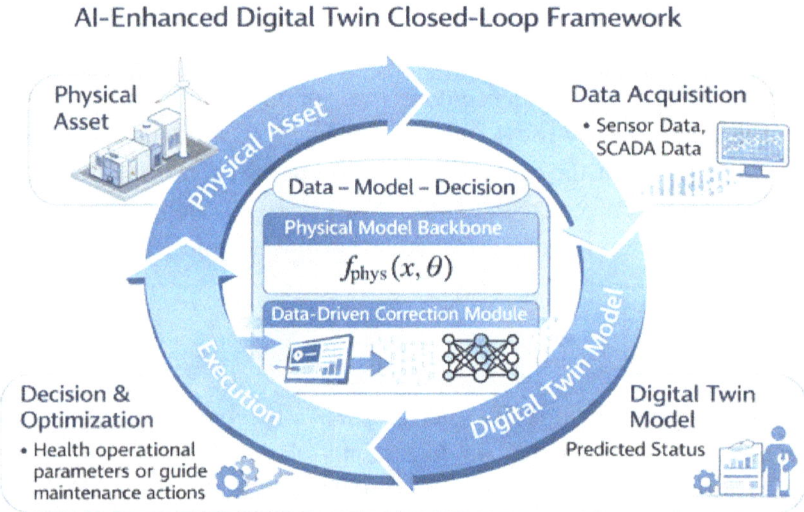

AI-enhanced digital twins fuse real-time measurements with physics-based models and data-driven modules to construct a continuously updated virtual counterpart of the wind turbine and its converters, supporting condition assessment, anomaly detection, predictive maintenance, and operational optimization (Subedi et al., 2025). As depicted in Figure 5.7, this "AI-enhanced digital twin" is not a static offline replica but a closed-loop architecture that explicitly couples the Physical Asset (turbine and converters), Data Acquisition (sensors/SCADA), Digital Twin Model (physics backbone plus data-driven correction), and Decision & Optimization (O&M actions), thereby keeping the model usable under changing operating conditions and device aging. Compared with approaches relying solely on offline models or offline data analytics, the value of a digital twin lies in closing the loop of "field data–model update–decision output" (and its execution back to the field), enabling the twin to remain valid in long-term operation.

Engineering implementations typically adopt a closed-loop "data–model–decision" framework, consistent with the three-layer core in Figure 5.7, so that the digital twin can both track system states and produce actionable engineering conclusions. The data layer synchronizes the physical asset and its twin through SCADA/measurement streams; the model layer generally uses control-friendly state-space models or averaged models as a physics backbone, augmented by data-driven residual correction to compensate for unmodeled effects, and coupled with uncertainty quantification to flag mismatch and rare events; the decision layer outputs health indices, alarms, and operational recommendations for O&M and dispatch (Subedi et al., 2025), which correspond to the "Decision & Optimization" block in Figure 5.7 and are applied through the "Execution" pathway.

The long-term effectiveness of a digital twin depends critically on online calibration and consistency maintenance; otherwise, the model will drift and become inaccurate, breaking the closed-loop premise in Figure 5.7. This can be summarized by the following parameter update form indicated in (26).

$$\hat{\phi}_{k+1} = \hat{\phi}_k + K_k\left(z_k - h\left(\hat{x}_k, \hat{\phi}_k\right)\right) \tag{26}$$

Here, ϕ denotes slowly varying equivalent parameters (e.g., effective capacitance, thermal resistance, and aging indicators), z_k is the measurement, and $h(\cdot)$ is the twin's measurement model. Incorporating physical constraints and uncertainty awareness can suppress noise overfitting and improve long-term consistency under changing conditions and aging (Xiao et al., 2021).

For grid-connected and wind-farm-scale use, the twin should also capture network interaction effects; otherwise, unit-level consistency is insufficient for stability and interoperability assessment. (Lyu et al., 2023) shows that integrating unit-level data-driven models with the system topology enables online impedance modeling and stability assessment without internal controller parameters—well suited to heterogeneous wind farms. From an O&M standpoint, SCADA-driven anomaly detection and fault diagnosis are the most scalable entry point and can be extended to predictive maintenance and optimization (Xu et al., 2025); sustained performance relies on online calibration, physical constraints, and uncertainty quantification to address mismatch over time (Xiao et al., 2021).

It is also important to mention that co-simulation frameworks combining electromagnetic transient (EMT) models with AI-Enhanced control system dynamics provide comprehensive platforms for analyzing wind generator performance under fault conditions (Makolo, 2013). These integrated simulation approaches are essential for validating converter control strategies and protection schemes prior to field deployment, particularly for complex FRT scenarios involving asymmetric faults and voltage recovery transients.

Grid Integration and Compliance

Overview of International Grid Codes (LVRT/HVRT)

International grid codes increasingly require converter-interfaced wind turbines to stay connected and support the grid during voltage sags and swells, defined through LVRT and HVRT profiles (Hansen & Michalke, 2007; IRENA, 2019) as in Table 6.1. Compliance demands that turbines withstand specified fault conditions without tripping, while providing reactive current injection and controlled active power recovery to support system voltage and frequency (Hansen & Michalke, 2007; IRENA, 2019).

Table 5. Different integration codes and compliances

Region/Standard	LVRT Profile	HVRT Profile	Reactive Support	Duration
Europe (ENTSO-E RfG)	0% V for 150ms; 20% for 625ms	110% V for 100ms; 130% for 20ms	100% Iq at 50% V dip	1.5s recovery
USA (FERC 661)	0% V for 150ms; 15% for 625ms	110% V continuous; 120% for 1s	1.05 pu Iq during faults	0.625s min
China (GB/T 19963)	0% V for 0.625s; 20% for 20s	120% V for 0.1s; 110% for 10s	1.05 pu Iq at 20% V	3s full recovery
Australia (NER)	0% V for 250ms; 25% for 500ms	130% V for 0.2s continuous	Dynamic Iq ramp	0.5s post-fault
Brazil (ANEEL)	0% V for 100ms; 15% for 625ms	110% V continuous	Max Iq during sag	2s recovery

Fault Ride-Through (FRT) Capabilities and Implementation

Fault ride-through capabilities are implemented using advanced converter controls, crowbar and chopper circuits in DFIGs, series and shunt compensation devices, and coordinated protective functions that limit over-currents and over-voltages (Hansen & Michalke, 2007). For example, improved PMSG-based systems often combine multilevel inverters and intelligent voltage controllers, such as type-2 fuzzy logic, to achieve LVRT performance with low total harmonic distortion and improved grid compatibility.

Provision of Ancillary Services: Frequency and Voltage Support

Wind converters also contribute to ancillary services such as frequency containment, voltage regulation, and power oscillation damping by modulating active and reactive power within converter and mechanical limits (IRENA, 2019; Muyeen, 2012). Power quality issues, including harmonics and flicker, are mitigated through appropriate switching strategies, filters, and advanced control schemes, often supported by standards specifying harmonic emission limits and dynamic voltage behavior (Muyeen, 2012; Teodorescu et al., 2011). Modern wind converters increasingly provide synthetic inertia to support grid frequency stability as conventional synchronous generation is displaced. Recent advances in flexible synthetic inertia optimization enable wind plants to adaptively modulate their inertial response based on real-time grid conditions, significantly enhancing frequency containment capabilities in inverter-dominated power systems (Makolo et al., 2024).

Mitigation of Power Quality Issues (Harmonics, Flicker)

Wind converters mitigate harmonics and flicker through multilevel topologies, advanced PWM strategies, and filters to meet grid codes limiting THD to <5% and flicker severity (P_st <1.0). Harmonics arise from PWM switching; flicker from wind variability and tower shadow effects.

Passive filters (LCL, tuned 5th/11th) at turbine/system levels detune resonances, while active compensation via grid-side converters damps low-order harmonics using dc-link energy. Multilevel converters (NPC/MMC) inherently reduce dv/dt and THD.

7. EMERGING TRENDS AND FUTURE DIRECTIONS

Wide-Bandgap Semiconductors (SiC, GaN) in Wind Converters

Wide-bandgap semiconductor devices such as silicon carbide (SiC) and gallium nitride (GaN) are reshaping wind converter design by enabling higher switching frequencies, higher junction temperatures, and significantly lower losses than conventional silicon IGBTs (Blaabjerg et al., 2024; Millan et al., 2014). In multi-MW and offshore turbines, SiC-based devices are particularly attractive for medium- and high-voltage back-to-back converters and HVDC export interfaces, where their high breakdown strength and superior thermal performance support compact, high-power-density designs and improved reliability under harsh cyclic loading (Blaabjerg et al., 2024; Millan et al., 2014).

Figure 19. Application space of Si, SiC

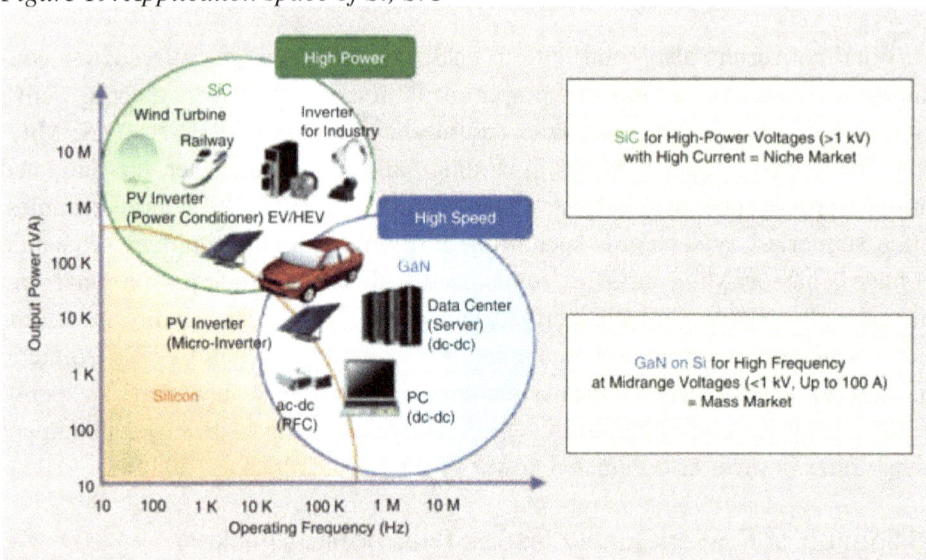

By contrast, GaN—particularly GaN-on-Si—occupies the high-speed (high-frequency) region at midrange voltages (<,1 kV, up to ~100 A), which aligns with mass-market applications such as AC–DC front ends (PFC), server/data-center DC–DC, and auxiliary/intermediate converters. The overlap region (e.g., EV/HEV and PV inverter applications in Figure 5.8) motivates hybrid device partitioning across stages, rather than a single-technology replacement strategy across the entire wind-energy conversion chain.

Despite these advantages, large-scale WBG adoption in wind power converters remains constrained by engineering, cost, and reliability considerations. Reported system-level studies indicate that performance improvements are typically realized through coordinated converter and system co-design, rather than by straightforward device substitution. Key constraints include device cost and packaging maturity, increased complexity in gate driving and protection, and electromagnetic compatibility (EMC) and insulation-coordination challenges associated with high voltage slew rate ($\frac{dv(t)}{dt}$) given in (27).

$$i_{cm}(t) \approx C_{par}\frac{dv(t)}{dt} \tag{27}$$

where C_{par} denotes the effective parasitic capacitance to ground (or across insulation barriers), highlighting that higher $\frac{dv(t)}{dt}$ directly drives larger displacement/common-mode currents and thus elevates EMC and insulation stress. In particular, rapid voltage transitions and elevated common-mode stress can intensify partial discharge risk and accelerate insulation aging in high-frequency magnetic components and related insulation systems, posing challenges to long-term reliability (Castellazzi et al., 2019). Consequently, WBG deployment is evolving from device-level replacement toward holistic system-level co-design that jointly considers semiconductor selection, converter topology, modulation and switching strategies, filtering/insulation design, and thermal management. Reported implementations further suggest that gains in efficiency and power density are often achieved via moderately increased switching frequency combined with multilevel topologies and optimized magnetic design, rather than through high-frequency operation alone. These trends are accelerating hybrid-device and hybrid-topology solutions, where SiC and GaN are allocated across conversion stages according to electrical and thermal stress profiles instead of relying on a single device type across the entire system.

Hybrid Converter Topologies

Hybrid converter topologies aim to exploit the complementary strengths of different semiconductor devices and structural concepts, which is especially relevant for wind plants increasingly co-located with PV, storage, and HVDC infrastructure (Teodorescu et al., 2011). Examples include converters that place SiC devices in high-stress positions while retaining silicon or GaN switches in less demanding legs, as well as multiport and multi-input DC–DC and DC–AC structures that interface wind, PV, and storage on a common DC bus, thereby reducing the number of conversion stages and simplifying coordinated MPPT and power management (Teodorescu et al., 2011).

From a system-level viewpoint, recent review studies reveal that hybrid power conversion topologies are transitioning from interim design solutions to core architectural elements underpinning integrated wind–solar–storage energy systems (Najafzadeh et al., 2021). By facilitating the coordinated integration of diverse energy sources via multiport structures or shared DC buses, hybrid topologies improve overall system efficiency, enhance operational flexibility, and enable effective power coordination and scalability among heterogeneous generation units. It can be said that the integration of wind power converters with distributed generation architectures represents a key pathway toward smarter, more resilient power systems (Adefarati & Bansal, 2019). Hybrid converter topologies facilitate coordinated energy management across heterogeneous renewable sources, enabling wind plants to participate in broader microgrid and distributed generation frameworks.

At the same time, hybrid and multiport power conversion architectures, while enhancing system integration, introduce a new set of engineering challenges. Disparities in dynamic behavior, power fluctuation levels, and control time scales among wind, photovoltaic, and energy storage units strengthen internal coupling, which in turn elevates the complexity of control and energy management. Moreover, shared DC bus architectures place stricter demands on protection design and fault isolation, since DC-side disturbances can propagate across multiple energy sources. Consequently, the design and selection of hybrid power conversion topologies necessitate a holistic trade-off among efficiency, reliability, control complexity, and energy management objectives, moving beyond optimization at the device or individual converter level alone (Najafzadeh et al., 2021).

Table 6 is a table that provides a comparison of most common hybrid converter topologies.

Table 6. Comparison of most common hybrid converter topologies

Topology	Key Features	Wind Energy Applications	Advantages
SiC-Silicon Hybrid	SiC devices in high-stress positions; silicon/GaN in low-stress legs	Multi-MW offshore turbines, HVDC export	Higher efficiency, reduced losses, compact design
Multiport DC-DC/DC-AC	Common DC bus interfacing wind, PV, storage	Hybrid wind-PV farms, multi-input MPPT	Fewer stages, simplified control, better power sharing
Cuk-SEPIC MIC	Series/parallel integration of Cuk and SEPIC converters	PV-WECS hybrids with single controller	Reduced switches, simultaneous source management
Transformer-Based MIC	Low-voltage side with multiple ports; high-voltage rectifier	Scalable m-input wind hybrids with storage	High gain, isolation, fault tolerance

Grid-Forming Control for Inverter-Dominated Grids

Grid-forming control is emerging as a key technology for operating inverter-dominated systems, shifting wind-turbine converters from passive grid-following units to active providers of voltage, frequency, and inertia-like services (IRENA, 2019). Grid-forming inverters behave as controlled voltage sources with internal frequency and voltage dynamics—often implemented through virtual synchronous machine, droop-based power synchronization, or advanced model-predictive schemes—so that large wind plants can contribute to frequency containment, black start, and stable islanded or weak-grid operation even when synchronous generation is scarce (IRENA, 2019).

8. KEY RECOMMENDATIONS FOR INDUSTRY AND RESEARCH

Technology Adoption Pathways

Wide-Bandgap Semiconductor Integration: The deployment of silicon carbide (SiC) and gallium nitride (GaN) devices offers substantial improvements in switching frequency, junction temperature capability, and power loss reduction compared to conventional silicon IGBTs. However, successful implementation requires holistic system-level co-design addressing gate drive complexity, electromagnetic compatibility challenges, and insulation coordination for high dv/dt operation. Industry should prioritize hybrid device strategies allocating WBG semiconductors to high-stress converter stages while retaining silicon devices where appropriate, rather than pursuing wholesale technology replacement (Adefarati & Bansal, 2019; Makolo et al., 2024).

Modular Multilevel Converters for Offshore Applications: MMC topologies provide superior voltage scalability, redundancy, and harmonic performance essential for offshore wind farms interfacing with HVDC transmission. Despite higher component count and control complexity, their benefits in terms of power quality, fault tolerance, and HVDC compatibility justify adoption in large-scale offshore installations exceeding 500 MW capacity.

Hybrid and Multiport Converter Architectures: As wind generation increasingly co-locates with solar PV and energy storage, multiport power conversion topologies enable efficient energy coordination through shared DC buses. These architectures reduce conversion stages, improve system efficiency, and simplify power management, though they introduce additional challenges in protection coordination and DC-side fault isolation that require careful engineering attention.

Advanced Control Strategy Evolution

Transition to Grid-Forming Control: As synchronous generation retires and inverter-based resources dominate, grid-forming control strategies become essential for maintaining system stability. Wind converters must evolve from passive grid-following units to active voltage sources providing frequency and voltage regulation through virtual synchronous machine, droop-based power synchronization, or advanced model-predictive control schemes. This transition enables large wind plants to contribute to frequency containment, voltage support, and even black-start capability in low-inertia grids (Gidwani et al., 2013; Makolo, 2013).

AI-Assisted Adaptive Control: Intelligent supervisory control layers leveraging machine learning can adaptively tune controller parameters, optimize current limits, and adjust operating strategies based on real-time grid conditions, wind variability, and converter aging. Implementation should position AI as a bounded supervisory layer augmenting—not replacing—certified control and protection functions, ensuring verifiability and safe fallback under out-of-distribution conditions.

Coordinated Wind Farm Control: Plant-level coordinated control optimizing power output, reactive power distribution, and ancillary service provision across multiple turbines enhances grid support capability and economic performance. Advanced optimization algorithms balancing energy production, voltage support, and mechanical stress mitigation should be deployed at the wind farm supervisory control level (Makolo et al., 2018).

Digital Transformation and Operational Intelligence

AI-Enhanced Digital Twins: Closed-loop digital twin frameworks combining physics-based models with data-driven corrections enable real-time condition monitoring, predictive maintenance, and performance optimization. Effective implementation requires online calibration mechanisms, physical constraint enforcement, and uncertainty quantification to maintain model validity as systems age and operating conditions change.

Data-Driven Surrogate Modeling: Rapidly callable surrogate models trained on high-fidelity EMT simulation data accelerate design optimization, parameter screening, and multi-scenario analysis. Physics-informed learning approaches incorporating energy balance, dq-frame dynamics, or impedance consistency constraints improve generalization across operating regimes and reduce dependence on extensive training datasets.

Embedded AI for Real-Time Perception: Deployment of lightweight AI models on existing DSP/MCU platforms enables low-latency sensor fault detection, parameter estimation, and health monitoring without compromising hard real-time control deadlines. Model compression, pruning, and quantization techniques make embedded inference feasible within typical 50-100 μs control cycles.

Future Trends

Grid Integration and Ancillary Services

Synthetic Inertia and Fast Frequency Response: Wind plants will increasingly provide synthetic inertial response emulating the frequency-stabilizing characteristics of synchronous generators. Flexible synthetic inertia optimization algorithms adaptively modulating inertial support based on wind availability, grid conditions, and economic incentives will become standard features (Makolo et al., 2024).

Dynamic Voltage Support and Power Oscillation Damping: Beyond basic reactive power provision, advanced control strategies enable wind converters to actively damp inter-area oscillations, provide fast voltage support during disturbances, and contribute to system stability margins previously maintained exclusively by synchronous condensers and FACTS devices.

Black-Start and Islanded Operation: Grid-forming wind plants equipped with energy storage will increasingly participate in system restoration following blackouts, either energizing local grids in island mode or supporting black-start procedures for conventional generation, enhancing overall grid resilience.

System Architecture Evolution

Software-Defined Power Electronics: Modular converter platforms with reconfigurable hardware and software-defined control will enable rapid deployment of new grid support functions, seamless firmware updates, and adaptive responses to evolving grid codes without hardware modifications. This paradigm shift mirrors software-defined networking in telecommunications.

Integrated Wind-Solar-Storage Systems: Co-optimization of wind generation with collocated solar PV and battery storage through unified power conversion and energy management systems will improve capacity factors, reduce transmission infrastructure requirements, and provide firm renewable capacity with minimal curtailment.

HVDC Integration for Offshore Wind: Multi-terminal HVDC networks collecting power from multiple offshore wind farms and delivering it to onshore grids at diverse connection points will become standard infrastructure. Advanced converter control coordinating offshore AC collection systems with HVDC transmission enables optimal power flow management and enhanced system reliability.

9. SUMMARY AND CONCLUSION

Modern wind turbines have evolved from simple fixed-speed mechanical-electrical energy converters into highly sophisticated, fully controllable power-electronic-based generation units whose operational behavior is predominantly determined by converter topology, semiconductor technology, and advanced control strategies rather than by the generator alone. This chapter has examined the key technological pillars that enable this transformation: variable-speed architectures with partial- and full-scale converters, multilevel and modular converter topologies, coordinated machine- and grid-side control strategies, AI-assisted modeling and simulation frameworks, and increasingly stringent grid integration requirements.

The comparative analysis of DFIG-based partial-scale converter systems and PMSG-based full-scale converter architectures revealed distinct trade-offs between cost, efficiency, control complexity, and grid support capability. While DFIG systems remain competitive in onshore applications where lower initial investment is prioritized, full-scale converter topologies offer superior fault ride-through performance, wider speed ranges, and enhanced reactive power support—characteristics that are increasingly valued as wind penetration grows and grid codes become more demanding. The emergence of advanced multilevel topologies such as neutral-point-clamped, flying-capacitor, cascaded H-bridge, and modular multilevel converters has enabled higher voltage operation, reduced harmonic distortion, and improved power density, particularly in offshore and high-power applications where conventional two-level converters reach their practical limits.

10. IMPLICATIONS FOR FUTURE POWER SYSTEMS

The convergence of advanced power electronics, intelligent control, and digital technologies is fundamentally reshaping the role of wind energy in electric power systems. Wind plants are transitioning from passive energy sources that must be accommodated by flexible conventional generation to active grid participants providing essential stability services, voltage and frequency regulation, and even system restoration capabilities. This transformation is critical for enabling high renewable penetration scenarios where wind and solar collectively supply 60-80% of annual electricity generation.

The integration of wide-bandgap semiconductors, modular multilevel converters, grid-forming control strategies, and AI-enhanced operational intelligence will enable wind energy systems to operate as integral, service-oriented components of future carbon-neutral grids. These systems will be characterized by minimal synchronous generation, high inverter penetration, bidirectional power flows, and active consumer participation—operating conditions fundamentally different from the centralized, synchronous-machine-dominated grids of the past century.

Success in this transition requires continued innovation across multiple dimensions: power semiconductor materials and packaging, converter topologies optimized for grid support rather than merely power conversion, control algorithms balancing energy production with system stability services, and digital platforms enabling predictive operation and proactive maintenance. Equally important is the evolution of grid codes, market mechanisms, and regulatory frameworks that properly value the flexibility, controllability, and grid support capabilities that modern wind power converters can provide.

Looking forward, the power electronics technologies and control strategies examined in this chapter represent not merely incremental improvements to existing wind turbine designs, but rather the foundational elements enabling a fundamental reimagining of electric power systems. The modular, software-defined, service-oriented wind energy conversion systems emerging today will be essential building blocks of the resilient, carbon-neutral, and economically efficient electricity infrastructure required to address climate change while meeting growing global energy demands.

REFERENCES

Ackermann, T. (2012). Wind Power in Power Systems. In Ackermann, T. (Ed.), *Wind Power in Power Systems* (2nd ed.). Wiley., DOI: 10.1002/9781119941842

Adefarati, T., & Bansal, R. C. (2019). Energizing Renewable Energy Systems and Distributed Generation. In Tascikaraoglu, A., & Erdinç, O. (Eds.), *Pathways to a Smarter Power System* (pp. 29–65). Academic Press, Elsevier.

Atallah, A. M., & El Tantawy, E. S. F. (2015). Direct torque control of machine side multilevel converter for variable speed wind turbines. *Energy*, *90*, 1091–1099. DOI: 10.1016/j.energy.2015.06.065

Blaabjerg, F., Chen, M., & Huang, L. (2024). Power electronics in wind generation systems. *Nature Reviews Electrical Engineering*, *1*(4), 234–250. DOI: 10.1038/S44287-024-00032-X

Blaabjerg, F., Teodorescu, R., Liserre, M., & Timbus, A. V. (2006). Overview of control and grid synchronization for distributed power generation systems. *IEEE Transactions on Industrial Electronics*, *53*(5), 1398–1409. DOI: 10.1109/TIE.2006.881997

Castellazzi, A., Gurpinar, E., Wang, Z., Hussein, A. S., & Fernandez, P. G. (2019). Impact of wide-bandgap technology on renewable energy and smart-grid power conversion applications including storage. *Energies*, *12*(23). Advance online publication. DOI: 10.3390/EN12234462

Chattejee, D., Chakraborty, C., & Dalapati, S. (2023). Pulse-width modulation techniques in two-level voltage source inverters - State of the art and future perspectives. *Power Electronics and Drives*, *8*(1), 335–367. DOI: 10.2478/PEAD-2023-0023

Davari, M., & Mohamed, Y. A. R. I. (2017). Robust vector control of a very weak-grid-connected voltage-source converter considering the phase-locked loop dynamics. *IEEE Transactions on Power Electronics*, *32*(2), 977–994. DOI: 10.1109/TPEL.2016.2546341

Debnath, S., Qin, J., Bahrani, B., Saeedifard, M., & Barbosa, P. (2015). Operation, control, and applications of the modular multilevel converter: A review. *IEEE Transactions on Power Electronics*, *30*(1), 37–53. DOI: 10.1109/TPEL.2014.2309937

Desalegn, B., Gebeyehu, D., & Tamirat, B. (2022). Wind energy conversion technologies and engineering approaches to enhancing wind power generation: A review. *Heliyon*, *8*(11), e11263. DOI: 10.1016/J.HELIYON.2022.E11263 PMID: 36345520

Errami, Y., Maaroufi, M., Cherkaoui, M., & Ouassaid, M. (2012). Maximum power point tracking strategy and direct torque control of permanent magnet synchronous generator wind farm. *Proceedings of 2012 International Conference on Complex Systems, ICCS 2012*. https://doi.org/DOI: 10.1109/ICOCS.2012.6458520

Errami, Y., Maaroufi, M., & Ouassaid, M. (2013). Maximum power point tracking of a wind power system based on the PMSG using sliding mode direct torque control. *Proceedings of 2013 International Renewable and Sustainable Energy Conference, IRSEC 2013*, 218–223. https://doi.org/DOI: 10.1109/IRSEC.2013.6529706

Gidwani, L., Tiwari, H., & Bansal, R. C. (2013). Improving power quality of wind energy conversion system with unconventional power electronic interface. *International Journal of Electrical Power & Energy Systems*, *44*(1), 445–453. DOI: 10.1016/j.ijepes.2012.07.037

Hansen, A. D., & Michalke, G. (2007). Fault ride-through capability of DFIG wind turbines. *Renewable Energy*, *32*(9), 1594–1610. DOI: 10.1016/J.RENENE.2006.10.008

Hazra, S., Madhusoodhanan, S., Moghaddam, G. K., Hatua, K., & Bhattacharya, S. (2016). Design considerations and performance evaluation of 1200-V 100-A SiC MOSFET-based two-level voltage source converter. *IEEE Transactions on Industry Applications*, *52*(5), 4257–4268. DOI: 10.1109/TIA.2016.2587098

Hu, J., Shang, L., He, Y., & Zhu, Z. Q. (2011). Direct active and reactive power regulation of grid-connected DC/AC converters using sliding mode control approach. *IEEE Transactions on Power Electronics*, *26*(1), 210–222. DOI: 10.1109/TPEL.2010.2057518

IRENA. (2019). *Future of wind: Deployment, investment, technology, grid integration and socio-economic aspects*. International Renewable Energy Agency.

Lezana, P., Silva, C. A., Rodríguez, J., & Pérez, M. A. (2007). Zero-steady-state-error input-current controller for regenerative multilevel converters based on single-phase cells. *IEEE Transactions on Industrial Electronics*, *54*(2), 733–740. DOI: 10.1109/TIE.2007.891994

Lyu, J., Rao, Y., Wang, Z., Dai, J., & Cai, X. (2023). Data-driven impedance identification and stability online assessment of wind farm connected with MMC-HVDC. *IEEE Transactions on Industry Applications*, *60*(2), 2567–2576. DOI: 10.1109/TIA.2023.3333760

Makolo, P. (2013). *Wind generator co-simulation with fault case analysis* [Master's thesis, Chalmers University of Technology]. Department of Energy and Environment.

Makolo, P., Justo, J. J., Mwasilu, F., & Zamora, R. (2018). Fault ride through technique for DFIG-based wind turbines under grid three-phase faults. In *2018 Australasian Universities Power Engineering Conference (AUPEC)* (pp. 1–5). IEEE. https://doi .org/DOI: 10.1109/AUPEC.2018.8757993

Makolo, P., Zamora, R., Perera, U., & Lie, T. T. (2024). Flexible synthetic inertia optimization in modern power systems. *Inventions (Basel, Switzerland)*, *9*(1), 18. DOI: 10.3390/inventions9010018

Millan, J., Godignon, P., Perpina, X., Perez-Tomas, A., & Rebollo, J. (2014). A survey of wide bandgap power semiconductor devices. *IEEE Transactions on Power Electronics*, *29*(5), 2155–2163. DOI: 10.1109/TPEL.2013.2268900

Muyeen, S. M. (2012). Wind energy conversion systems: Technology and trends. *Green Energy and Technology*, *78*. Advance online publication. DOI: 10.1007/978-1-4471-2201-2

Najafzadeh, M., Ahmadiahangar, R., Husev, O., Roasto, I., Jalakas, T., & Blinov, A. (2021). Recent contributions, future prospects and limitations of interlinking converter control in hybrid AC/DC microgrids. *IEEE Access : Practical Innovations, Open Solutions*, *9*, 7960–7984. DOI: 10.1109/ACCESS.2020.3049023

Okedu, K. E., & Muyeen, S. M. (2022). Comparative performance of DFIG and PMSG wind turbines during transient state in weak and strong grid conditions considering series dynamic braking resistor. *Energies*, *15*(23). Advance online publication. DOI: 10.3390/EN15239228

Polinder, H., Van Der Pijl, F. F. A., De Vilder, G. J., & Tavner, P. J. (2006). Comparison of direct-drive and geared generator concepts for wind turbines. *IEEE Transactions on Energy Conversion*, *21*(3), 725–733. DOI: 10.1109/TEC.2006.875476

Rodríguez, J., Bernet, S., Wu, B., Pontt, J. O., & Kouro, S. (2007). Multilevel voltage-source-converter topologies for industrial medium-voltage drives. *IEEE Transactions on Industrial Electronics*, *54*(6), 2930–2945. DOI: 10.1109/TIE.2007.907044

Shukla, S., & Singh, B. (2019). MPPT control technique for solar powered direct torque control of induction motor drive with a robust speed and parameters adaptation scheme for water pumping. *IET Renewable Power Generation*, *13*(2), 273–284. DOI: 10.1049/iet-rpg.2018.5390

Subedi, S., Gui, Y., & Xue, Y. (2025). Applications of data-driven dynamic modeling of power converters in power systems: An overview. *IEEE Transactions on Industry Applications*, *61*(2), 2434–2456. DOI: 10.1109/TIA.2025.3529797

Teodorescu, R., Liserre, M., & Rodriguez, P. (2011). *Grid converters for photovoltaic and wind power systems* (1st ed.). Wiley.

Vukosavić, S. N. (2018). *Grid-side converters control and design: Interfacing between the AC grid and renewable power sources*. Springer.

Xiao, C., Liu, Z., Zhang, T., & Zhang, X. (2021). Deep learning method for fault detection of wind turbine converter. *Applied Sciences (Basel, Switzerland)*, *11*(3), 1–22. DOI: 10.3390/APP11031280

Xu, Y., Zhou, B., Sun, X., Tian, Y., & Jiang, X. (2025). Dynamic modeling and online updating of full-power converter wind turbines based on physics-informed neural networks and Bayesian neural networks. *Electronics (Basel)*, *14*(15). Advance online publication. DOI: 10.3390/ELECTRONICS14152985

Zarei, M. E., Ramirez, D., Veganzones, C., & Rodriguez, J. (2020). Predictive direct control of SPMS generators applied to the machine side converter of an OWC power plant. *IEEE Transactions on Power Electronics*, *35*(7), 6719–6731. DOI: 10.1109/TPEL.2019.2956738

Zhang, Z., Zhao, Y., Qiao, W., & Qu, L. (2014). A space-vector-modulated sensorless direct-torque control for direct-drive PMSG wind turbines. *IEEE Transactions on Industry Applications*, *50*(4), 2331–2341. DOI: 10.1109/TIA.2013.2296618

Zhang, Z., Zhao, Y., Qiao, W., & Qu, L. (2015). A discrete-time direct torque control for direct-drive PMSG-based wind energy conversion systems. *IEEE Transactions on Industry Applications*, *51*(4), 3504–3514. DOI: 10.1109/TIA.2015.2413760

Chapter 4
Navigating Grid Code Requirements With Advanced Monitoring and Control

Dieudonne Musongya Bisimwa
http://orcid.org/0009-0008-3991-1043
University of Dar es Salaam, Tanzania

Ibrahim Abdallah Mwammenywa
http://orcid.org/0000-0003-2735-643X
University of Dar es Salaam, Tanzania

ABSTRACT

The modern power grid is progressively being released from traditional plants in response to environmental concerns. This is possible due to advanced technologies in managing and controlling renewable energy resources, which enable these to participate in complex power systems. Wind and solar PV power stations are increasingly becoming significant on a global scale, continuously expanding their presence due to their ability to deliver environmentally friendly, sustainable energy. Moreover, integrating renewable energy sources, especially wind power, poses challenges related to management and control. This chapter offers detailed information regarding the requirements of grid codes, as well as the latest monitoring and control strategies that ensure stability, reliability, and the effective operation of wind power system integration. Additionally, it gives some practical worked examples and illustrations to emphasize the necessity for updated policies. The execution of these regulatory frameworks is essential for enabling the effective incorporation of renewable energy into the grid.

DOI: 10.4018/979-8-2600-0106-6.ch004

1 INTRODUCTION

Many countries are endowed with rich natural resources, such as wind, solar, geothermal energy, and rivers, which can be harnessed for electricity generation to reduce dependence on imported fossil fuels. In 2018, the most appealing sectors for renewable energy (RE) investments were solar and wind power, commanding significant shares of 48.9% and 47.2% of total new investments, respectively, with other sectors lagging significantly behind. In the same year, RE resources accounted for 26.2% of global electricity generation. The countries that led in electricity consumption from RE resources included Canada, Sweden, Denmark, Portugal, Chile, Germany, Spain, Italy, Ireland, and the United Kingdom (Kurbatova & Perederii, 2020). Additionally, Grid codes (GCs) establish the technical regulations and behaviors for all participants in the power system, including power generators, adjustable loads, and storage assets. Their implementation ensures that system operators are confident these connected assets will not compromise the security of the electricity supply (Vrana et al., 2018, IRENA, 2022). It is internationally acknowledged that RE policies are fundamental to fostering the advancement of RE. Going forward, the current trends in RE indicate that a robust policy framework will be crucial for the large-scale development of RE (Zeng et al., 2013).

Over the last 15 years, technical standards for RE grid integration in Europe, North America, and China have primarily been established by electricity associations or power grid companies. As RE generation has augmented and its effects on the power grid have become more pronounced, more countries are legitimizing RE grid integration standards through national legislation. Countries that have been early adopters of wind power have also developed specific wind power grid-connected codes tailored to their national situation. In 2000, ELTRA in Denmark established regulations for wind farms connected to power systems. Ireland's National Grid Corporation followed in 2002 with its own wind grid connection standards. Scotland's Transmission and Distribution Company, alongside the Scottish Hydropower Company, proposed regulations for wind grid integration. In 2003, E.ON in Germany issued a standard for connecting wind power to high-voltage grids. By 2005, the US Federal Energy Regulatory Commission enacted Decree No. 661 to outline Wind Power Integration Procedures. These developments reflect ongoing improvements in wind power grid codes worldwide (Yan et al., 2019)

Besides this, on December 12, 2005, China established the "Technical Rule for Connecting with Wind Farm to Power Systems" (GB/Z 19963-2005) to support large-scale wind power integration. This marked the introduction of technical requirements for connecting wind farms to power systems. Since 2009, revisions have been made to national standards, incorporating international best practices while considering China's unique wind power context. The updated guidelines include provisions for active/reactive control, output forecasting, low voltage ride through capability, and requirements for large wind power bases of 1 million kilowatts. However, as RE integration scales up, new operational issues are emerging, posing significant challenges for the industry and research focus (Yan et al., 2019)

On the other hand, to address 20% of the energy demand, the US Department of Energy plans to integrate an additional 300 GW of wind power plants (WPPs) by 2030. The Federal Energy Regulatory Commission (FERC) establishes requirements for WPPs, focusing on ensuring reliable and sustainable energy services. FERC's GCs for WPP interconnection emphasize low voltage ride through (LVRT) capability, power factor design criteria, reactive power capability, and Supervisory Control and Data Acquisition (SCADA) systems, reflecting ongoing global improvements in WPP GCs requirements (Islam et al., 2020). Moreover, by 2019, the global installed capacity of wind power reached 650.8 GW, with an addition of 59.7 GW in that year alone. Projections indicated that installed capacity would rise to 664.5 GW by the end of 2019, reflecting an increase of 65.4 GW, or approximately 17.4% from 2018. Geographic distribution of wind turbines shows that Asia holds the leading position, followed by North America, Europe, and the rest of the world, as depicted in Figure 1 (Abubakr et al., 2022)

Figure 1. Wind Capacity in 2018 by Region (Abubakr et al., 2022)

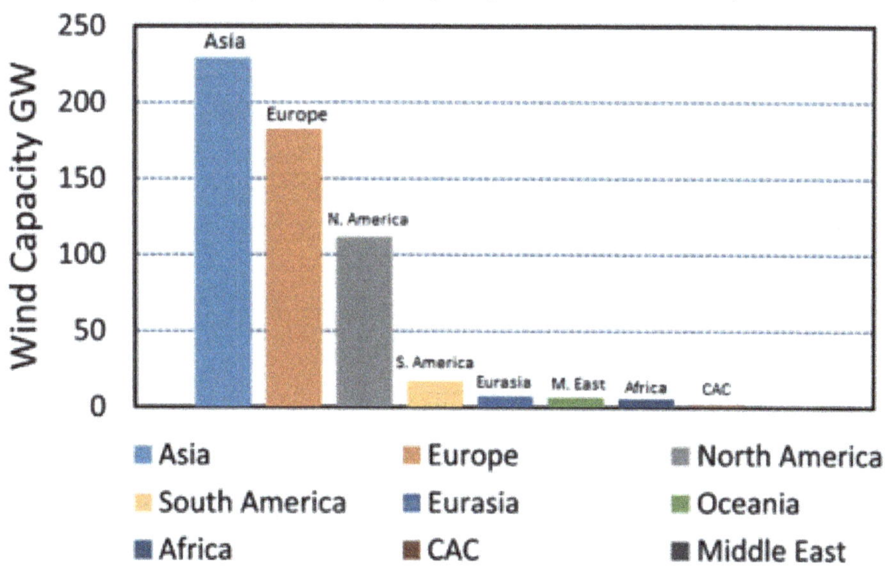

In this chapter, details and current information regarding the requirements and policies of GCs, as well as the latest monitoring and control strategies that ensure stability, reliability, and the effective operation of wind power system integration, are provided. The rest of the chapter is organized in the following sections: Section II is dedicated to global standards of RE. Section III offers a concise overview of the recent GCs for wind power integration into the power grid. GCs for wind power advanced control methods, and recent monitoring techniques are developed in Section IV. Then, practical examples and illustrations are given in Section V and future trends in Section VI. Finally, Section VII concludes the study, summarizing the key insights and outcomes.

2 GLOBAL STANDARDS OF RENEWABLE ENERGY

Across the world, different countries have developed RE plans tailored to their needs, establishing technical specifications and standards, notably influenced by IEEE and IEC standards (Zheng & Tang, 2012). The International RE Agency (IRENA) defines "RE" as energy consistently generated from renewable sources, including hydropower, bioenergy, ocean energy, geothermal, wave, tidal, wind, and solar power, as well as other sources like landfill gas and biogas (Tovstukha, 2017).

2.1 Non-Technical Standards of Renewable Energy

At present, RE development currently relies on various factors, including policy, geographical, technological, economic, and social. It should be noted that energy production from renewables is currently more expensive than from fossil fuels, making it difficult for "green" energy to compete with traditional energy generation. Thus, the growth of renewable energy relies heavily on government support through effective mechanisms that encourage green energy generation and consumption (Kurbatova & Perederii, 2020).

Nowadays, many countries throughout the world use diversified support schemes to promote RE expansion. The most popular are presented below:

- The feed-in tariff is a pricing mechanism that guarantees government purchases of renewable energy, securing returns for investors. It is the predominant global strategy for promoting renewable energy, adopted by 111 countries or regions by the end of 2018. Various adaptations have emerged, including feed-in premium, fixed feed-in tariff, front-end loaded feed-in tariff, and adjusting feed-in tariff (Carley et al., 2017).
- The renewable portfolio standard mandates certain quotas for electricity generation from RE sources, affecting all energy market participants. Producers sell electricity at conventional market prices, while additional costs are com-

pensated through the sale of green certificates. This mechanism ensures sufficient revenue for covering generation costs and profits, utilized in countries like Norway, the USA, Poland, Denmark, Romania, and Australia (Carley et al., 2017).

- The net metering system supports households with renewable energy capabilities by using two-sided meters to measure electricity flow in both directions. This system incentivizes private households to install renewable facilities, ensuring that any excess electricity can be sold back to the grid at retail prices. It is implemented in countries like Denmark, the USA, Japan, Italy, Singapore, Cyprus, and Greece (Kurbatova & Perederii, 2020).
- The tender schemes involve announcing tenders for constructing RE plants, where the authority specifies desirable characteristics such as installed capacity, implementation timing, and environmental impact. Potential investors can then submit competitive proposals, with the winner receiving partial state financial compensation for their investment project, as seen in countries like the United Kingdom, Israel, Belgium, Spain, Slovenia, and Hungary.
- Tax and customs initiatives often provide full or partial exemptions from income tax, value-added tax, and customs duties on equipment and components for RE plant construction, as seen in countries like Ukraine, Canada, Switzerland, the Czech Republic, and Austria (Kurbatova & Perederii, 2020).

Creating effective policies for renewable energy markets is complex, as there is currently no ideal support scheme to ensure the successful deployment of renewable energy capacities. Consequently, many countries utilize multiple mechanisms concurrently. As of the end of 2018, according to Figure 2, 135 countries have implemented special incentive mechanisms in the power sector, compared to 70 in the transport sector and 20 in heating and cooling, highlighting the power sector as the most developed in terms of renewable energy incentives (Kurbatova & Perederii, 2020).

Figure 2. Several Countries with RE Policy by Sector 2012- 2018 (Kurbatova & Perederii, 2020).

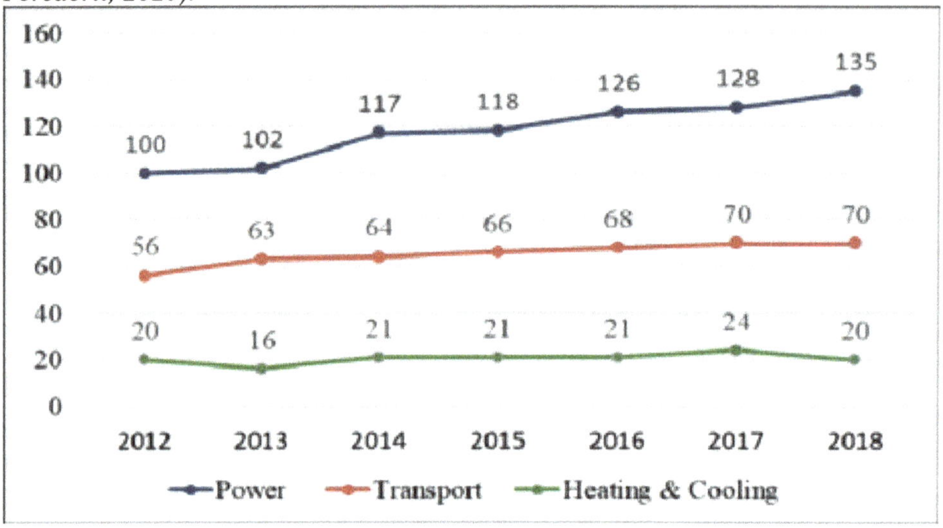

2.2 Technical Standards of Renewable Energy Grid Integration

GCs for RE integration vary by country but share common elements, emphasizing the need for active power control, reactive power management, fault ride-through (FRT) capability, and the ability to tolerate voltage and frequency deviations in the power system (Yan et al., 2019). Innovative solutions like grid-forming inverters, dynamic line rating (DLR), and advanced energy storage systems have been developed to address the complexities introduced by the intermittent nature of solar and wind energy, improving grid stability and maximising renewable energy utilization (Cavus, 2025).

2.2.1 Active Power Regulations

To effectively manage the increasing active power from RE generation systems, it is crucial to control their output as a function of active power. The primary operation of the power grid involves supplying sufficient power to the market to balance load and generate electricity, while also considering grid stability and transmission capacity limitations. The power station's scheduling mechanism regulates the generation of RE by adjusting the generator's output between its rated and minimum levels based on resource availability. As RE generation relies on external conditions, there are limited minimum operating levels. When sufficient resources are present, the generator will modulate its output to align with available capacity (Yan et al., 2019).

2.2.2 Reactive Power Regulations

The power system operator regulates generator reactive power to maintain voltage levels. Conventional large power plant generators provide substantial reactive power, whereas RE generators have limited capabilities. Enhancing reactive power capacity in RE generators can improve their integration by reducing voltage fluctuations and meeting reactive power demands, as they replace traditional power plants. The reactive power requirements should align with the largest renewable power plants to minimize dependency on conventional voltage control capacity (Yan et al., 2019).

2.2.3 Frequency Regulations

Frequency is a critical global parameter in synchronous AC power systems, maintained by balancing generation and load at the transmission level. Stability is threatened by sudden imbalances, such as faults leading to increased load or disconnections of power plants. To address frequency variations, operators keep power reserves within the control area for continuous balancing. Therefore, it emphasizes improving energy permeability while keeping frequency stability. Generator response to frequency interference differs depending on the imbalance type (Yan et al., 2019).

2.2.4 Faults Ride Through

A fault in a power system, such as a short circuit, causes high current that can rapidly damage equipment. Effective fault detection is achieved through relays that disconnect faulty components while keeping the generator connected for proper fault detection and maintaining power balance. The behavior analysis of RE generators shows that their requirements are akin to conventional generators but with specific conditions for grid reconnection post-voltage drops. RE generators must support current for fault detection, constrained by technology, and unlike traditional generators, they must deliver excess current beyond their rated capacity, affecting overall costs (Yan et al., 2019).

2.2.5 Power Quality

Power quality aspects include deviations in voltage and current from nominal values alongside amplitude and frequency. These deviations, which relate to waveform characteristics and short-term fluctuations, should be measured with appropriate methods. These aspects of power quality are particular to a specific position in the grid observed over a designated time interval (Yan et al., 2019). Technical requirements for frequency and voltage for renewable energy are given in Table 1.

Table 1. Technical requirements for frequency and voltage control in power systems with renewable energy sources. (Murdan et al., 2023)

Organization	European Union ENTSO-E	United States Federal Energy Regulatory Commission	China National Energy Administration	India Central Electricity Regulatory Commission
Frequency control	Technical requirements for frequency control in power systems integrated with wind and solar energy sources	Technical requirements for frequency control in hybrid power systems, and guidelines for integrating RE sources	Technical requirements for frequency and voltage control in hybrid power systems	Technical requirements for frequency and voltage control in power systems integrated with wind and solar energy sources
Voltage control	Requirements for voltage control in hybrid power systems outlined in the European Network Code on Requirements for Grid Connection of Generators	Requirements for reactive power support, power factor control, and compliance regulations for wind and solar energy sources-FERC Order 845	Requirements for reactive power and power factor control, voltage stability, testing and certification requirements for voltage control equipment	Requirements for reactive power control, power factor control, voltage stability, testing and certification requirements for voltage control equipment

All generators in their vicinity affect power quality, with impacts varying based on generator technology. Conventional generators differ from RE (generators in this regard. The RE Grid Code establishes rules addressing various aspects of a generator's impact to guarantee that all equipment and users connected comply with a specific voltage quality standard (Yan et al., 2019).

3 GRID CODES FOR WIND POWER INTEGRATION INTO THE POWER GRID

Wind energy is an inexhaustible and environmentally friendly alternative to fossil fuels that reduces greenhouse gases and diversifies the electricity supply. The conversion process includes four stages: wind turbine, gearbox (used in some WPPs), electric generator, and power electronics converter. Wind turbines convert wind's kinetic energy into mechanical energy, while the gearbox increases the turbine shaft's speed (Salameh et al., 2018). The generator then converts mechanical power to electrical power, regulated by power electronics before integration into the grid.

Wind turbine designs are classified into five types, and harnessing wind energy can occur both onshore and offshore (Ullah et al., 2025; Ogunlana & Ezeoha, 2021).

In recent years, the integration of large-scale WPPs into utility grids poses challenges for grid stability during disturbances, as disconnecting them rapidly can be unfavorable. Consequently, global power system operators are revising GCs to support the high penetration of WPPs. Compliance with these updated GCs is essential for WPP owners to ensure continuous connection to the grid while maintaining stability during both steady-state and dynamic operations (Ullah et al., 2025).

The interconnection standards for RE power established by institutions like IEEE and IEC focus on two main aspects: the structure and performance of the generation system, and the operational, control, and protection impacts on the power system once the generation system connects to the grid. While the specific content of wind power interconnection standards varies across different countries, the fundamental categories remain consistent, as indicated in Table 2 (Zheng & Tang, 2012).

Table 2. Contents of Interconnection for Wind Power (Zheng & Tang, 2012)

	Contents
General requirements	Power control, voltage and frequency regulation
Power quality	Operating voltage and frequency, harmonic and waveform distortion, voltage unbalance
Safety and protection	Abnormal condition requirement, LVRT
Test	Basic requirements, test contents
Communication signal	Normal and fault signal
Model information	Simulation model and parameter

3.1 Grid Code Requirements Under Steady State Conditions

The GCs for WPP integration in steady state focus on frequency control, voltage control, and power quality, varying by country due to factors such as network operator requirements and wind power levels. Larger WPPs must comply with stricter requirements, especially in countries with high renewable energy shares, which necessitate support for system inertia and frequency fluctuations. Compliance is mandatory at the point of common coupling (PCC), irrespective of the plant's size or type. (Ullah et al., 2025).

3.1.1 Frequency Control Requirements

A wind power generation unit lacks the inherent ability to manage frequency variations, necessitating a GC to align power output with these changes. Wind power operators commonly use spinning reserve and supplementary reserves to address frequency deviations, which require sophisticated control strategies (Fernández-Guillamón et al., 2019). Under the British GC, mandatory frequency control applies to WPPs with capacities of 50 MW or more. Similarly, in Denmark, WPPs connected to high-voltage networks must implement specific frequency support functions to adjust active power production in response to frequency variations (Wu, 2018; Ullah et al., 2025).

Frequency variation guidelines and operational requirements for WPPs vary internationally, with most countries (excluding the USA and Canada) adopting a rated frequency of 50 Hz. Germany's WPPs operate between 46.5 and 53.5 Hz, affected by grid voltage and location. In Spain, the frequency operating range is 48–51 Hz within normal voltage, while Australia allows a 2.5% deviation from nominal frequency for connectivity. Denmark's requirements relate frequency operation (FOR) to the installed capacity, contrasting with Canada's stricter range of 55.5–61.7 Hz (Li et al., 2023). The implications of these requirements on WPP frequency stability are explored in greater detail by Díaz-González et al. (2014).

3.1.2 Voltage Control Requirements

WPPs must adjust their reactive power output to respond to voltage changes at the PCC. Their connection to the grid through long subsea cables results in significant transmission losses (Oleka et al., 2016). Fluctuating wind speeds lead to irregular power output from wind turbine generators, complicating voltage stability. Modern wind turbine generators use power electronics to manage voltage through reactive power regulation, though this is constrained by converter limitations. Regulatory bodies have mandated that WPPs provide reactive power support, characterized by curves that link active power to PCC voltage (Khan et al., 2022). Table 3 indicates that the operating voltage range of PCC in certain standards, particularly for countries with significant wind power capacity, is narrower in China compared to Germany and Ireland. This suggests that China's power transmission and distribution networks are less robust in handling the intermittency of wind power (Zheng & Tang, 2012; Ullah et al., 2025).

Table 3. Operating voltage by countries (Ullah et al., 2025).

Denmark		**Germany**		**Ireland**		**Scotland**	
400kV	-10% ~5%	400kV	-8% ~ 10%	400kV	-13% ~ 5%	400kV	±5%
150kV	-3% ~13%	220kV	-13% ~ 12%	220kV	-9 ~ 12%	275kV	±10%
132kV	-5% ~ 10%	110kV	-13% ~ 12%	110kV	-10% ~ 12%	132kV	±10%
UK		**Australia**		**China**		**USA**	
400kV	±5%	All Voltages	±10%	220kV	-10% ~ 10%	≥161kV	±10%
275Kv	±10%			110kV	-10% ~ 10%	<161kV	±2.5%
132kV	±10%						

3.1.3 Power Quality Standards

Key power quality disturbances in steady-state wind power integration include voltage sags, swells, flicker, harmonic distortion, voltage unbalance, and resonance. GCs and power quality standards set limits on voltage magnitude, frequency stability, harmonic emissions, and flicker severity to ensure safe operations. Flicker emission limits vary by country; Denmark has 0.35 for P_{st} and 0.25 for P_{lt}, while Germany allows 0.8 for P_{st} and 0.9 for P_{lt}. The UK and US provide detailed flicker requirements based on system voltage levels. Total harmonic distortion (THD) for voltage and current must remain below 5% according to standards such as IEE Std 51-201, IEEE 1547, and IEC standards, with specific UK limits of 3% for odd and 5% for even THD at the PCC (Cleveland, 2008). Australia's GC mirrors these THD requirements, while sag standards show similarities between Denmark and Germany, with the USA enforcing the strictest swell limits and more flexibility for sag limits (Ullah et al., 2025).

Voltage unbalance factor is a key metric for assessing voltage unbalance in three-phase systems, with different GCs setting specific limits for its acceptable values at PCC. For instance, the German GC allows up to 2.5% unbalance at voltage levels between 60 and 150 kV, while the Danish Grid Code enforces a stricter limit of 0.2% for all voltage levels. The U.K. standards establish two distinct limits: 2% for voltages up to 150 kV and 1.5% for those above 50 kV (Islam et al., 2020; Ullah et al., 2025).

3.2 Grid Code Requirements Under Dynamic Operation

The quick penetration of large-scale WPPs into power grids poses challenges, particularly regarding dynamic performance and system stability during grid faults (Alsmadi et al., 2018). These plants must remain connected to the grid during such disturbances to support stability. The GC includes voltage ride-through (VRT) requirements to ensure safe operation under disturbances, expecting WPPs to endure significant voltage fluctuations relative to nominal values for a defined period (Nguyen et al., 2023).

3.2.1 Low-Voltage Ride-Through and High-Voltage Ride-Through Requirements

The requirements for LVRT and high voltage ride-through (HVRT) in international GCs are essential for ensuring power system stability during voltage disturbances. These requirements mandate that renewable energy generators, particularly wind turbines, have the capability to withstand and manage voltage fluctuations (Murdan et al., 2023). LVRT specifically allows generators to remain grid-connected under low-voltage conditions. Various countries have established GCs that outline voltage levels, durations, and required generator responses; for instance, Germany requires turbines to endure voltage dips to 0.85 p.u., while Denmark mandates a 0.9 p.u. limit. Similar regulations are in place across the US, Australia, and other nations, defining acceptable voltage dip levels and response times for wind turbines (Wu, 2018). Figure 3 depicts the LVRT curves of different European countries.

Figure 3. LVRT Curves for Different European Countries (Wu, 2018).

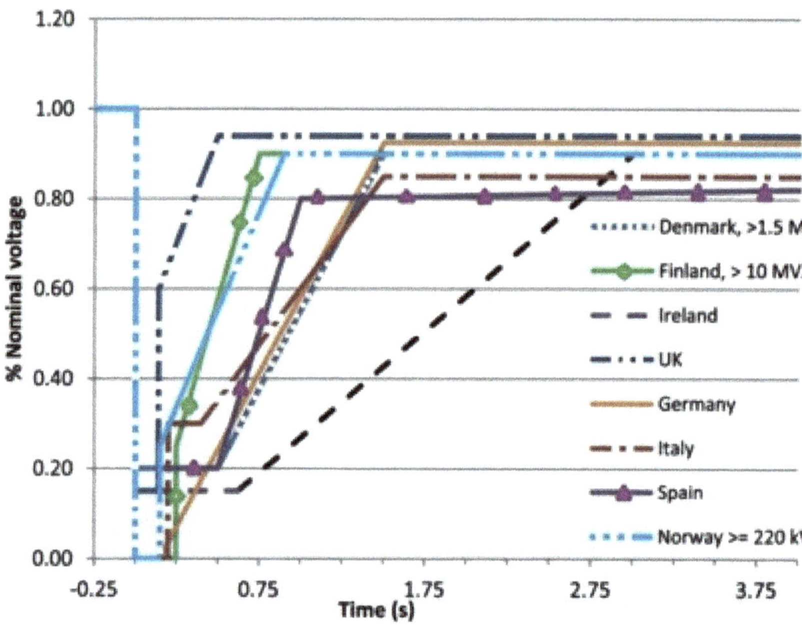

Conversely, HVRT refers to the ability of renewable generators to remain operational during high voltage levels without disconnecting from the grid. This is critical during voltage surges. Grid Codes dictate the acceptable voltage levels and durations for generator endurance, such as the requirement for Spanish wind farms to withstand voltage surges up to 1.3 p.u. for a specific duration, compared to UK wind turbines that must endure levels up to 1.1 p.u, as illustrated in Table 4. These regulations ensure ongoing connectivity and functionality during high voltage events (Murdan et al., 2023).

Table 4. HVRT Requirements in Various Grid Codes (Ali et al., 2021).

	HVRT		**Power factor**	
	Vmax (p.u.)	**Tmax (s)**	**Leading**	**Lagging**
Australia	1.3	0.05	0.93	0.93
Denmark (for more than 1.5 MW)	1.2	0.1	0.975	0.975
Germany onshore	1.26	0.1	0.95	0.925
India	-	-	0.95	0.95
Ireland	1.1	-	0.95	0.95
Spain	1.3	0.5	0.99	0.99
UK	1.1	-	0.95	0.95
China	-	-	0.95	0.95
USA ERCOT	1.175	0.2	0.95	0.95

In addition to these requirements, some GCs require reactive current or power support functions to enhance grid voltage stability during faults. This support is vital during transient voltage changes, as it helps maintain the grid's voltage profile, typically through advanced control algorithms that adjust generators' reactive power output dynamically (Celik et al., 2022). Grid Codes enforce LVRT and HVRT requirements to maintain stable power system operations and enhance grid reliability while supporting the integration of RE generation. These requirements enable renewable generators to contribute to grid stability and overall power system resilience.

4 GRID CODES FOR WIND POWER CONTROL METHODS AND MONITORING TECHNIQUES

4.1 Frequency Regulations in Grid Codes

Frequency control plays a crucial role in international grid codes, which are established to manage deviations in grid frequency. These codes specify the acceptable operational frequency range and delineate the responsibilities of generators in addressing frequency fluctuations resulting from disturbances, thereby encouraging swift actions to stabilize frequency. A key component of many grid codes is the requirement for primary frequency response (PFR), which is vital for maintaining frequency stability, although the complexity of these requirements can differ significantly between countries (Murdan et al., 2023; Li et al., 2023).

Different countries have varying definitions of "droop." In China, it is defined as the ratio of the per-unit change in power to the per-unit change in frequency. In contrast, Denmark and Germany interpret droop as the percentage change in nominal output corresponding to a 1 Hz change in frequency. In the UK and Spain, droop is characterized as the percentage change in frequency for a 100% change in power. Furthermore, the output of WPPs remains constant when frequency fluctuates within the deadband. The smallest deadband is found in the USA's Electric Reliability Council of Texas (ERCOT) Interconnection at 60 ± 0.017 Hz, while larger deadbands exist in Canada's Quebec Interconnection (59~60.5 Hz) and Australia's east coast electricity grid (50±1 Hz). Most other countries maintain a frequency deadband of fn ±0.2 Hz. Additionally, a decrease in system inertia leads to a greater rate of change of frequency (RoCoF) following disturbances. Canada's Quebec Interconnection mandates that WPPs endure a RoCoF of 4 Hz/s, whereas most countries have RoCoF specifications between 1.5–2.5 Hz/s, and as of now, China and Germany have not defined a specific RoCoF value (Li et al., 2023).

The significance of fast frequency services is heightened in countries heavily reliant on renewable energy sources, like Australia, Ireland, New Zealand, and the UK, as shown in Table 5, which face unique challenges due to reduced system inertia. As these systems are often disconnected from larger grids, they lack the stabilizing benefits of conventional generators, placing the burden of frequency stability on new technologies. The deployment of fast frequency response (FFR) services is crucial, as they enable swift adjustments to active power output, preventing issues like wind farm tripping and customer load shedding. Advanced control systems, demand response strategies, and energy storage technologies are essential for ensuring rapid responses and providing the necessary ancillary support for stable grid operation (Murdan et al., 2023).

Table 5. International New Fast Frequency Services (Murdan et al., 2023).

Country	System services	Type	Response time	Duration	Target
Great Britain (National grid)	Enhanced frequency control capability (EFCC)	Post-fault	0.5 s	N/A	Renewable technologies, demand side resources, and other new technologies
Ireland (EirGrid)	FFR	Post-fault	2.0 s	8.0 s	Wind and other emerging technologies
Australia (AEMO)	FFR	Post-fault	1.0 s	N/A	Wind and other technologies
Canada (Hydro-Quebec)	FFR	Post-fault	1.5 s	9.0 s	Wind turbine (6% rated capacity boost)
USA (ERCOT)	FFR	Post-fault	0.5 s	FFR1: up to 10 min FFR2: indefinite (longer than 10 min)	Wind cannot provide response for 10 min Demand response and Battery energy storage

4.2 Reactive Power Regulations

Reactive power regulation is essential in GCs for voltage stability and power factor management, specifying guidelines for reactive power exchanges between generators and the grid. Control strategies include voltage control, voltage-reactive power droop control, and power factor control, allowing generators to adjust output based on grid needs. Some GCs require generators to adhere to specific power factor thresholds to stabilize voltage, while also defining acceptable power factor ranges for adaptability to changing system conditions (Murdan et al., 2023).

In the Danish GC, wind generators exceeding 25 MW must provide a reactive power capability of ±0.3 p.u. within an active power range of 0.2 to 0.8 p.u., decreasing to ±0.2 p.u. as active power increases from 0.8 p.u. to 1 p.u. In Turkey, the plants should be capable of operating in the range of 0.835 leading and lagging power factor (PF) while operating above 50% of the nominal power. This specification reflects a reasonable adjustment in reactive power requirements at higher active power levels, balancing cost and technical limits of wind generators. These requirements are typically illustrated in Figure 4 using P-Q diagrams which show the relationship between available active and reactive power (Sourkounis & Tourou, 2013; Sarkar et al., 2018; Celik et al., 2022).

Figure 4. Grid Code Reactive Power Requirements Specified for Wind Generators in some European Countries (Sarkar et al., 2018).

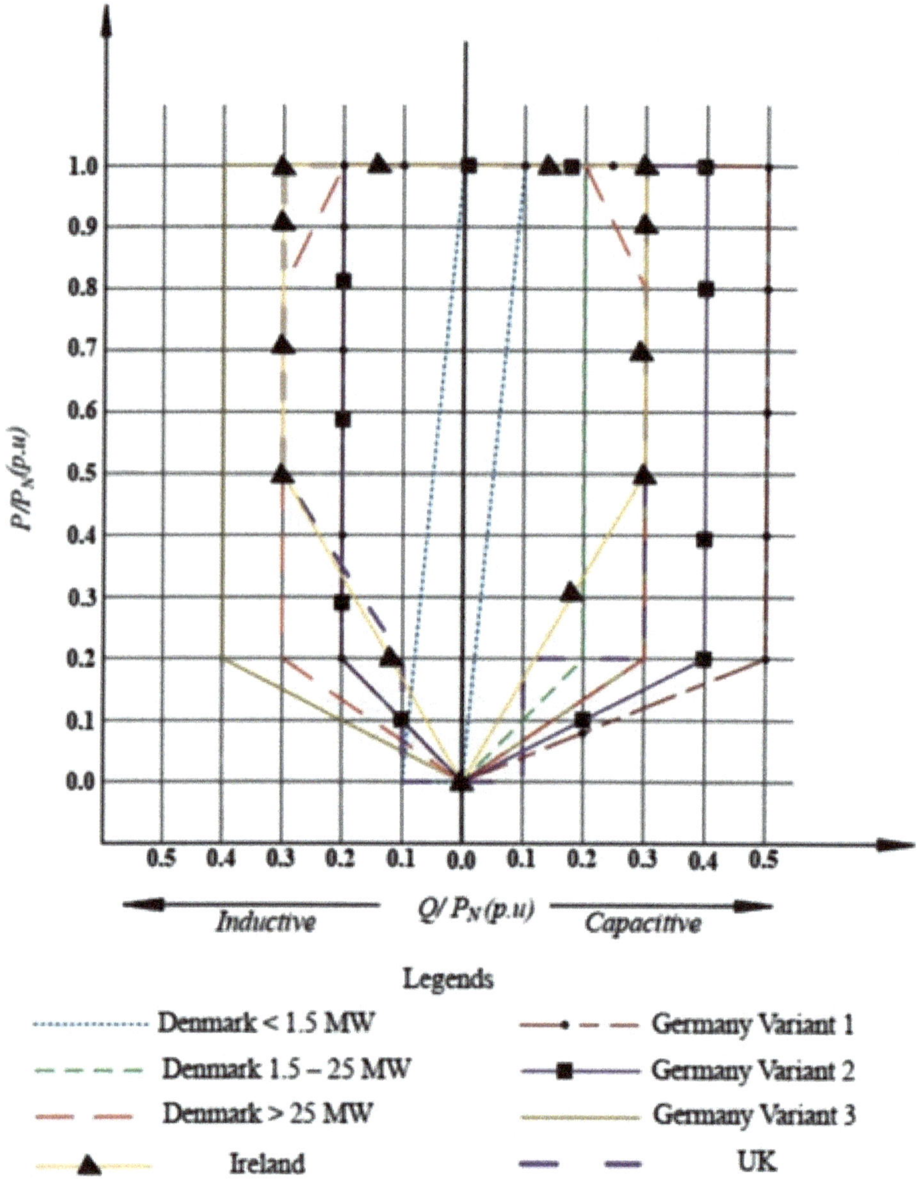

The National Electricity Rules (NER) in Australia mandate that wind generators must have reactive power control capability of +/-0.93 power factor at full output across the entire operating range, and maintain +/-10% of nominal voltage. The minimum access standard allows for zero reactive power capability. Specifically, for

South Australia, wind farms are required to have this +/-0.93 power factor capability at full output and also provide 50% dynamic reactive power capability as a fraction of their rated power (Sarkar et al., 2018).

4.3 Requirements Under Disturbances

System transient stability refers to a power system's ability to maintain synchronism after disturbances, with critical fault clearing time being vital for stability. Research shows that grid-forming converters can significantly extend this time. By improving transient stability in grid-following converters, the Remote Irish electrical network can reliably operate with less than 30% grid-forming contribution, down from 40%, even in three-phase fault scenarios (Nycander & Soder, 2018; Vrana et al., 2021). Additionally, Grid-forming converters enhance power system robustness during and after faults by stabilizing oscillations and providing swift voltage and frequency support. Their synchronization stability is vital for maintaining transient stability, allowing these converters to autonomously set critical parameters and ensuring the resilience and reliability of the power system (Murdan et al., 2023). Therefore, GCs must specify requirements for the FRT capabilities of grid-forming converters to enhance performance and system transient stability, which contributes to a more resilient power system. Additionally, aspects could be added to existing codes to improve ride-through capabilities.

These aspects include:

- *Voltage sag/swell characteristics:* Including specifications for the duration and characteristics of voltage sags and swells could enhance guidelines for renewable generators to effectively manage varying voltage conditions, alongside voltage dip and surge levels.
- *Response time:* The response time of WPPs during voltage disturbances is vital for grid stability. Current codes may define the maximum permissible response time for generators to normalize after voltage dips or surges.
- *Advanced regulation strategies:* The incorporation of advanced control strategies and technologies in grid converters can improve the ride-through capabilities of renewable generators, utilizing advanced control algorithms, fault detection, isolation mechanisms, and coordinated control schemes to enhance dynamic response during voltage disturbances.
- *Fault clearance requirements*: Existing codes on renewable generators address their behavior during voltage disturbances. To improve ride-through capabilities, incorporating fault clearance requirements is essential. This entails defining the generators' actions during faults to ensure safe and efficient grid reconnection post-fault clearance.

- *Harmonization of requirements:* Harmonizing the LVRT and HVRT requirements across various jurisdictions can enhance consistency and interoperability, facilitating the integration of renewable energy generation into power systems and improving grid resilience.

Incorporating aspects into existing grid codes will enhance the ride-through capabilities of WPPs and their contribution to grid stability. Continuous updates and refinements by power system authorities can ensure effective WPP integration and promote sustainable power system development.

4.4 New Coordinated Control Scheme for DFIG-Based WTGs

Chang et al. (2022) introduce a coordinated control strategy that adheres to reactive current requirements specified in GCs for both positive- and negative-sequence systems. This strategy allows WPPs to contribute additional positive-sequence reactive current (I1R) to support system voltage and enhance power system transient stability. Recent GCs from Germany and China also specify requirements for negative-sequence reactive current (I2R) from WPPs.

Authors investigate the I2R characteristics of Doubly-fed induction generator (DFIG)-based wind turbine generators (WTGs) under existing control strategies, including balanced positive-sequence control (BPSC) and positive- and negative-sequence control (PNSC). The analysis reveals that these strategies do not meet I2R requirements under severe voltage unbalance caused by asymmetrical faults. A new coordinated control strategy, PNSC-I12R, proposed by Chang et al. (2022), aims to address these deficiencies. The conventional BPSC allocates most rotor-side converter (RSC) current capacity to positive-sequence currents, but the coupled control introduces a low-impedance path affecting performance. Consequently, DFIGs contribute both I2R and negative-sequence active current (I2A) to the grid, with significant negative sequence rotor current. These components can be predicted using the proposed negative-sequence equivalent impedance ZDFIG. Under practical voltage unbalance, rotor current may exceed the RSC capacity (1.2 p.u.), triggering protections. The PNSC-Tem and existing coordinated control manage the RSC current capacity to mitigate double grid frequency oscillations in electromagnetic torque (Tem) while fulfilling I1R requirements (Chang et al., 2022).

In the context of grid management, eliminating double grid frequency oscillations inadvertently leads to an I2R (current squared resistance) value that is contrary to the desired outcome. Consequently, during practical voltage imbalances caused by asymmetrical short circuits, the existing coordinated control can produce a positive I2R via the grid-side converter (GSC). However, this output remains inadequate in mitigating overvoltages on the healthy phases, thereby exacerbating the overvoltage issues during unbalanced fault conditions. Additionally, under these control schemes, the voltage requirements for the rotor-side converter (RSC) control can surpass their maximum limits, leading to RSC over-modulation and loss of control over the rotor current (Chang et al., 2022).

The PNSCI12R coordinated control strategy maximizes current capacity for I1R and I2R components utilizing a decoupled double synchronous reference frame phase lock loop (DDSRF-PLL) to evaluate active and reactive components in positive- and negative-sequence systems. It generates eight current references, effectively reducing rotor voltage demand for RSC control and mitigating negative-sequence voltages and healthy phase voltages at the grid. This approach addresses voltage imbalances from asymmetrical faults and shows lower oscillation amplitudes in Tem compared to BPSC (Chang et al.,2022).

4.5 Relevance for Marine Energy Installations

Future ocean energy projects will require modifications to national grid codes similar to those for wind energy. Ocean farms must manage voltage and reactive power at grid connections, employing techniques from wind energy. Regulations currently mandate wind farms to adjust active power for grid frequency management, a challenge for ocean farms due to their limited control over energy from variable ocean resources. Countries like Denmark, Ireland, and Finland specify regulatory requirements for wind power plants in their grid codes, while Germany has unique regulations for offshore wind farm connections (Robles et al., 2019).

WPPs must now manage local and remote control of active power variations, except those from sudden wind speed drops. A demonstration during the Twenties EU project tested controlled shutdowns of wind farms in storm conditions, enabling delays for secondary controllers to engage. If advanced, this technology could allow WPPs to shut down at wind speeds above cut-out thresholds in future GCs. Presently, WPPs provide steady state frequency control by adjusting active power output, while there is a growing need for dynamic frequency control, particularly inertia emulation, currently recommended in the GCs of Spain and Norway, with potential applicability to other marine energy installations soon. (Robles et al., 2019).

4.6 Time-Frequency Analysis Technique for Wind Power Grid Codes

WPPs are classified as intermittent power sources, necessitating the use of time-frequency analysis (TFA) to surpass the constraints of the conventional Fast Fourier Transform (FFT) method when dealing with non-stationary signals. The time-frequency distribution (TFD) exhibits advantageous characteristics that address these limitations, leading Islam et al. (2020) to select the reduced interference distribution (RIDB) for analyzing wind power GCs under conditions of unbalance and distortion within the grid. This choice is due to RIDB's effective capability in minimizing interference terms (Islam et al., 2015)

The IEEE Std. 1459-2010 provides definitions for various power components like effective RMS voltage and current, effective power factor, and others, particularly under non-sinusoidal and unbalanced conditions, utilizing the conventional FFT method (Blodt et al., 2006). However, FFT may lead to inaccuracies in assessing these power indices in challenging grid environments. To address this, Islam et al. (2020) have redefined these indices using the TFA method, which aims to deliver a more accurate evaluation for analyzing wind power grid connections, focusing on TFA-based effective RMS voltage, power factor, and data acquisition for SCADA systems.

In a WPP, the SCADA system is essential for real-time data access and monitoring, facilitating comparative analyses to ensure the reliability of power systems. For effective analysis and monitoring, parameters such as voltage and current THDs, active power, apparent power, reactive power, and harmonic pollution information are acquired through the TFA method proposed by Islam et al. (2020).

4.7 Limitations of Current Grid Codes and the Need for Enhanced Codes

Historically, GCs have focused on synchronous generators and conventional power plants. However, the rise of RE generators necessitates updates to GCs by power system authorities and stakeholders. These revisions aim to establish a cohesive operational framework that enhances interaction among system operators, suppliers, and consumers while maintaining their responsibilities. While some countries are revising their GCs, many continue to use outdated codes that hinder the integration of renewable energy. To address these shortcomings and improve the integration of renewable energy sources, it is crucial to consider several key areas for reform (Murdan et al., 2023).

Flexibility in the electric power system is the capacity to adapt to changes in supply and demand, ensuring safe operations. It encompasses resources like flexible consumption, generation, storage, and grid interconnections. Traditionally, this flexibility was managed by adjusting conventional power plant outputs. However, in a decarbonized system, the phase-out of these plants poses major challenges in meeting flexibility needs (Murdan et al., 2023).

To improve grid flexibility, integrating advanced control algorithms and demand-side management into GCs is proposed, allowing for real-time adjustments that accommodate renewable energy variability. Europe is working on a specific network code to clarify demand side flexibility and define stakeholder roles. Despite progress, barriers hinder effective scaling of this flexibility. The European association for smart energy emphasizes the network code's role in enhancing transparency, facilitating market interactions, and promoting a market-oriented flexibility procurement approach (Murdan et al., 2023).

4.8 Cybersecurity Challenges

The integration of digital communication networks in smart grids introduces notable cybersecurity challenges, particularly due to the digitization of renewable energy sources like wind farms. This transition raises issues such as increased vulnerabilities linked to to insecure legacy systems, SCADA systems and Internet of Things (IoT)-enabled devices. Emphasizing regulatory compliance, thorough workforce training, and collaboration among stakeholders is crucial for safeguarding grid integrity (Ogunmoye et al., 2025). In Europe, the Network and Information Security (NIS) Directive is designed to enhance cybersecurity for essential service operators like energy providers, mandating strict security measures (Adeoba, Odjegba and Pandelani, 2025). Conversely, the U.S. employs the North American Electric Reliability Corporation's Critical Infrastructure Protection (NERC CIP) standards, which enforce cybersecurity protocols for bulk electric system overseers, focusing on risk management and incident response to safeguard the electric grid from cyber and physical threats (Tuyen et al., 2022; Ekechukwu and Simpa, 2024). Cyber threats are evolving, presenting challenges for regulatory bodies, which emphasizes the need for proactive measures. Power plant stakeholders, including owners, engineers, and developers, must adopt a comprehensive strategy that entails technology upgrades, personnel training, policy enforcement, and collaboration with regulators to effectively mitigate these risks (Mikac, 2023; Kumar et al., 2025).

5 PRACTICAL EXAMPLES AND ILLUSTRATION

5.1 Overview of Wind Power Grid Codes in the US

The electrical power grid in the US is divided into three main regions: the Western Interconnection, the Eastern Interconnection, and the Texas Interconnection. Within these regions, seven independent system operators (ISOs) and regional transmission operators (RTOs) manage the grid in accordance with standards set by the North American Electric Reliability Corporation (NERC) to ensure the reliability of bulk electric power systems. Following the Northeast blackout of 2003, NERC was designated as the Electric Reliability Organization (ERO) for the US by an order from the Federal Energy Regulatory Commission (FERC), which established that adherence to NERC's standards was previously encouraged but not enforced (Islam et al., 2020).

In response to the rising integration of wind energy in the United States, the Federal Energy Regulatory Commission (FERC) has established Grid Code requirements that apply to both small WPPs with a capacity of 20 megawatts (MW) or less, as well as large WPPs exceeding 20 MW. These requirements are in addition to the compliance mandates set by the North American Electric Reliability Corporation (NERC). The GCs issued by FERC for WPPs in the US are summarized as follows:

5.1.1 Low-Voltage Ride-Through Capability

The first GC addresses the LVRT capability of a WPP, which refers to the ability of wind turbine generators to function during periods of reduced grid voltage. There are two critical requirements: a WPP must maintain LVRT capability down to 15% of the rated line voltage for a duration of 0.625 seconds, and it must also be capable of continuous operation at 90% of the rated line voltage, as measured on the high voltage side of the WPP substation transformer(s) (Wu, 2018).

5.1.2 Power Factor Design Criteria

According to the second GC, the power factor of both synchronous and non-synchronous wind generating facilities must be maintained within a range of 0.95 leading to 0.95 lagging, measured at POC. This requirement is essential for ensuring the safety and reliability of the electric power system. Additionally, the power factor range is dynamic, and power electronic devices, such as fixed and switched capacitors, or a combination of these, can be utilized to achieve the necessary reactive capability (Islam et al., 2020).

5.1.3 Supervisory Control and Data Acquisition

To ensure system reliability, the WPP must implement a SCADA system for data transmission and instruction reception from the transmission provider. Compliance with Federal Energy Regulatory Commission (FERC) Guidance Committee requirements necessitates adherence to distinct protocols established by seven ISOs, which are mandatory for all WPPs. The WPP studied is situated in Texas and operates under the Electric Reliability Council of Texas (ERCOT). It is essential for all WPPs to furnish a real-time SCADA point to communicate with ERCOT regarding the number of available wind turbine generators (WTGs) for power injection, both active and reactive. Additionally, two further real-time SCADA points must report the number of WTGs that are non-communicative and out of service, ensuring comprehensive monitoring and reliability of power flow.

5.1.4 Harmonic

ERCOT mandates that all WPPs adhere to IEEE Std. 519, as outlined in the technical report released by the National Renewable Energy Laboratory (NREL) in the United States.

5.1.5 Reactive Power

All WPPs with a capacity above 20 MVA are required to manage reactive power within a range of -32.8% to +32.8% of their net active power output at POC, corresponding to a power factor of 0.95. This regulation by the ERCOT aims to uphold the voltage profile, which varies by location and transmission voltage but generally stays between 0.95 and 1.05 p.u. under steady-state conditions. Furthermore, ERCOT mandates that WPPs be capable of delivering reactive power at all output levels equal to or exceeding 10% of their nameplate capacity. If a WPP operates below this threshold and cannot provide necessary voltage support, ERCOT reserves the right to disconnect the WPP to ensure system reliability (Islam et al., 2020).

5.1.6 Cybersecurity Requirements

The United States Department of Energy plays a crucial role in enhancing cybersecurity within the energy sector through its Office of Cybersecurity, Energy Security, and Emergency Response. This office aims to strengthen the U.S. energy infrastructure's resilience against cyber threats by collaborating with energy companies. Emergency Response focuses on developing cybersecurity technologies, sharing training resources, and fostering partnerships between the public and private sectors. Among its initiatives are the Cybersecurity for Energy Delivery Systems program and the Cyber Testing for Resilient Industrial Control Systems program, which are designed to improve the cybersecurity framework of energy systems and assess critical devices for vulnerabilities (Kumar et al., 2025).

5.2 China's Wind Power Grid Codes

In 2018, China emerged as the leading investor in the RE industry, with an investment of $88.5 billion. Notably, investments from developing countries in the RE sector have surpassed those from developed nations in recent years, largely driven by a significant increase in electricity demand, particularly in China and India (Kurbatova & Perederii, 2020).

The development of RE has been a crucial aspect of China's policy for an extended period; however, the formal policy framework did not begin to take shape until 2005, when the RE Law was ratified during the 4th meeting of the 10th National People's Congress. This legislation marks the first significant legal foundation for RE in China and outlines the strategic importance of RE growth within the country. It established essential policy measures aimed at fostering RE development and forming the foundational legal and policy architecture for China's RE initiatives. Since the enactment of the RE Law, a series of policies have been introduced, including laws enacted by the National People's Congress, regulations and development plans from the State Council and its ministries, as well as various administrative measures, pricing policies, financial and tax strategies, technological standards, and a range of comprehensive and specialized measures (Zeng et all., 2013).

5.2.1 Frequency Modulation Requirements

Mandatory GB 38755-2019 guidelines stipulate that wind farms accessing power systems at voltage levels of 35kV and above must have primary frequency modulation, rapid voltage regulation, and peak regulation capabilities. In centralized access areas, they should also support moment capacity and short circuit. Additionally, electric power industry standard DL/T 1870-2018 mandates that wind farms and photovoltaic stations at 110 (66) kV and above must implement primary frequency modulation, requiring startup within 3 seconds, a response time of no more than 12 seconds, and an adjustment time of no more than 15 seconds, with active power accuracy capped at 1% P_n when output exceeds 10% P_n (Bao & Zhang, 2022).

5.3 Ethiopia's Wind Power Grid Codes

Ethiopia's renewable energy sector is dominated by hydropower, with wind power contributing about 7.65% (324 MW) of the total electricity generated. The Ethiopian National Distribution System Grid Code (ENDS-GC), regulated by the GC, sets technical standards for grid integration and performance requirements during network disruptions, including voltage control, power factor limits, reactive power requirements, failure responses, frequency variations, and ride-through capabilities for minor interruptions (Khan et al., 2022).

5.3.1 Technical Specifications for Integration at Distribution Level

A renewable power plant (RPP) must meet future grid requirements, with varying strictness based on its size and grid connectivity. As per the Grid Code, RPPs are categorized according to capacity. The code outlines technical requirements like voltage and frequency bands, and disconnection times that RPPs must adhere to depending on their category. At the point of consumption, RPPs must handle rapid temperature changes of up to 20 degrees and resume normal production within 5 seconds after conditions stabilize. Voltage at the POC must remain within limits; RPPs are required to inject reactive current if voltage drops below 80%, peaking at 50%. If voltage exceeds 110%, RPPs must absorb reactive power. Compliance with the frequency response curve mandated by the system operator is essential, with active power reductions required during high frequency conditions. Metering for grid frequency must maintain an accuracy of at least 10 *mHz*, and RPPs must reduce active power when frequency exceeds 50.5 Hz and disconnect if it surpasses 51.5 Hz for over 4 seconds to protect the ENDS (Khan et al., 2022).

The RPP is mandated to execute voltage and reactive power control at the POC, but can only select one function—voltage, power factor, or reactive power—at a time due to their mutual exclusivity. Control parameters are developed by the distribution network system planner and regional control center, documented in the operating agreement. Reactive power control focuses on the management of reactive power supply and absorption at the POC. The RPP must adjust its echo analog set point within two seconds of changes made by the planner or control center, with a 30-second compliance window. Control performance must maintain a precision tolerance of 2% of the set point or 0.5% of the maximum reactive power, requiring set point accuracy of 1 kVAr. Power Factor Control adjusts reactive power in relation to active power at POC, with similar responsiveness and accuracy requirements for adjustments dictated by the control center or planner (Khan et al., 2022).

5.3.2 High Wind Curtailment

The wind power generating plant must remain connected to the distribution system even when average wind speeds drop below the cut-out threshold of at least 25 m/s. To mitigate interruptions in Active Power generation near this threshold, the plant needs an automatic downward regulation mechanism, allowing the RPP Active Power to adjust continuously between 100% and at least 40% of its rated capacity. Individual Wind Turbine Generating Plants may shut down during downward regulation to match load characteristics. Two methods of downward control exist: continuous and discrete, with discrete regulation limited to 25% steps of rated power. The downward regulatory band will be set in collaboration with the system planner upon commissioning. Figure 5 provide an example of a high wind downward control chart (Khan et al., 2022).

Figure 5. Illustrative High Wind Downward Regulation Chart (Khan et al., 2022).

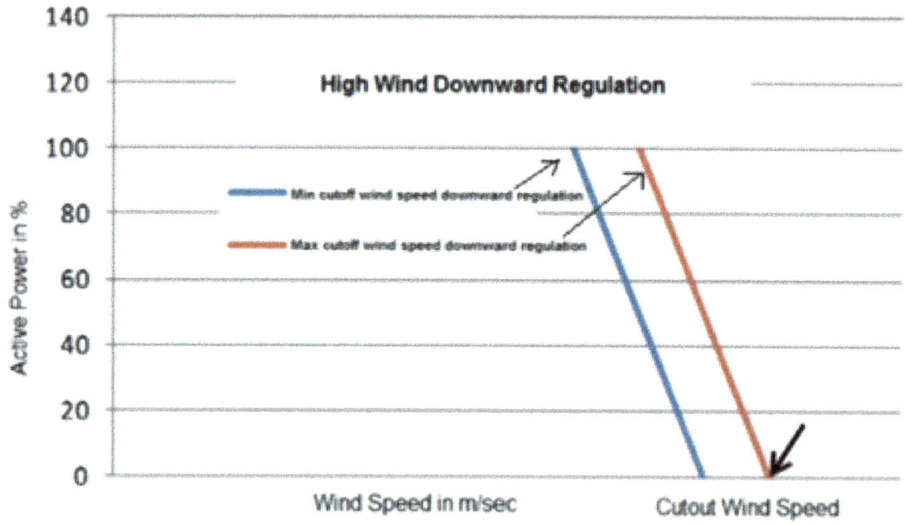

6 FUTURE TRENDS

Many solutions are being developed by researchers to face emerging renewable integration challenges like higher renewable penetration, voltage and frequency deviations, coordination challenges, etc. This section covers emerging technologies which need requirements from policymakers for their effective implementation.

6.1 Dynamic Line Rating (DLR) technologies

DLR technologies provide real-time monitoring and optimization of transmission line capacity by adapting to environmental conditions and load demands. DLR adjusts line ratings according to factors such as ambient temperature and wind conditions, affecting the conductor's heat dissipation ability. However, for broader adoption in modern power systems, DLR must overcome technical and regulatory challenges for seamless deployment (Cavus, 2025).

6.2 High-Voltage Direct Current (HVDC) technologies

HVDC technology minimizes power loss and enables long-distance electricity transmission, linking remote renewable energy sources, like offshore wind farms, to urban areas. Its capacity to interconnect asynchronous power systems enhances grid reliability and flexibility, essential for integrating renewable energy and meeting climate goals (Cavus, 2025). However, regulatory frameworks must adapt to support HVDC deployment, necessitating policymakers to align it with overall energy transition objectives through appropriate legislation and incentives (Pragati et al., 2023).

6.3 Energy Storage System Technologies

The energy storage system is highly effective for frequency regulation due to its rapid response and operational flexibility, particularly in wind power plants. Although energy storage is essential for managing system frequency, research on configuring its capacity for optimal frequency regulation alongside wind power remains limited (Pontes et al., 2023). The potential of energy storage to regulate frequency through differential droop control highlights the need for supportive market mechanisms and policies to promote its deployment. This includes discussions on ancillary service markets, capacity markets, and regulatory frameworks to enhance energy storage technology adoption in power systems (Ullah et al., 2024).

6.4 Grid-forming Inverters

Grid-forming (GFM) controls enhance the stability of wind turbines in weak grid conditions, unlike grid-following (GFL) techniques. GFM wind turbines, functioning as controllable voltage sources, create voltage amplitude and frequency independently of the main grid, allowing black start capability with adequate energy backup (Nguyen et al., 2022). Current standards like IEEE 1547 focus on GFL inverters and need updates to accommodate GFM behaviors, such as eliminating the anti-islanding requirement due to their ability to operate in stand-alone mode (Lin et al., 2020). The recently approved IEEE P2800 standard outlines interconnection requirements for inverter-based resources but does not restrict inverter technology adoption; however, existing grid-code requirements must be improved to support GFM advancements (Nguyen et al., 2022).

7 CONCLUSION

This chapter investigates the GC requirements for the integration of WPPs. It discusses steady-state operational requirements set by grid codes, notably focusing on reactive power requirements, continuous voltage operating ranges, frequency operating requirements, and voltage quality requirements. An analysis is presented on the correlation between reactive power requirements and voltage operating requirements, alongside proposed methods to fulfill both criteria. Additionally, the chapter covers the dynamic LVRT requirement. As the penetration of wind power increases, WPPs are expected to function similarly to conventional power plants, ensuring the provision of frequency and voltage control support, while also being capable of riding through low voltage conditions with defined residue voltages and durations.

The following conclusions are extracted from this study:

1) WPPs should show increased tolerance for abnormal operating conditions to improve grid stability during disturbances. They must maintain a broad FOR and support extended operational durations, with Germany and Canada leading in having FOR parameters of 46.5–53.5 Hz and 59–60.5 Hz, respectively.
2) Increased wind power generation reduces system inertia, resulting in significant RoCoF and low frequency nadir. To counteract this reduced inertia, synthetic inertia control is crucial for WPPs, allowing for rapid inertial response (IR) to stabilize frequency after power imbalances. However, current IR requirements are often inadequate or absent, highlighting the need for enhanced grid codes. Countries like China, Canada, and Spain, as well as organizations like IEEE, have developed specific IR requirements for WPPs.
3) Technologies in the wind energy sector are evolving rapidly and bringing new challenges. Regulators must also have the capacity to evolve regulations at the same speed to facilitate their application.

When integrating onshore and offshore WPPs into the grid, active power control and frequency regulation are critical due to the inherent variability of wind energy. The significant fluctuations in power generation primarily impact frequency, presenting challenges for grids that are dominated by renewable energy sources. Many new technologies are being developed and regulations must be established as quick as possible.

REFERENCES

Abubakr, H., Vasquez, J. C., Mahmoud, K., Darwish, M. M. F., & Guerrero, J. M. (2022). Comprehensive review on renewable energy sources in Egypt—Current status, grid codes and future vision. *IEEE Access : Practical Innovations, Open Solutions*, *10*, 4081–4101. DOI: 10.1109/ACCESS.2022.3140385

Ali, S. W., Sadiq, M., Terriche, Y., Naqvi, S. A. R., Hoang, L. Q. N., Mutarraf, M. U., Hassan, M. A., Yang, G., Su, C.-L., & Guerrero, J. M. (2021). Offshore wind farm-grid integration: A review on infrastructure, challenges, and grid solutions. *IEEE Access : Practical Innovations, Open Solutions*, *9*, 102811–102827. DOI: 10.1109/ACCESS.2021.3098705

Alsmadi, Y. M., Xu, L., Blaabjerg, F., Ortega, A. J. P., Abdelaziz, A. Y., Wang, A., & Albataineh, Z. (2018). Detailed investigation and performance improvement of the dynamic behavior of grid-connected DFIG-based wind turbines under LVRT conditions. *IEEE Transactions on Industry Applications*, *54*(5), 4795–4812. DOI: 10.1109/TIA.2018.2835401

Bao, X., & Zhang, L. (2022). Review of grid codes and implementation for wind farm frequency regulation. In *E3S Web of Conferences*, 352. https://doi.org/DOI: 10.1051/e3sconf/202235202012

Blodt, M., Chabert, M., Régnier, J., & Faucher, J. (2006). Mechanical load fault detection in induction motors by stator current time-frequency analysis. *IEEE Transactions on Industry Applications*, *42*(6), 1454–1463. DOI: 10.1109/TIA.2006.882631

Carley, S., Baldwin, E., MacLean, L. M., & Brass, J. N. (2017). Global expansion of RE generation: An analysis of policy instruments. *Environmental and Resource Economics*, *68*(2), 397–440. DOI: 10.1007/s10640-016-0025-3

Cavus, M. (2025). Advancing Power Systems with Renewable Energy and Intelligent Technologies: A Comprehensive Review on Grid Transformation and Integration. *Electronics (Basel)*, *14*(6), 1159. DOI: 10.3390/electronics14061159

Çelik, Ö., Yalman, Y., Tan, A., Bayındır, K. Ç., Çetinkaya, Ü., Akdeniz, M., Chaudhary, S. K., Høyer, M., & Guerrero, J. M. (2022). Grid code requirements–A case study on the assessment for integration of offshore wind power plants in Turkey. *Sustainable Energy Technologies and Assessments*, *52*, 102137. DOI: 10.1016/j.seta.2022.102137

Chang, C., Kocar, I., Hu, J., Karaagac, U., Chan, K. W., & Mahseredjian, J. (2022). Coordinated control of DFIG converters to comply with reactive current requirements in emerging grid codes. *Journal of Modern Power Systems and Clean Energy*, *9*(1), 1–14. DOI: 10.35833/MPCE.2021.000191

Cleveland, F. M. (2008). IEC 61850-7-420 communications standard for distributed energy resources (DER). *IEEE Power Energy*, *1–4*, 1–4. Advance online publication. DOI: 10.1109/PES.2008.4596553

Díaz-González, F., Hau, M., Sumper, A., & Gomis-Bellmunt, O. (2014). Participation of wind power plants in system frequency control: Review of grid code requirements and control methods. *Renewable & Sustainable Energy Reviews*, *34*, 551–564. DOI: 10.1016/j.rser.2014.03.040

Ekechukwu, D., & Simpa, P. (2024). The importance of cybersecurity in protecting renewable energy investment: A strategic analysis of threats and solutions. *Engineering Science and Technology Journal*, *5*(6), 1845–1883. DOI: 10.51594/estj.v5i6.1186

Fernández-Guillamón, A., Gómez-Lázaro, E., Muljadi, E., & Molina-García, Á. (2019). Power systems with high RE sources: A review of inertia and frequency control strategies over time. *Renewable & Sustainable Energy Reviews*, *115*, 109369. DOI: 10.1016/j.rser.2019.109369

IRENA. (2022). *Grid codes for renewable powered systems*. International RE Agency.

Islam, M., Mohammadpour, H. A., Ghaderi, A., & Zare, M. (2015). Time-frequency-based instantaneous power components for transient disturbances according to IEEE standard 1459. *IEEE Transactions on Power Delivery*, *30*(3), 1288–1297. DOI: 10.1109/TPWRD.2014.2361203

Islam, M. M., Hossain, E., Padmanaban, S., & Brice, C. W. (2020). A new perspective of wind power grid codes under unbalanced and distorted grid conditions. *IEEE Access : Practical Innovations, Open Solutions*, *8*, 15931–15944. DOI: 10.1109/ACCESS.2020.2966907

Khan, B., Josep, M., Guerrero, J. M., Chaudhary, S., Vasquez, J. C., Frederiksen, K. H. B., & Wu, Y. (2022). A review of grid code requirements for the integration of renewable energy sources in Ethiopia. *Energies*, *15*(14), 5197. DOI: 10.3390/en15145197

Kumar, K., Pandey, N. K., Verre, F., & Shah, R. (2025). Cybersecurity challenges in the digitization and integration of renewable energy systems: A review. *IEEE Transactions on Engineering Management*, *72*, 3042–3054. Advance online publication. DOI: 10.1109/TEM.2025.3586557

Kurbatova, T., & Perederii, T. (2020, October). *Global trends in RE development. In 2020 IEEE KhPI Week on Advanced Technology (KhPIWeek), 260-263.*, DOI: 10.1109/KhPIWeek51551.2020.9250098

Li, L., Zhu, D., Zou, X., Hu, J., Kang, Y., & Guerrero, J. M. (2023). Review of frequency regulation requirements for wind power plants in international grid codes. *Renewable & Sustainable Energy Reviews*, *187*, 113731. DOI: 10.1016/j.rser.2023.113731

Lin, Y., Eto, J. H., Johnson, B. B., Flicker, J. D., Lasseter, R. H., Villegas Pico, H. N., ... & Ellis, A. (2020). *Research roadmap on grid-forming inverters* (No. NREL/TP-5D00-73476). National Renewable Energy Lab.(NREL), Golden, CO (United States). https://doi.org/DOI: 10.2172/1721727

Mikac, R. (2023). Protection of the eu's critical infrastructures: Results and challenges. *Applied Cybersecurity and Internet Governance*, *2*(1), 1–5. DOI: 10.60097/ACIG/162868

Murdan, A., Jahmeerbacus, I., & Hassen, S. Z. S. (2023). Challenges of existing grid codes and the call for enhanced standards. *Clean Technologies and Recycling*, *3*(4), 241–256. DOI: 10.3934/ctr.2023015

Nguyen, T. K., Chang, G. W., Liu, Y. J., & Li, G. Y. (2023). Improving Voltage Ride-Through Capability of Grid-Tied Microgrid With Harmonics Mitigation. *IEEE Transactions on Power Delivery*, *38*(1), 738–747. DOI: 10.1109/TPWRD.2022.3199081

Nguyen, T. T., Vu, T., Paudyal, S., Blaabjerg, F., & Vu, T. L. (2022). Grid-forming inverter-based wind turbine generators: Comprehensive review, comparative analysis, and recommendations. *arXiv preprint arXiv:2203.02105*. https://doi.org//arXiv.2203.02105DOI: 10.48550

Nycander, E., & Söder, L. (2018). Review on European grid codes for wind farms and their implications for wind curtailments. *17th int'l wind integration Workshop*. URN: urn:nbn:se:kth:diva-236565

Ogunlana, O., & Ezeoha, M. (2021). The future of RE in Nigeria: charting the path for RE in Nigeria. *Available at SSRN* 3960313.

Ogunmoye, K. A., Agupugo, C. P., Ejichukwu, E., Barros, P., & Hayden, M. D. (2025). Protecting the Grid: Cybersecurity strategies for renewable energy integration. https://doi.org/DOI: 10.30574/wjarr.2025.26.3.2234

Oleka, E. U., Ndubisi, S. N., & Ijemaru, G. K. (2016). Electric power transmission enhancement: A case of Nigerian electric power grid. *Amer. J. Elect. Electron*, *4*(1), 33–39. DOI: 10.12691/ajeee-4-1-5

Pontes, L., Costa, T., Souza, A., Dantas, N., Vasconcelos, A., Rissi, G., Dias, R., Mohamed, M. A., Siano, P., & Marinho, M. (2023). Operational data analysis of a battery energy storage system to support wind energy generation. *Energies*, *16*(3), 1468. DOI: 10.3390/en16031468

Pragati, A., Mishra, M., Rout, P. K., Gadanayak, D. A., Hasan, S., & Prusty, B. R. (2023). A comprehensive survey of HVDC protection system: Fault analysis, methodology, issues, challenges, and future perspective. *Energies*, *16*(11), 4413. DOI: 10.3390/en16114413

Robles, E., Haro-Larrode, M., Santos-Mugica, M., Etxegarai, A., & Tedeschi, E. (2019). Comparative analysis of European grid codes relevant to offshore renewable energy installations. *Renewable & Sustainable Energy Reviews*, *102*, 171–185. DOI: 10.1016/j.rser.2018.12.002

Salameh, J. P., Cauet, S., Etien, E., Sakout, A., & Rambault, L. (2018). Gearbox condition monitoring in wind turbines: A review. *Mechanical Systems and Signal Processing*, *111*, 251–281. DOI: 10.1016/j.ymssp.2018.03.052

Sarkar, M. N. I., Meegahapola, L. G., & Datta, M. (2018). *Reactive power management in renewable rich power grids: A review of grid-codes, renewable generators, support devices, control strategies and optimization algorithms*. IEEE., DOI: 10.1109/ACCESS.2018.2838563

Sourkounis, B., & Tourou, P. (2013). Grid code requirements for wind power integration in Europe. *Energy*, *2013*, 1–9. DOI: 10.1155/2013/437674

Tovstukha, I. (2017). Modern approaches to defining alternative sources of energy. In *International Scientific and Practical Conference* (p. 124).

Tuyen, N., Quan, N., Linh, V., Vu, T., & Fujita, G. (2022). A comprehensive review of cybersecurity in inverter based smart power system amid the boom of renewable energy. *IEEE Access : Practical Innovations, Open Solutions*, *10*, 35846–35875. DOI: 10.1109/ACCESS.2022.3163551

Ullah, F., Zhang, X., Khan, M., Mastoi, M. S., Munir, H. M., Flah, A., & Said, Y. (2024). A comprehensive review of wind power integration and energy storage technologies for modern grid frequency regulation. *Heliyon*, *10*(9), e30466. Advance online publication. DOI: 10.1016/j.heliyon.2024.e30466 PMID: 38756608

Ullah, M., Guan, Y., Yu, Y., Chaudhary, S. K., Vasquez, J. C., & Guerrero, J. M. (2025). Dynamic Performance and Power Quality of Large-Scale Wind Power Plants: A Review on Challenges, Evolving Grid Code, and Proposed Solutions. *IEEE Open Journal of Power Electronics*, *6*, 1148–1173. DOI: 10.1109/OJPEL.2025.3582012

Vrana, T., Attya, A., & Trilla, L. (2021). Future-oriented generic grid code regarding wind power plants in Europe. *International Journal of Electrical Power & Energy Systems*, *125*, 106490. DOI: 10.1016/j.ijepes.2020.106490

Vrana, T. K., Flynn, D., Gómez-Lázaro, E., Kiviluoma, J., Marcel, D., Cutululis, N., & Smith, J. C. (2018). Wind power within European grid codes: Evolution, status and outlook. *Wiley Interdisciplinary Reviews. Energy and Environment*, *7*(3), 1–21. DOI: 10.1002/wene.285

Yan, L., Yongning, C., Haiyan, T., Xinshou, T., Zhankui, Z., & Jianqing, J. (2019, November). Common focus and new requirement on technical standards of RE grid integration. In *2019 Chinese Automation Congress (CAC)*, 3719-3723. DOI: 10.1109/CAC48633.2019.8996943

Zeng, M., Li, C., & Zhou, L. (2013). Progress and prospective on the police system of RE in China. *Renewable & Sustainable Energy Reviews*, *20*, 36–44. DOI: 10.1016/j.rser.2012.11.048

Zheng, H., & Tang, Y. (2012, December). *Comparison of interconnection standards of RE generation. In 2012 10th International Power & Energy Conference*. IPEC., DOI: 10.1109/ASSCC.2012.6523254

APPENDIX

Glossary of Acronyms

I1R: positive-sequence reactive current
I2A: negative-sequence active current
I2R: negative-sequence reactive current
P_{lt}: Long-term flicker perceptibility
P_n: rated power
P_{st}: Short-term flicker perceptibility

Chapter 5
AI-Digital Twin Synergy for Smart and Autonomous Wind Energy Systems

Cherala Vikram
BVRIT HYDERABAD College of Engineering for Women, India

Rampelli Manojkumar
http://orcid.org/0000-0002-8141-0321
BVRIT HYDERABAD College of Engineering for Women, India

ABSTRACT

This chapter explores the synergy between Artificial Intelligence (AI) and Digital Twin (DT) technologies for advancing smart and autonomous wind-energy systems. It examines how AI-driven analytics and learning models enhance forecasting, control, and predictive maintenance, while DTs provide real-time virtual replicas for simulation and optimization. Their integration enables self-learning, adaptive, and resilient wind farms capable of real-time decision-making and performance optimization. The chapter also addresses implementation challenges, including data reliability, scalability, and cybersecurity, highlighting future research directions for sustainable, intelligent energy infrastructures.

DOI: 10.4018/979-8-2600-0106-6.ch005

INTRODUCTION

Overview of Wind Energy Systems and Their Role in Renewable Integration

The twenty-first century has witnessed an unprecedented global movement toward sustainable energy. Among all renewable sources, wind energy stands out as one of the most mature, cost-effective, and scalable technologies for decarbonizing power generation. Wind turbines convert the kinetic energy of moving air masses into electrical energy through aerodynamic blades and electromechanical conversion systems. Modern wind farms, whether onshore or offshore, are now key contributors to national electricity portfolios and are projected to supply more than one-third of global electricity by 2050. This growth is driven by declining levelized costs of energy, technological innovation, and strong policy commitments to carbon neutrality (Shafik, 2024). Wind energy plays a pivotal role in supporting the integration of renewables within electric power networks. Unlike conventional fossil-fuel plants, wind turbines do not emit greenhouse gases during operation, helping nations meet their emission-reduction targets under frameworks such as the Paris Agreement. The modularity of wind installations from community-scale turbines to multi-gigawatt offshore clusters enables flexible deployment in diverse geographic settings. Hybridization with solar and energy-storage systems further enhances system resilience and reliability. Integrated wind-solar networks, coordinated through intelligent frequency and voltage-control algorithms, can mitigate intermittency and stabilize grid frequency during rapid load variations (Alam et al., 2023).

The growing digitalization of energy systems has also reshaped wind-power operations. Supervisory control and data acquisition (SCADA) systems collect massive volumes of data on wind speed, turbine temperature, vibration, and power output. These data streams support predictive analytics, enabling operators to adjust turbine parameters in real time. The synergy between wind power and advanced computational models is therefore central to achieving sustainable grid integration. As Hannan et al. (2023) emphasize, future energy transitions depend not only on increasing renewable capacity but also on optimizing the efficiency, reliability, and intelligence of existing assets.

In emerging economies, wind energy contributes to energy access and economic development. Small-scale wind turbines can electrify remote communities, reduce dependency on imported fuels, and promote decentralized generation models aligned with circular-economy principles. Moreover, wind energy integration has a multiplier effect across industries, creating green jobs, stimulating manufacturing of turbine components, and fostering research in aerodynamics and materials science.

The cumulative environmental and socio-economic benefits make wind energy a cornerstone of sustainable development goals.

Nevertheless, wind power systems also face technological and operational complexities that challenge their performance and dependability. As installed capacity expands, issues of efficiency, reliability, and maintenance become more pronounced, necessitating advanced monitoring, control, and optimization approaches, discussed in the following section. An illustration of the overview of artificial intelligence (AI), machine learning (ML), and digital-twin (DT) roles in wind-energy systems is given in Figure 1.

Figure 1. Overview of AI, ML, and digital-twin roles in wind-energy systems.

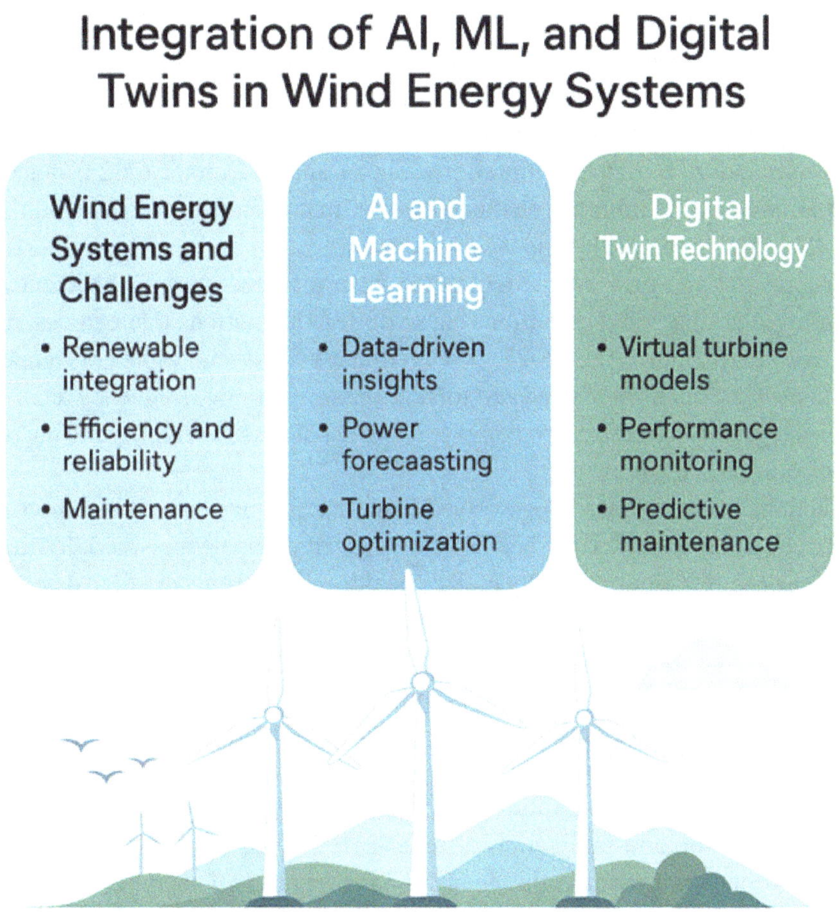

Challenges in Efficiency, Reliability, and Maintenance of Wind Turbines

While wind energy is environmentally benign, ensuring high operational performance remains a formidable task. The efficiency of a wind turbine is constrained by the Betz limit, aerodynamic losses, mechanical friction, and suboptimal control settings. Turbines must operate efficiently across a wide range of wind speeds; yet, under turbulent or low-wind conditions, the power coefficient deteriorates, leading to reduced energy yield. Wake effects, where downstream turbines experience reduced wind velocity due to upstream extraction, further diminish overall farm efficiency (Elkelawy, Atta, & Seleem, 2024).

Reliability is another critical issue faced by wind energy systems. A modern wind turbine comprises thousands of components: gearboxes, bearings, blades, generators, and converters subjected to cyclic mechanical and thermal stresses. Failures in these subsystems can result in extended downtime and significant financial loss. Offshore turbines operate in harsh marine environments characterized by high humidity, corrosion, and wave-induced vibrations. Sheng and O'Connor (2023) noted that gearbox and bearing failures constitute most of maintenance costs over a turbine's life cycle, with accessibility and weather windows often limiting repair operations. Environmental and grid-interaction factors also influence turbine efficiency and reliability. Variable wind conditions cause power fluctuations that can destabilize grid frequency. Alam et al. (2023) observed that hybrid renewable configurations with coordinated control of wind and solar generation improve frequency regulation. Yet these systems introduce new complexity in dispatch scheduling, requiring robust optimization frameworks.

Maintenance strategies have evolved from reactive and preventive approaches toward condition-based and predictive paradigms. Traditional time-scheduled maintenance can be inefficient, as it may replace healthy components or fail to detect early degradation. Recent studies advocate data-driven maintenance, where continuous monitoring identifies anomalies long before catastrophic failure occurs (Afridi, Ahmad, & Hassan, 2022). Such predictive maintenance relies heavily on sensors, SCADA data, and AI algorithms capable of pattern recognition and fault prognosis.

Energy infrastructure in coastal and port regions faces similar challenges. Sadiq et al. (2021) highlighted that future "green ports" demand energy-efficient designs, predictive control, and integration of renewable microgrids. Lessons from these industrial contexts reinforce the need for intelligent asset management in wind farms. From a lifecycle perspective, the economic implications of maintenance are significant. The operation and maintenance (O&M) costs can account for up to 30 percent of the total cost of energy. Minimizing unscheduled downtime, improving spare-part logistics, and utilizing autonomous inspection technologies, such as drones for blade assessment, are becoming essential to maintaining competitiveness. In sum, enhancing the efficiency, reliability, and maintainability of wind turbines requires a holistic framework combining advanced sensing, real-time analytics, and adaptive control. This need has paved the way for the integration of AI, ML, and digital-twin (DT) technologies, which promise to revolutionize how wind-energy assets are monitored and optimized.

Motivation for Integrating AI, ML, and DT in Wind Energy Systems

The growing complexity of modern wind farms has made traditional engineering models insufficient for optimal management. The motivation for integrating AI, ML, and DT arises from the necessity to convert vast operational data into actionable insights. These technologies represent the digital transformation of the wind-energy sector, enabling systems that are not only automated but also intelligent and self-learning.

AI and ML

AI encompasses computational methods that mimic human cognitive functions, learning, reasoning, and decision-making, while ML, a subset of AI, focuses on algorithms that improve performance through experience. In wind-energy systems, ML algorithms analyze SCADA and condition-monitoring data to forecast power output, detect faults, and optimize turbine control parameters. For example, supervised learning models can predict short-term wind speed based on meteorological variables, while unsupervised clustering identifies abnormal patterns in vibration signals. Deep-learning networks further enhance accuracy in detecting subtle faults such as blade cracks or bearing degradation (Rosca & Stancu, 2025).

AI-driven control systems also improve energy capture by continuously adapting yaw and pitch angles to maximize the aerodynamic coefficient. Reinforcement-learning agents have been deployed to learn optimal control policies under uncertain and fluctuating wind conditions, reducing fatigue loads and extending component life. These applications illustrate the transformative potential of data-centric intelligence in renewable-energy optimization (Ohalete et al., 2023). These AI/ML applications are shown in Figure 2.

Figure 2. AI/ML applications in wind assessment, power prediction, and turbine control.

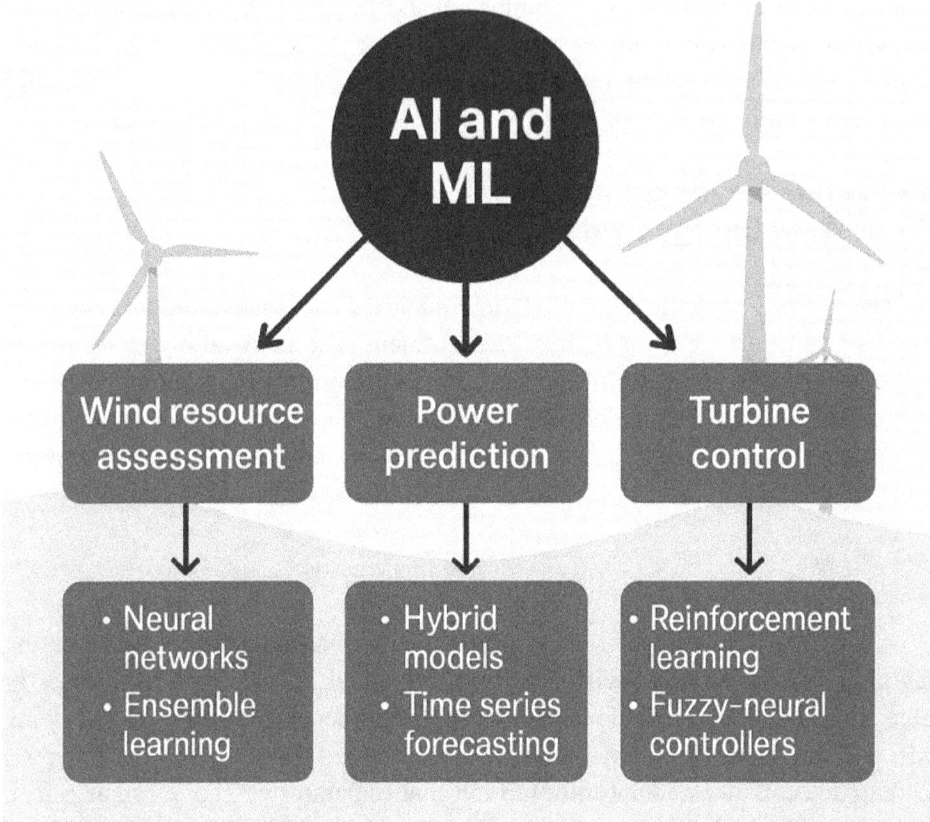

Digital-Twin Technology

A digital twin (DT) is a dynamic virtual replica of a physical asset that mirrors its behavior through real-time data synchronization, simulation, and analytics. Within wind-energy systems, DTs integrate sensor data, physics-based models, and AI algorithms to provide a holistic view of turbine performance. Kandemir et al. (2024) defined predictive digital twins as systems capable of self-updating based on sensor feedback and capable of simulating potential failure scenarios to guide maintenance decisions.

The practical benefits are manifold. DTs can predict the remaining useful life of components, evaluate the impact of control strategies, and simulate grid-interaction behavior under extreme conditions. They bridge the gap between design, operation, and maintenance phases, transforming static monitoring into dynamic, prognostic insight. Guerreiro et al. (2024) emphasized that DT-based validation methods, including hardware-in-the-loop and software-in-the-loop testing, ensure that wind-turbine subsystems comply with evolving grid codes and reliability standards.

Integration Into Smart Grids and Energy Ecosystems

Digital-twin technologies extend beyond individual turbines to encompass entire wind farms and their interfaces with smart grids. Through interconnected DTs, operators can coordinate multiple energy assets, turbines, batteries, and even electric vehicles, enabling advanced energy-management functions such as load forecasting and vehicle-to-grid integration. Al-Shetwi et al. (2025) demonstrated how DTs, when combined with AI-driven optimization, facilitate real-time balancing of generation and demand while enhancing grid stability.

The convergence of AI, ML, and DTs supports self-healing, adaptive energy systems capable of responding autonomously to faults and external disturbances. For instance, predictive models can anticipate extreme weather impacts and adjust operational parameters proactively. Similarly, AI-assisted DTs contribute to the design of next-generation turbines by simulating aerodynamic and structural responses under varying scenarios before physical prototypes are built.

Socio-Economic and Environmental Motivation

Beyond technical efficiency, these digital technologies align with broader sustainability objectives. Optimized operation reduces energy waste, lowers maintenance-related emissions, and extends asset lifespan, contributing to life-cycle sustainability. Intelligent automation also decreases human exposure to hazardous maintenance tasks, particularly in offshore environments. Economically, predictive analytics minimize downtime, increasing the return on investment for renewable projects.

From a policy perspective, governments and industry stakeholders recognize AI-enabled DT frameworks as enablers of Industry 4.0 in renewable energy. They bridge interdisciplinary domains like mechanical engineering, data science, and information technology, while fostering collaboration across academia and industry. This transformation promises not only technical advancement but also knowledge innovation, digital-skill development, and new business models for service-based energy analytics. Collectively, the integration of AI, ML, and digital-twin technologies marks a paradigm shift from reactive to proactive, intelligent renewable-energy systems, setting the foundation for the research and applications explored throughout this work.

AI AND MACHINE LEARNING IN WIND ENERGY SYSTEMS

Role of AI and Machine Learning in Wind Resource Assessment, Power Prediction, and Turbine Control

AI and ML have revolutionized the field of wind-energy engineering by enabling intelligent forecasting, adaptive control, and data-driven decision-making. Wind energy systems are inherently complex, characterized by nonlinearity, high variability, and dynamic interactions among aerodynamics, mechanical components, and electrical subsystems. Traditional analytical and control methods often fall short of managing this complexity effectively. By contrast, AI and ML techniques provide adaptive frameworks capable of learning from data, predicting system behavior, and optimizing performance in real time (Farrar, Ali, & Dasgupta, 2023).

Wind-Resource Assessment and Power Prediction

Accurate wind-resource assessment is fundamental to wind-farm planning, site selection, and investment decisions. Conventional statistical methods, such as Weibull and Rayleigh distributions, estimate wind potential but fail to capture the full variability of wind dynamics, particularly in complex terrains. ML-based approaches, especially those leveraging artificial neural networks (ANNs), support vector machines (SVMs), and ensemble learning (EL), enable precise estimation of wind-speed distributions and power curves based on multi-dimensional meteorological data (Anushalini & Sri Revathi, 2024).

Hybrid ML frameworks combine time-series decomposition with AI regression models to improve forecast accuracy. For example, empirical mode decomposition (EMD) separates fluctuating wind signals into intrinsic components, which are then modeled by ANNs or random forest (RF) regressors. These models handle short-term and ultra-short-term forecasting with higher precision than physics-only or statistical approaches. Accurate prediction supports grid stability, demand forecasting, and economic dispatch. According to Bošnjaković, Martinović, and Đokić (2025), EL approaches like gradient-boosted regression trees and random forests outperform single-model predictors by reducing variance and improving generalization.

Turbine Control and Optimization

AI-driven control systems are central to ensuring that wind turbines operate optimally under rapidly changing conditions. Conventional proportional-integral-derivative (PID) controllers, though robust, cannot adapt efficiently to nonlinear dynamics or sudden disturbances. AI and ML introduce intelligent control paradigms that learn adaptive policies to maximize power extraction while minimizing structural stress and fatigue.

Hassan, Ahmad, Shafiullah, Islam, and Alam (2025) reviewed various AI-based pitch-control strategies, concluding that reinforcement learning (RL) and fuzzy-neural networks achieve faster convergence and better load reduction than traditional controllers. These intelligent frameworks continuously optimize blade pitch angles to maintain rated power during high wind speeds, improving both efficiency and reliability. Similarly, Farrar et al. (2023) demonstrated that model-free RL agents can autonomously adjust yaw and torque controls, enabling turbines to adapt to stochastic wind inflows and turbulence.

At the system level, AI-enabled multi-agent control architectures coordinate turbine clusters for optimal wake steering and energy capture. Optimization algorithms, including particle swarm optimization (PSO) and genetic algorithms (GA), are commonly used to manage wake interactions in large farms. Ioannides, Stamelos, Papazis, Stamataki, and Stamatakis (2024) emphasized that integrating AI-based controllers with Internet of Things (IoT) infrastructure, using distributed sensors and edge computing, enables low-latency monitoring, decentralized control, and predictive fault detection. The convergence of AI and IoT thus creates a foundation for autonomous, self-optimizing wind farms.

Data-Driven Modeling Versus Physics-Based Modeling in Wind Energy

Modern wind-energy analytics rely heavily on predictive modeling to forecast power output and optimize control. Two primary paradigms dominate this landscape: physics-based modeling and data-driven modeling. Each has distinct advantages and limitations, and hybrid approaches are increasingly preferred for comprehensive system analysis.

Physics-Based Models

Physics-based models are derived from first principles, representing wind-turbine aerodynamics, structural mechanics, and electrical dynamics through analytical equations. These models rely on established laws of fluid dynamics and electromechanical theory to describe power generation. They provide transparency, interpretability, and robustness under known conditions. However, their performance deteriorates in turbulent flows, unmodeled disturbances, or when accurate parameterization is not feasible (Erge & van Oort, 2022). Computational fluid dynamics (CFD)-based simulations are powerful but computationally intensive, limiting their real-time applicability.

Data-Driven Models

In contrast, data-driven models learn directly from empirical data using ML algorithms. They can approximate complex nonlinear relationships without requiring explicit physical knowledge of system parameters. For instance, ANNs and SVMs learn mappings between meteorological inputs (wind speed, temperature, humidity) and turbine responses (rotor speed, output power). RF and deep-learning models further improve prediction accuracy by leveraging large-scale SCADA data collected from operational turbines (Wang, Chen, Zhu, & Sun, 2022).

Data-driven approaches are particularly effective for fault diagnosis, short-term forecasting, and real-time control. However, they can be vulnerable to overfitting and lack interpretability, making their standalone use risky in safety-critical systems.

Hybrid Modeling Approaches

Recent research focuses on hybrid physics–data frameworks, combining the interpretability of physics-based models with the flexibility of ML. Chinnappan (2025) proposed an ML–PINN–Simulink architecture integrating physics-informed neural networks (PINNs) that embed governing fluid dynamics equations directly into the neural network structure. This hybrid approach preserves physical consistency while correcting unmodeled nonlinearities and sensor errors.

Similarly, hybrid methods enhance the resilience of forecasting systems under sparse or noisy data conditions. They serve as an essential foundation for digital twin technology, which depends on both data-driven learning and physical simulation fidelity. The fusion of data and physics enables wind-energy systems that are both accurate and trustworthy, ensuring safe, explainable AI deployment in real-world scenarios.

Overview of Key AI and ML Algorithms Used in Wind Energy Systems

AI and ML frameworks in wind energy rely on several core algorithms, each offering unique strengths for prediction, control, and optimization tasks. The key algorithms and their respective roles are discussed in the following subsections and shown in Figure 3.

Figure 3. AI and ML algorithms and their respective roles.

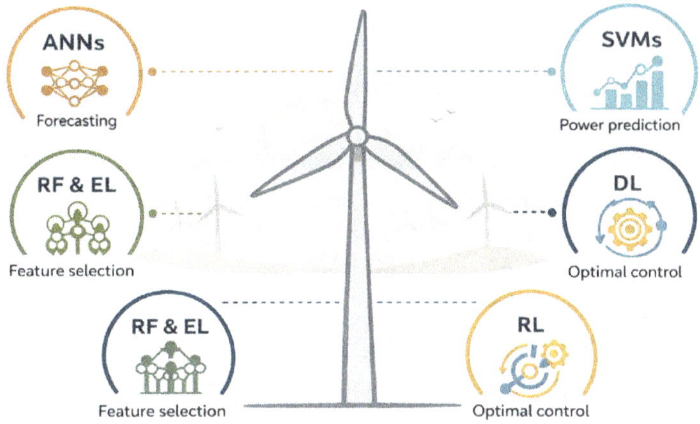

ANNs

ANNs are widely used for wind-speed and power prediction because of their ability to model complex, nonlinear relationships. Multilayer perceptrons (MLPs) with backpropagation remain popular for short-term forecasts. Advanced variants, such as recurrent neural networks (RNNs) and long short-term memory (LSTM) networks, capture the temporal dependencies that are critical to wind-speed time-series analysis (Anushalini & Sri Revathi, 2024). Hybrid ANN models combined with fuzzy logic or wavelet transforms achieve superior performance for highly volatile datasets. Despite their accuracy, ANNs require large datasets and may lack transparency in internal decision processes.

SVMs

SVMs perform regression and classification using kernel-based transformations to separate data into high-dimensional feature spaces. They excel in moderate-sized datasets and noisy environments where ANN models may overfit. In wind-energy forecasting, SVM regression models yield robust and interpretable predictions of power curves and turbine performance (Farrar et al., 2023). However, they may struggle with scalability when processing large SCADA datasets.

Random Forest (RF) and EL

RF and ensemble algorithms combine multiple decision trees to improve accuracy and reduce variance. RF models are effective for feature selection and uncertainty quantification. Bošnjaković et al. (2025) noted that RF algorithms are computationally efficient, require minimal hyperparameter tuning, and are well-suited for online forecasting in operational wind farms. EL methods such as gradient boosting and extreme gradient boosting (XGBoost) further enhance predictive robustness.

DL Architectures

DL architectures, including CNNs and hybrid CNN–LSTM models, automatically extract hierarchical features from raw data. CNNs are applied for analyzing vibration and image data from turbine inspections, while LSTM networks manage sequential data such as wind-speed time series (Lipu et al., 2021). Autoencoders and generative adversarial networks (GANs) enable unsupervised anomaly detection and data augmentation, improving fault diagnosis reliability. Although computationally demanding, DL models achieve unmatched accuracy for large-scale, high-dimensional datasets.

RL

RL enables turbines and controllers to learn optimal strategies through interaction with the environment. Unlike supervised learning, RL does not rely on labeled data but instead maximizes a reward signal, such as energy capture or reduced fatigue. Deep reinforcement learning (DRL), which combines RL with deep neural networks, supports adaptive pitch, yaw, and torque control (Hassan et al., 2025). Farrar et al. (2023) emphasized that RL frameworks achieve self-optimization in turbulent wind conditions, reducing mechanical loads while improving efficiency. These systems represent the future of intelligent, autonomous wind-energy operation.

Case Studies Highlighting AI-Based Wind Forecasting and Operational Optimization

Hybrid Forecasting Models

AI-based hybrid forecasting models have achieved remarkable success in improving prediction accuracy for wind power generation. Lipu et al. (2021) proposed a hybrid architecture combining ARIMA models with ANNs, yielding significant reductions in root-mean-square error (RMSE) compared to single-method predictors.

This approach decouples deterministic and stochastic wind components, enabling robust short-term forecasting. Smart, Olanrewaju, Usman, Otaru, and Umar (2025) applied ensemble AI models for renewable-energy optimization and found that hybrid ML–statistical frameworks enhanced both forecasting accuracy and computational efficiency. Such models are vital for integrating wind energy into smart grids, where prediction errors can compromise stability and cause power-quality disturbances.

Operational Optimization and Predictive Maintenance

Operational optimization through AI has expanded beyond forecasting to include condition monitoring and predictive maintenance. AI algorithms process SCADA, vibration, and temperature data to identify anomalies that may indicate mechanical wear or failure. Autoencoder-based anomaly detection and hybrid classification models now achieve near-real-time fault diagnosis with high accuracy (Ukoba, Olatunji, Adeoye, Jen, & Madyira, 2024). These predictive maintenance systems minimize downtime and extend component life.

In practice, AI-optimized control strategies have reduced maintenance costs and improved capacity factors. Farrar et al. (2023) demonstrated that AI-driven controllers enhance wind-turbine grid compatibility and power-quality indices, preventing voltage fluctuations and frequency deviations. Similarly, Ioannides et al. (2024) showed that IoT-connected control systems using ML can adapt to environmental and load variations autonomously, ensuring stable operation without continuous human supervision.

Grid Integration and Energy Management

AI has become essential for large-scale grid integration of wind power. Forecasting models based on AI help system operators to plan to reserve capacities and manage frequency control. Smart et al. (2025) demonstrated that AI-supported dispatch optimization reduces curtailment and improves reliability in hybrid wind-solar grids. In addition, RL-based scheduling agents can dynamically adjust energy storage and grid-support services, contributing to stable operation in variable renewable scenarios.

Outlook

AI integration is expected to extend into self-learning wind farms, where turbines communicate and coordinate in real time using federated learning (FL) and distributed optimization algorithms. These systems will allow collaborative learning across geographically dispersed wind farms while preserving data privacy. With advances in edge computing, AI models can be deployed directly on turbine controllers, enabling rapid response and reducing cloud dependence. The transition toward intelligent, autonomous, and adaptive wind-energy systems marks a defining moment in the global pursuit of sustainable energy.

AI and ML have emerged as transformative enablers in the design, control, and operation of wind-energy systems. From wind forecasting to turbine control and maintenance, these technologies provide adaptive, data-driven solutions that address longstanding challenges in efficiency, reliability, and cost reduction. The evolution from purely physics-based to hybrid AI–physics models has strengthened model robustness and interpretability. With continued progress in deep learning (DL), RL, and IoT integration, future wind farms are poised to become self-optimizing, resilient systems forming a critical part of the intelligent energy grid.

DIGITAL TWIN TECHNOLOGY FOR WIND TURBINES

Concept and Architecture of Digital Twins in Wind Energy

The DT concept represents a transformative leap in the management of renewable-energy assets. A DT is a dynamic, virtual counterpart of a physical system that mirrors its real-time state through continuous data exchange, analytics, and simulation (Song et al., 2023). Within wind energy, it serves as a cyber-physical bridge linking sensors, controllers, and computational models to deliver predictive insights and optimize performance.

Mahmoud et al. (2024) designed and prototyped a DT architecture that integrates thermal, mechanical, and electrical subsystems of a turbine within a unified simulation framework. Their model demonstrated how accurate replication of real-world conditions enables engineers to analyze failures before they occur and refine control parameters proactively. For offshore environments, where accessibility is limited, Haghshenas et al. (2023) developed a predictive DT that combines sensor data with machine-learning (ML) models to forecast degradation and minimize maintenance costs. Song et al. (2023) noted that DTs extend beyond individual turbines to encompass entire power systems, enabling grid-level resilience and interoperability. Embedded AI algorithms simulate variable wind conditions, component stress, and grid contingencies to inform operational decisions. Incorporating ML converts the DT from a static monitoring tool into an adaptive intelligence layer. Fahim et al. (2022) implemented an ML-driven DT using SCADA data and neural networks to model nonlinear turbine behavior accurately. This evolution positions DTs as self-learning, continuously optimizing systems at the core of digitalized wind energy. The key components and benefits of digital-twin technology in wind turbines are shown in Figure 4 and discussed as follows.

Figure 4. Key components and benefits of digital-twin technology in wind turbines.

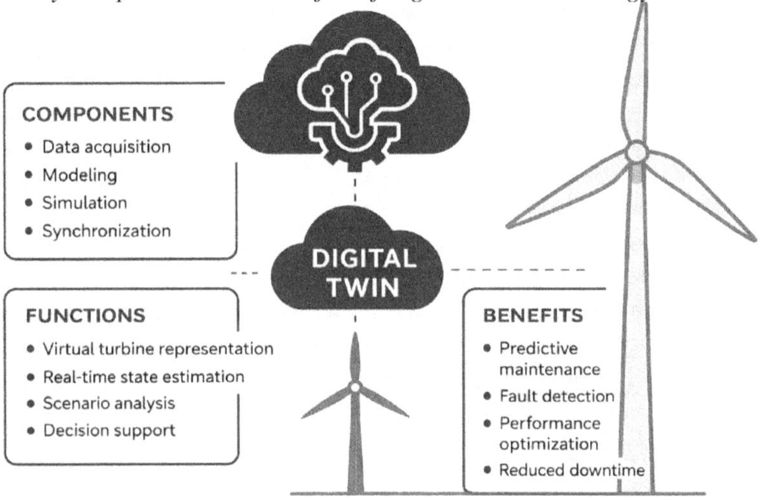

Components

A functional DT comprises four interdependent components: data acquisition, modeling, simulation, and synchronization with the real system.

Data Acquisition

Data acquisition supplies the lifeblood of the DT. Sensors capture wind speed, rotor torque, vibration, and temperature, transmitting these data streams through IoT and SCADA networks. Qays et al. (2022) demonstrated that IoT-aided SCADA systems enhance scalability and responsiveness by enabling real-time communication among distributed assets. Adaptive sampling and event-triggered monitoring further improve bandwidth efficiency and diagnostic precision.

Modeling and Simulation

Modeling transforms raw measurements into representative mathematical or computational constructs. Yan et al. (2023) introduced a hybrid physical–data-driven model that merges aerodynamic equations with ML estimators to simulate dynamic turbine operation. Their approach captures nonlinearities often neglected in purely physics-based analyses. For offshore installations, Liu et al. (2023) created a DT-based framework integrating hydrodynamic and structural models for floating turbines, supporting reliability assessment under complex marine loads. Simulation enables what-if experimentation and controller validation without physical risk. Nguyen et al. (2022) combined DTs with power-hardware-in-the-loop (PHIL) testing to synchronize virtual models and real equipment, thus accelerating design verification while maintaining safety.

Synchronization and Feedback Loops

Synchronization maintains fidelity between the physical turbine and its digital counterpart through continuous bidirectional communication. State-estimation and filtering algorithms align measured and simulated variables even amid noise or latency (Liu et al., 2023). Feedback from analytics, such as predicted wear or efficiency drift, is relayed to on-site controllers, enabling cyber-physical optimization. These adaptive loops keep the DT a living, evolving entity that mirrors turbine health across its lifecycle.

Benefits: Predictive Maintenance, Fault Detection, and Performance Optimization

Predictive Maintenance

Predictive maintenance is among the most mature DT applications in wind energy. By merging real-time data with probabilistic models, DTs forecast component failures and schedule maintenance before costly breakdowns. Zhang et al. (2026) proposed a probabilistic DT for offshore turbines that integrates uncertainty quantification to determine cost-optimal service intervals. Zhong et al. (2023) reviewed DT-based maintenance strategies and found higher detection sensitivity and fewer false alarms compared with traditional systems.

Fahim et al. (2022) applied ML within their DT to predict gearbox and bearing wear using SCADA features, achieving early anomaly identification. Udo et al. (2024) further confirmed that combining AI and DTs can reduce maintenance costs by 20 to 30 percent while improving overall efficiency.

Fault Detection

DT-enabled fault detection leverages continuous comparison between expected and observed behavior. Deviations trigger diagnostic routines supported by ML classification of vibration or electrical signatures. Zhong et al. (2023) reported that evolving DTs refine their fault libraries over time, enhancing diagnostic accuracy for remote offshore assets where manual inspection is limited. Automated alarm generation and root-cause tracing streamline maintenance response and cut downtime.

Performance Optimization

Beyond fault prevention, DTs continuously optimize turbine performance. Fahim et al. (2022) demonstrated that their ML-enhanced DT dynamically adjusted pitch and yaw to maintain optimal power coefficients, improving energy yield. Udo et al. (2024) integrated RL within the DT to mitigate wake losses and turbulence impacts across entire farms. Zhang et al. (2026) added lifecycle simulations that guide refurbishment decisions, ensuring long-term efficiency and sustainability.

Example Frameworks: Real-Time Integration with SCADA and IoT Platforms

The fusion of DTs with SCADA and IoT architecture enables real-time observability and control across large wind farms. Kabir et al. (2024) surveyed IoT-driven energy systems and concluded that DT–IoT integration delivers persistent synchronization between assets and their digital models, empowering decentralized analytics and decision-making.

Mansour et al. (2023) highlighted interoperability among SCADA databases, edge-computing devices, and cloud services as vital for scalable DT deployment. Embedding DT functions within IoT frameworks allows operators to manage geographically dispersed turbines while ensuring data consistency. In offshore contexts, Haghshenas et al. (2023) implemented a predictive DT that assimilated SCADA streams and AI algorithms for real-time adaptation to environmental fluctuations. Issa, Hamad, and Abdel-Geliel (2023) provided a practical demonstration using the Microsoft Azure IoT platform, achieving sub-second latency between sensors and cloud-based digital replicas. Their Azure dashboards enabled remote visualization, anomaly alerts, and command execution. Key advantages of such integrated frameworks include

1. Scalability: Cloud-hosted DTs manage hundreds of turbines concurrently.
2. Interoperability: Standard protocols (e.g., MQTT, OPC UA) unify heterogeneous devices.
3. Cybersecurity: Encryption and anomaly detection protect data integrity.
4. Edge Intelligence: Local analytics maintain responsiveness even with limited connectivity.

Collectively, these systems transform conventional farms into self-aware, intelligent energy ecosystems capable of adaptive optimization.

Digital-twin technology has redefined operational intelligence in wind energy by merging sensing, data analytics, and simulation into a cohesive framework. From the conceptual architectures proposed by Mahmoud et al. (2024) and Haghshenas et al. (2023) to IoT-based deployments demonstrated by Issa et al. (2023), DTs have proven indispensable for predictive maintenance, fault detection, and efficiency enhancement. Integration with SCADA and IoT ensures continuous synchronization, while AI and ML empower autonomous decision-making. As renewable systems evolve toward Industry 4.0 paradigms, DTs will underpin smart, sustainable, and resilient wind farms. They close the loop between design and operation, enabling every turbine to function not merely as equipment but as a data-driven, adaptive entity contributing to a carbon-neutral future.

INTEGRATED AI–DIGITAL TWIN FRAMEWORK FOR WIND TURBINE OPERATION

Synergy Between AI/ML Models and Digital Twins

The integration of AI and DT technologies marks a new era in smart wind-energy systems. Both paradigms complement each other, AI brings predictive and decision-making intelligence, while DT provides real-time virtual representation and simulation capabilities. Together, they form an intelligent ecosystem that transforms wind turbines into self-learning, adaptive assets capable of optimizing performance and reliability.

Ranawaka, Alahakoon, Sun, and Hewapathirana (2024) emphasized that AI and DT convergence is essential for achieving sustainability in modern power grids. They identified that DTs enable continuous data flow between physical and digital environments, while AI algorithms process this data to make predictive and prescriptive decisions. Similarly, Alghamdi et al. (2025) described this relationship as mutual reinforcement, AI enhances DT accuracy through learning and optimization, whereas DTs provide high-fidelity data environments that accelerate AI training and validation. Rahmani-Sane, Azad, Ameli, and Haghani (2025) highlighted that the combined AI–DT framework has revolutionized predictive maintenance, system design, and operational intelligence in renewable energy systems. Within wind energy, DTs replicate turbine behavior, while AI algorithms analyze operational patterns to detect anomalies, optimize control actions, and forecast power generation. By closing the loop between sensing, learning, and simulation, this synergy supports autonomous decision-making and real-time optimization.

Maksimović, Jokić, and Bošković (2025) expanded the concept further by linking 5G communication networks with AI-driven DTs to create ultra-responsive, low-latency systems for smart grid and turbine coordination. They proposed that future wind-energy ecosystems will rely on the seamless interaction of DTs, AI agents, and advanced connectivity to realize self-organizing and self-healing power systems. The synergy between AI and DTs fundamentally shifts operational paradigms from reactive monitoring to proactive intelligence. In wind farms, this means anticipating wind fluctuations, predicting component failures, and dynamically adjusting turbine settings, tasks that were once manual and data-intensive, are now automated and optimized in real time.

Workflow

The integrated AI–DT workflow in wind-turbine operation consists of four interlinked stages: data collection, model training, real-time monitoring, and feedback optimization as shown in Figure 5.

Figure 5. AI–digital-twin synergy workflow for smart turbine operation.

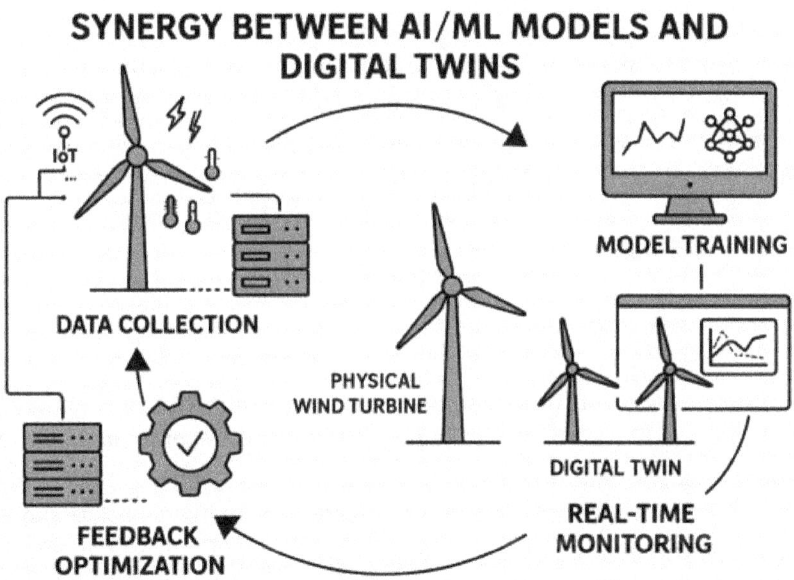

Data Collection

The foundation of any AI–DT framework lies in comprehensive, high-quality data. Sensors installed on turbines continuously measure wind speed, vibration, torque, temperature, and voltage. These data streams are transferred through IoT gateways and SCADA systems to cloud or edge servers for processing and storage. Fahim, Sharma, Cao, Canberk, and Duong (2022) demonstrated how extensive SCADA datasets fuel DT-driven ML models, enabling precise replication of turbine dynamics and environmental responses.

Model Training

In this phase, ML and deep-learning algorithms are trained using historical and real-time data to predict operational parameters and identify patterns associated with faults or inefficiencies. Kavousi-Fard, Dabbaghjamanesh, Sheikh, and Jin (2025) developed a deep-learning-based DT to mitigate wake effects in large wind farms. Their model learned complex flow interactions between turbines and adjusted yaw control to optimize power capture. This data-driven adaptation ensures that DTs evolve with changing environmental conditions.

Stadtmann et al. (2023) identified model training as the critical step where data-driven insights merge with physical simulations. By calibrating AI models with sensor feedback, DTs maintain alignment with the physical turbine, ensuring predictive accuracy. Yuan, Wang, and Zhao (2025) demonstrated how integrating DTs with Building Information Modeling (BIM) and IoT systems streamlines project workflows during wind-farm construction, facilitating dynamic quality management and continuous learning.

Real-Time Monitoring

Once trained, the AI–DT system continuously monitors the turbine's performance in real time. The DT mirrors the physical turbine's operational state, while AI algorithms analyze data streams to detect deviations or inefficiencies. Fahim et al. (2022) implemented neural networks to identify anomalies in rotational speed and blade loading, enabling preventive maintenance scheduling. The integration of AI analytics within DTs transforms static visualization into intelligent surveillance, capable of autonomously diagnosing issues before they escalate.

Feedback Optimization

The final stage involves feedback from the DT to the physical turbine through closed-loop control. Using RL and optimization algorithms, the system adjusts turbine parameters such as pitch angle, yaw orientation, and generator torque to maintain maximum efficiency. Kavousi-Fard et al. (2025) highlighted that feedback optimization based on predictive DT simulations improved farm-level power output by reducing wake losses and load imbalances. This continuous cycle, data collection, model training, monitoring, and feedback, creates an intelligent self-improving loop. It allows the turbine to evolve dynamically, learning from every operational event, thus ensuring resilient and sustainable performance across varying environmental conditions.

Cloud and Edge Computing for Real-Time Analytics

AI–DT integration depends heavily on computational infrastructure that supports real-time analytics. Cloud computing provides scalable resources for large-scale data storage and processing, while edge computing offers low-latency analytics close to the data source. Together, they create a hybrid ecosystem essential for high-speed wind-energy applications. Liu (2025) explored cloud-edge collaboration for intelligent monitoring and control, proposing an investment framework for digital energy management. His research demonstrated that hybrid deployment minimizes latency and reduces bandwidth consumption while maintaining data integrity. In wind-energy systems, this architecture allows time-sensitive operations, like fault detection and control optimization, to be executed locally at the turbine (edge), while long-term analytics and model training occur in the cloud. Abdullahi, Perinpanayagam, and Hamidu (2022) proposed a fog computing-based architecture that bridges cloud and edge layers for asset management in wind-turbine plants. Their design improved responsiveness and fault tolerance by decentralizing computational workloads, which is particularly vital for remote offshore farms. By distributing intelligence across the network, turbines can operate semi-autonomously, even with intermittent connectivity.

Feng, Qiao, and Lv (2023) developed a soft computing-enabled cloud optimization framework for DTs within IoT systems. Their study demonstrated that soft computing, using fuzzy logic and evolutionary algorithms, enhances resource allocation, energy efficiency, and service reliability in cloud–DT interactions. This system architecture enables scalable deployment of AI–DT frameworks, supporting predictive maintenance and energy forecasting in near real time. The synergy of cloud, fog, and edge computing enables real-time analytics, scalability, and reliability, key prerequisites for next-generation wind-energy management. By leveraging distributed intelligence, AI–DT ecosystems achieve self-healing capabilities and continuous optimization without human intervention.

Emerging Research: Hybrid Digital Twins Combining Physics-Based and AI-Based Models

Hybrid digital twins integrate physics-based modeling with data-driven AI learning, achieving both interpretability and predictive precision. These models leverage the deterministic accuracy of physical laws while incorporating the adaptability of ML algorithms. Shittu and Shittu (2024) emphasized that hybrid DTs overcome the limitations of purely data-driven systems by embedding domain knowledge and engineering principles. In renewable energy, AI-powered hybrid DTs can simulate turbine degradation, predict failures, and recommending optimal control actions based on both historical data and physical constraints.

Pujana, Esteras, Perea, Maqueda, and Calvez (2023) developed a hybrid-model-based DT for wind-turbine drivetrains that uses synthetic data generation for failure modeling. Their DT combines mechanical equations with AI estimators to detect faults under limited real-world data availability. The resulting model produces high-fidelity simulations that assist in preventive maintenance planning and digital validation of new turbine designs. Rahmani-Sane et al. (2025) noted that hybrid DTs provide the backbone for physics-informed AI in power systems. These systems leverage real-time sensor data while adhering to energy-conservation laws, thereby improving model reliability and reducing computational errors. Such hybrid DTs will be instrumental in developing self-learning turbines that adapt autonomously to complex environmental and operational challenges. As AI and simulation technologies advance, hybrid DTs will dominate future wind-energy research, supporting explainable, trustworthy, and data-efficient intelligence.

Implementation Challenges and Solutions: Data Reliability, Scalability, and Cybersecurity

Despite their advantages, implementing integrated AI–DT frameworks present several challenges, primarily concerning data reliability, system scalability, and cybersecurity.

Data Reliability

AI and DT systems rely on vast amounts of high-quality data. However, inconsistencies, sensor drift, or data loss can compromise model accuracy. Data preprocessing, redundancy, and sensor calibration are critical to maintaining reliability. Rahmani-

Sane et al. (2025) highlighted that data-fusion algorithms combining SCADA, IoT, and simulation data enhance model robustness. Incorporating blockchain-based audit trails could further ensure traceability and data integrity throughout the lifecycle of the DT.

Scalability

Scalability remains a key concern, particularly for large wind farms deploying multiple DTs. As each turbine generates terabytes of data daily, system architecture must support high-throughput analytics. Liu (2025) suggested that cloud–edge orchestration enables scalable analytics without compromising latency. Furthermore, containerized microservices and distributed AI agents allow each turbine's DT to operate semi-independently while contributing to farm-level optimization.

Cybersecurity

Cybersecurity represents one of the most pressing issues in DT-enabled energy systems. The interconnected nature of IoT, AI, and cloud services introduces vulnerabilities to data theft, system intrusion, and control manipulation. Kumar, Pandey, Verre, and Shah (2025) conducted a comprehensive review of cybersecurity challenges in renewable-energy digitization, identifying malware, ransomware, and data spoofing as emerging threats. They recommended adopting zero-trust architecture, multi-factor authentication, and real-time intrusion detection systems for enhanced protection. Rekeraho et al. (2024) examined cybersecurity threats in IoT-based smart renewables, emphasizing encryption and secure communication protocols such as TLS, AES128, MQTT-S, etc. Similarly, Al Abdulwahid (2024) proposed a secure cyber-physical architecture for Industry 4.0 power grids, incorporating redundant communication pathways and anomaly detection to safeguard DT–AI systems in wind farms. Together, these measures address the triad of challenges, reliability, scalability, and cybersecurity, ensuring that integrated AI–DT ecosystems remain secure, resilient, and future-ready.

The integration of AI and DT technologies marks a paradigm shift in wind-energy operations, creating intelligent, adaptive, and autonomous systems. The synergy between AI's learning capabilities and DT's real-time simulation enables predictive maintenance, fault detection, and dynamic optimization. From the foundational synergy studies by Ranawaka et al. (2024) and Rahmani-Sane et al. (2025) to hybrid frameworks and edge-cloud deployments (Liu, 2025; Abdullahi et al., 2022), research demonstrates that AI–DT collaboration enhances efficiency, sustainability, and resilience in modern wind-energy infrastructure. While challenges persist in data reliability, scalability, and cybersecurity, ongoing innovations, such as hybrid DTs, blockchain verification, and secure architecture are steadily resolving these barriers.

Ultimately, the integrated AI–DT framework serves as the intelligence core of next-generation wind farms. It enables real-time adaptation, system-wide learning, and operational excellence, ushering in the era of fully autonomous, self-optimizing renewable-energy systems aligned with global sustainability goals.

CASE STUDIES AND FUTURE TRENDS

Real-World Examples of AI and DT Enabled Wind Farms

The emergence of AI and DT technologies has transformed modern wind-energy operations, enabling predictive maintenance, real-time optimization, and intelligent monitoring. Multiple real-world implementations now illustrate how DTs and AI frameworks improve turbine reliability, energy yield, and cost efficiency. Chen, Liu, Yu, and Li (2024) presented a comprehensive review on enhancing reliability in floating offshore wind turbines through DT technology. Offshore environments pose unique challenges due to harsh weather and limited maintenance accessibility. The researchers highlighted real-world pilot projects in Europe and Asia where DTs monitor structural integrity, fatigue loads, and mooring-line stresses. By integrating sensor data with physics-based simulations, operators achieve near-continuous monitoring and reduce unplanned downtime by up to 25%. These DT systems are connected to cloud analytics platforms that automatically generate maintenance alerts based on predictive models.

Similarly, Habbouche, Amirat, and Benbouzid (2025) analyzed AI-driven DT applications for gearbox condition monitoring, a critical issue in turbine reliability. They demonstrated that neural networks embedded within DT frameworks can accurately predict gearbox faults weeks before physical symptoms appear. The study also showcased practical deployments where DT-enabled predictive maintenance decreased annual operational costs by 15%, improving the long-term sustainability of wind farms. Rodriguez (2025) provided insights into DT dynamics in smart energy grids, documenting industrial-scale implementations where DTs synchronize distributed energy assets, including wind farms. These systems utilize real-time digital replicas of turbines, substations, and grid nodes to manage power flow dynamically. By merging DT simulation capabilities with AI-based optimization, grid operators can balance generation and demand in milliseconds, improving overall grid resilience.

Furthermore, *A. W. Energy* (2024) reported on edge-driven DTs in distributed renewable systems. In these architectures, localized edge processors perform real-time analytics on turbine data, reducing latency and dependence on centralized cloud platforms. This approach has proven essential for remote wind farms where connectivity is intermittent. The edge-DT framework enhances operational continuity, supports fault detection, and allows turbines to make autonomous control adjustments without human intervention. Collectively, these case studies underscore that DT–AI integration is no longer theoretical but operationally transformative. From floating offshore turbines to distributed hybrid grids, DTs provide digital continuity and intelligence that fundamentally redefine reliability and performance optimization in the renewable sector. These examples are shown in Figure 6.

Figure 6. Real world examples of AI and DT enabled wind farms.

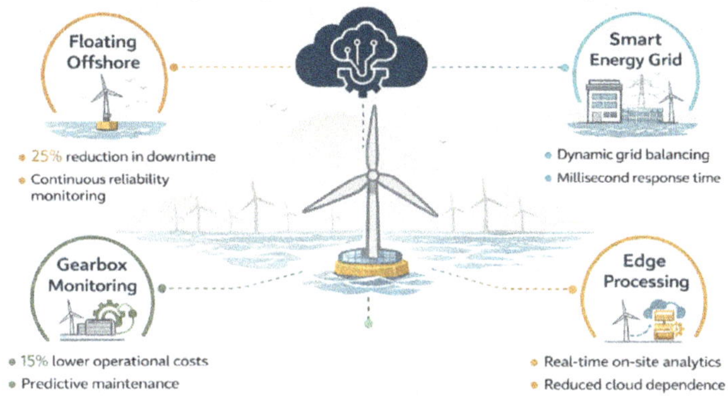

Real-Time Industrial Case Study: Long-Term SCADA-Based Performance Assessment

A real-time industrial case study based on ten years (2011–2020) of 10-minute SCADA data from seven 2-MW utility-scale wind turbines is presented by Astolfi et al., (2022). It provides a detailed assessment of long-term turbine performance evolution under the combined effects of aging and control optimization. The turbines operate within the same wind farm and were subjected to a farm-wide control optimization in 2018, offering a consistent and realistic dataset for turbine-level performance evaluation.

Preliminary SCADA-Based Performance Trends

Initial inspection using IEC-compliant power curves reveals substantial changes in the wind farm performance landscape over the decade. Turbines that were underperforming in early operation (e.g., T2 and T6 in 2011) differ from those exhibiting degradation in later years (e.g., T1 and T3 in 2020), while T7 remains consistently among the best-performing units. Vertical power-curve analysis further highlights contrasting aging trajectories, with T1 showing progressive deterioration and T7 demonstrating long-term improvement. These observations confirm that qualitative inspection alone is insufficient for understanding long-term performance evolution.

Detection of Structural Changes Using PCA and Mahalanobis Distance

Principal Component Analysis (PCA) and Mahalanobis-distance-based novelty detection are employed to assess the statistical comparability of SCADA datasets before regression-based analysis. Results indicate consistent turbine behavior from 2011 to 2016, followed by a clear and farm-wide deviation beginning in 2017 and becoming pronounced after 2018. This shift coincides with the documented control optimization, demonstrating that multivariate SCADA analytics can reliably detect real operational regime changes in industrial wind farms.

Impact of Control Optimization on Turbine Behavior

Analysis of the rotor-speed–power relationship in the partial-load operating region confirms a modification in control strategy starting in 2018, characterized by an altered rotational speed operating range. While this optimization delivers the expected efficiency gains for six turbines, one turbine (T1) exhibits further performance degradation, indicating that control upgrades may amplify latent mechanical or control-related deficiencies. This finding highlights the importance of turbine-specific health assessment prior to applying operational optimizations.

Time–Space Performance Assessment Using Data-Driven Models

Support Vector Regression models trained on SCADA data serve as simplified digital twins, enabling fixed-time, rolling-time, and spatial benchmarking analyses. The results show that four turbines achieve a net positive performance drift over the decade, two experience moderate aging consistent with literature-reported degradation rates, and one turbine undergoes significant worsening. Spatial benchmarking further reveals that turbines with negligible aging drift may still incur substantial cumulative energy losses if they consistently underperform relative to the best-performing turbine.

Implications for Real-Time Monitoring and Digital Twins

Overall, this case study demonstrates that continuous, turbine-specific SCADA analytics are essential for reliable long-term performance assessment. Aging, optimization, and component health interact in complex and non-uniform ways that cannot be captured by fleet-level averages or single aging coefficients. The findings strongly support the adoption of continuously updated, SCADA-driven digital twin frameworks as a foundation for AI-based monitoring, condition-aware optimization, and informed operational decision-making in modern wind farms.

Comparative Analysis of Traditional vs. Intelligent Monitoring Systems

Traditional wind-turbine monitoring systems rely primarily on threshold-based alarms and periodic inspections. Although effective for basic oversight, these systems lack adaptability to complex dynamic behaviors, often resulting in delayed fault detection and higher maintenance costs. In contrast, intelligent monitoring systems leverage AI and DTs to provide continuous learning, self-diagnosis, and proactive decision-making. Rababaah (2023) conducted a comparative study between deep-learning and traditional machine-learning models for intelligent health diagnosis. The findings revealed that DL models, such as CNNs and LSTM networks, achieved superior fault-classification accuracy compared with conventional algorithms. The DL models processed raw vibration and acoustic signals directly, eliminating the need for extensive feature engineering. As a result, intelligent DT-based systems can detect subtle anomalies that conventional SCADA systems often overlook.

Karad and Thakur (2021) discussed the limitations of early IoT-based monitoring frameworks, which focused more on data acquisition than analytics. Their review emphasized that, while IoT enabled real-time visibility, its lack of advanced AI integration limited predictive capabilities. However, integrating IoT networks with DTs and AI models now allows not only real-time observation but also dynamic optimization and lifetime estimation for critical components. Benbouzid et al. (2021) provided a detailed state-of-the-art review on intelligent condition monitoring. They noted that conventional methods depend on static parameter thresholds, which often lead to false positives or undetected early-stage failures. By contrast, intelligent monitoring systems incorporating DTs adaptively adjust thresholds based on contextual parameters such as temperature, humidity, and operational loads. These adaptive systems continuously refine themselves through machine learning, making them more resilient and accurate over time.

Comparative analyses demonstrate that intelligent monitoring reduces diagnostic latency and enhances decision accuracy. For example, while traditional SCADA alarms might trigger after critical failure symptoms, AI–DT systems can detect early degradation patterns, prompting maintenance interventions long before breakdowns occur. This paradigm shift from reactive to predictive maintenance defines the future of turbine health management.

Table 1. Comparison of Traditional and Intelligent Wind Turbine Monitoring Systems.

Aspect	Traditional Monitoring Systems	Intelligent (AI–DT-Based) Monitoring Systems
Monitoring approach	Threshold-based alarms	Data-driven, adaptive learning
Data utilization	Limited SCADA parameters	Multisource data (SCADA, IoT, sensors)
Analytics capability	Descriptive	Predictive and prescriptive
Fault detection	Reactive, post-failure	Early anomaly detection
Feature engineering	Manual	Automatic (DL-based)
Adaptability	Static thresholds	Context-aware, self-adjusting
Diagnostic latency	High	Low
Maintenance strategy	Preventive / reactive	Predictive maintenance
Decision support	Limited	Real-time decision support
Operational impact	Higher downtime and costs	Reduced downtime and O&M costs

Future Trends: Self-Learning Wind Turbines, FL, and AI-Based Swarm Optimization

The next generation of wind-energy systems will integrate self-learning, collaborative, and swarm-intelligent technologies driven by AI and DTs. These advances will enable wind farms to function as autonomous ecosystems that continuously adapt to environmental variability and operational demands. Safari, Daneshvar, and Anvari-Moghaddam (2024) provided a comprehensive review of "energy intelligence," highlighting how AI frameworks can manage energy production, consumption, and trading across distributed systems. Their review suggested that self-learning wind turbines, powered by RL, can autonomously adjust their control parameters to maximize energy yield. By learning from continuous environmental feedback, these turbines can respond to complex wind turbulence patterns with minimal human input.

Narayanan (2024) discussed the application of RL in wind-energy systems, demonstrating its potential for optimizing pitch, yaw, and torque control. The study revealed that RL algorithms outperform classical control strategies by dynamically learning optimal responses under uncertainty. RL-based controllers, when integrated within DTs, create autonomous feedback loops capable of real-time adaptation and performance tuning.

Another emerging trend is FL, a decentralized AI approach that allows multiple wind farms to collaboratively train models without sharing sensitive data. Jayachitra, Prasanth, Dhanaraj, and Pelusi (2025) emphasized that FL supports privacy-preserving model development across geographically distributed turbines. Each turbine learns locally from its data, while only model updates are aggregated centrally. This approach not only enhances model generalization but also ensures compliance with cybersecurity and data-protection requirements. AI-based swarm optimization represents an additional frontier in large-scale wind-farm coordination. Garcia Marquez and Peinado Gonzalo (2022) reviewed swarm-based algorithms, such as particle swarm optimization (PSO) and ant colony optimization (ACO), that collectively adjust turbine orientations to minimize wake interference and maximize power output. When combined with DT simulations, these algorithms can predict wake propagation and dynamically reconfigure turbine operations to maintain optimal efficiency.

Overall, the future of wind-energy systems lies in developing intelligent, self-organizing networks where turbines cooperate like agents in an ecosystem. These AI–DT hybrid frameworks will underpin digitalized, adaptive, and sustainable energy landscapes as shown in Figure 7.

Figure 7. Future Trends: Self-Learning Wind Turbines, FL, and AI-Based Swarm Optimization.

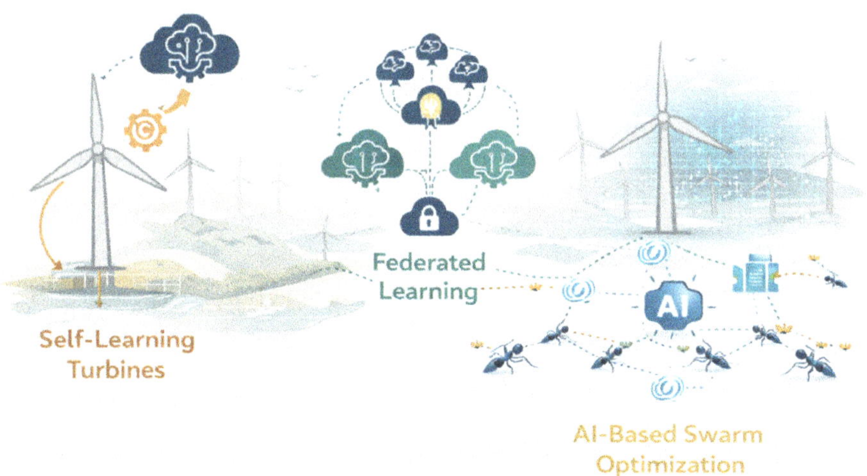

Policy, Standardization, and Interoperability Considerations for Digital Wind Energy Systems

The widespread deployment of AI–DT frameworks in wind energy requires standardized protocols, regulatory guidelines, and secure data-sharing mechanisms to ensure interoperability and safety. Without standardized frameworks, integration across diverse hardware, software, and regional infrastructures becomes fragmented, limiting scalability. Guerreiro, Martin, Dreyer, Yang, and Andresen (2024) analyzed grid compliance and interoperability issues, highlighting the importance of software- and hardware-in-the-loop (SIL/HIL) testing. Their research showed that DTs can simulate turbine–grid interactions to ensure that new control algorithms meet regulatory standards before deployment. Grid-compliant DT environments thus reduce certification time and enhance operational safety in high-voltage systems.

Clifton et al. (2023) identified grand challenges in the digitalization of wind energy, including data governance, cybersecurity, and AI model transparency. They stressed that regulatory bodies must address intellectual property, data ownership, and algorithmic accountability to maintain trust in digital infrastructure. Furthermore, they emphasized the need for open data standards to facilitate interoperability among turbine manufacturers, software vendors, and grid operators. Strasser et al. (2024) discussed the progress toward interoperability testing in smart energy systems, proposing testbeds for evaluating communication protocols and data exchange formats. They argued that achieving interoperability requires harmonizing standards like OPC UA, IEC 61850, and MQTT, enabling seamless connectivity across DT-enabled assets and control systems. Such harmonization is vital for scaling intelligent wind-farm networks across countries and utility grids.

Policy frameworks should also consider cybersecurity certification, ensuring that AI–DT architectures adhere to security-by-design principles. Standardization of model interfaces and communication layers will accelerate the integration of DTs across the renewable sector, fostering collaboration between academia, industry, and government.

CONCLUSIONS

The integration of AI, ML, and DT technologies represents a transformative leap toward intelligent, autonomous, and data-driven wind-energy systems. AI and ML serve as the analytical core, enabling forecasting, adaptive control, and predictive maintenance using vast SCADA and sensor datasets. Simultaneously, DTs act as dynamic virtual replicas of turbines and wind farms, offering real-time simulation, fault prediction, and performance optimization. Their synergy creates continuous learning loops, data collection, model training, real-time monitoring, and feedback optimization, resulting in self-learning, resilient, and efficient operations.

This integration enhances predictive maintenance, reducing downtime and costs by up to 30%, while optimizing aerodynamic efficiency and grid stability. However, challenges such as data reliability, scalability, and cybersecurity remain critical. Cloud–edge computing, blockchain, and zero-trust architectures are emerging as key enablers for secure and scalable implementation. Future research will advance hybrid digital twins, FL, and edge intelligence, fostering autonomous, cybersecure, and sustainable wind farms. Ultimately, AI–DT synergy signifies a paradigm shift from reactive management to intelligent, adaptive, and self-optimizing renewable systems, driving global progress toward carbon neutrality and resilient digital energy ecosystems.

REFERENCES

Abdullahi, I., Perinpanayagam, S., & Hamidu, I. (2022, September). A fog computing based approach towards improving asset management and performance of wind turbine plants using digital twins. In *2022 27th International Conference on Automation and Computing (ICAC)* (pp. 1–5). IEEE. DOI: 10.1109/ICAC55051.2022.9911170

Afridi, Y. S., Ahmad, K., & Hassan, L. (2022). Artificial intelligence based prognostic maintenance of renewable energy systems: A review of techniques, challenges, and future research directions. *International Journal of Energy Research*, *46*(15), 21619–21642. DOI: 10.1002/er.7100

Al Abdulwahid, A. (2024). Cyber-physical threat mitigation in wind energy systems: A novel secure architecture for industry 4.0 power grids. *Energy Informatics*, *7*(1), 141. DOI: 10.1186/s42162-024-00449-6

Al-Shetwi, A. Q., Atawi, I. E., El-Hameed, M. A., & Abuelrub, A. (2025). Digital twin technology for renewable energy, smart grids, energy storage and vehicle-to-grid integration: Advancements, applications, key players, challenges and future perspectives in modernising sustainable grids. *IET Smart Grid*, *8*(1), e70026. DOI: 10.1049/stg2.70026

Alam, M. S., Chowdhury, T. A., Dhar, A., Al-Ismail, F. S., Choudhury, M. S. H., Shafiullah, M., Hossain, M., Hossain, M., Ullah, A., & Rahman, S. M. (2023). Solar and wind energy integrated system frequency control: A critical review on recent developments. *Energies*, *16*(2), 812. DOI: 10.3390/en16020812

Alghamdi, W. Y., Alshamrani, R. M., Aloufi, R. K., Ba Lhamar, S. O., Altwirqi, R. A., Alotaibi, F. S., & Alazwari, A. S. (2025). Exploring the synergy between digital twin technology and artificial intelligence: A comprehensive survey. *International Journal of Advanced Computer Science and Applications*, *16*(3). Advance online publication. DOI: 10.14569/IJACSA.2025.0160399

Anushalini, T., & Sri Revathi, B. (2024). Role of machine learning algorithms for wind power generation prediction in renewable energy management. *Journal of the Institution of Electronics and Telecommunication Engineers*, *70*(4), 4319–4332. DOI: 10.1080/03772063.2023.2205838

Astolfi, D., Pandit, R., Celesti, L., Lombardi, A., & Terzi, L. (2022). SCADA data analysis for long-term wind turbine performance assessment: A case study. *Sustainable Energy Technologies and Assessments*, *52*, 102357. DOI: 10.1016/j.seta.2022.102357

Benbouzid, M., Berghout, T., Sarma, N., Djurović, S., Wu, Y., & Ma, X. (2021). Intelligent condition monitoring of wind power systems: State of the art review. *Energies*, *14*(18), 5967. DOI: 10.3390/en14185967

Bošnjaković, M., Martinović, M., & Đokić, K. (2025). Application of artificial intelligence in wind power systems. *Applied Sciences (Basel, Switzerland)*, *15*(5), 2443. DOI: 10.3390/app15052443

Chen, B. Q., Liu, K., Yu, T., & Li, R. (2024). Enhancing reliability in floating offshore wind turbines through digital twin technology: A comprehensive review. *Energies*, *17*(8), 1964. DOI: 10.3390/en17081964

Chinnappan, C. C. (2025). Integrating data-driven and physics-based approaches for robust wind power prediction: A comprehensive ML-PINN-Simulink framework. *Scientific Reports*, *15*(1), 1–29.

Clifton, A., Barber, S., Bray, A., Enevoldsen, P., Fields, J., Sempreviva, A. M., Williams, L., Quick, J., Purdue, M., Totaro, P., & Ding, Y. (2023). Grand challenges in the digitalisation of wind energy. *Wind Energy Science*, *8*(6), 947–974. DOI: 10.5194/wes-8-947-2023

Elkelawy, M., Atta, Z. A., & Seleem, H. (2024). Technological advances, efficiency optimization, and challenges in wind power plants: A comprehensive review. *Pharos Engineering Science Journal*, *1*(1), 57–65. DOI: 10.21608/pesj.2025.344906.1006

A. W. Energy. (2024). *Edge driven digital twins in distributed energy systems.*

Erge, O., & van Oort, E. (2022). Combining physics-based and data-driven modeling in well construction: Hybrid fluid dynamics modeling. *Journal of Natural Gas Science and Engineering*, *97*, 104348. DOI: 10.1016/j.jngse.2021.104348

Fahim, M., Sharma, V., Cao, T. V., Canberk, B., & Duong, T. Q. (2022). Machine learning-based digital twin for predictive modeling in wind turbines. *IEEE Access : Practical Innovations, Open Solutions*, *10*, 14184–14194. DOI: 10.1109/ACCESS.2022.3147602

Farrar, N. O., Ali, M. H., & Dasgupta, D. (2023). Artificial intelligence and machine learning in grid connected wind turbine control systems: A comprehensive review. *Energies*, *16*(3), 1530. DOI: 10.3390/en16031530

Feng, H., Qiao, L., & Lv, Z. (2023). Innovative soft computing-enabled cloud optimization for next-generation IoT in digital twins. *Applied Soft Computing*, *136*, 110082. DOI: 10.1016/j.asoc.2023.110082

Garcia Marquez, F. P., & Peinado Gonzalo, A. (2022). A comprehensive review of artificial intelligence and wind energy. *Archives of Computational Methods in Engineering*, *29*(5), 2935–2958. DOI: 10.1007/s11831-021-09678-4

Guerreiro, G. M. G., Martin, F., Dreyer, T., Yang, G., & Andresen, B. (2024). Advancements on grid compliance in wind power: Component & subsystem testing, software-/hardware-in-the-loop, and digital twins. *IEEE Access : Practical Innovations, Open Solutions*, *12*, 25949–25966. DOI: 10.1109/ACCESS.2024.3366091

Habbouche, H., Amirat, Y., & Benbouzid, M. (2025). Leveraging digital twins and AI for enhanced gearbox condition monitoring in wind turbines: A review. *Applied Sciences (Basel, Switzerland)*, *15*(10), 5725. DOI: 10.3390/app15105725

Haghshenas, A., Hasan, A., Osen, O., & Mikalsen, E. T. (2023). Predictive digital twin for offshore wind farms. *Energy Informatics*, *6*(1), 1. DOI: 10.1186/s42162-023-00257-4

Hannan, M. A., Al-Shetwi, A. Q., Mollik, M. S., Ker, P. J., Mannan, M., Mansor, M., Al-Masri, H. M. K., & Mahlia, T. I. (2023). Wind energy conversions, controls, and applications: A review for sustainable technologies and directions. *Sustainability (Basel)*, *15*(5), 3986. DOI: 10.3390/su15053986

Hassan, A., Ahmad, G., Shafiullah, M., Islam, A., & Alam, M. S. (2025). Review of the intelligent frameworks for pitch angle control in wind turbines. *IEEE Access : Practical Innovations, Open Solutions*, *13*, 29864–29885. DOI: 10.1109/ACCESS.2025.3540367

Ioannides, M. G., Stamelos, A. P., Papazis, S. A., Stamataki, E. E., & Stamatakis, M. E. (2024). Internet of things-based control of induction machines: Specifics of electric drives and wind energy conversion systems. *Energies*, *17*(3), 645. DOI: 10.3390/en17030645

Issa, R., Hamad, M. S., & Abdel-Geliel, M. (2023, February). Digital twin of wind turbine based on Microsoft® Azure IoT platform. In *2023 IEEE Conference on Power Electronics and Renewable Energy (CPERE)* (pp. 1–8). IEEE. DOI: 10.1109/CPERE56564.2023.10119576

Jayachitra, S., Prasanth, A., Dhanaraj, R. K., & Pelusi, D. (Eds.). (2025). *Applications of federated learning in technological advancements: Use cases and applications*. CRC Press. DOI: 10.1201/9781003532323

Kabir, M. R., Halder, D., & Ray, S. (2024). Digital twins for IoT-driven energy systems: A survey. *IEEE Access : Practical Innovations, Open Solutions*, *12*, 177123–177143. DOI: 10.1109/ACCESS.2024.3506660

Kandemir, E., Hasan, A., Kvamsdal, T., & Abdel-Afou Alaliyat, S. (2024). Predictive digital twin for wind energy systems: A literature review. *Energy Informatics*, *7*(1), 68. DOI: 10.1186/s42162-024-00373-9

Karad, S., & Thakur, R. (2021). Efficient monitoring and control of wind energy conversion systems using Internet of Things (IoT): A comprehensive review. *Environment, Development and Sustainability*, *23*(10), 14197–14214. DOI: 10.1007/s10668-021-01267-6

Kavousi-Fard, A., Dabbaghjamanesh, M., Sheikh, M., & Jin, T. (2025). A novel deep learning based digital twin model for mitigating wake effects in wind farms. *Renewable Energy Focus*, *53*, 100686. DOI: 10.1016/j.ref.2025.100686

Kumar, K., Pandey, N. K., Verre, F., & Shah, R. (2025). Cybersecurity challenges in the digitization and integration of renewable energy systems: A review. *IEEE Transactions on Engineering Management*, *72*, 3042–3054. DOI: 10.1109/TEM.2025.3586557

Lipu, M. H., Miah, M. S., Hannan, M. A., Hussain, A., Sarker, M. R., Ayob, A., Saad, M. H. M., & Mahmud, M. S. (2021). Artificial intelligence-based hybrid forecasting approaches for wind power generation: Progress, challenges, and prospects. *IEEE Access : Practical Innovations, Open Solutions*, *9*, 102460–102489. DOI: 10.1109/ACCESS.2021.3097102

Liu, Y., Zhang, J. M., Min, Y. T., Yu, Y., Lin, C., & Hu, Z. Z. (2023). A digital twin-based framework for simulation and monitoring analysis of floating wind turbine structures. *Ocean Engineering*, *283*, 115009. DOI: 10.1016/j.oceaneng.2023.115009

Liu, Z. (2025). Intelligent monitoring and control of farmland based on edge-cloud collaboration and digital twin for digital energy management: Investment benefit analysis. *Sustainable Energy Research*, *12*(1), 43. DOI: 10.1186/s40807-025-00179-7

Mahmoud, M., Semeraro, C., Abdelkareem, M. A., & Olabi, A. G. (2024). Designing and prototyping the architecture of a digital twin for wind turbine. *International Journal of Thermofluids*, *22*, 100622. DOI: 10.1016/j.ijft.2024.100622

Maksimović, M., Jokić, S., & Bošković, M. Č. (2025). Innovative horizons for sustainable smart energy: Exploring the synergy of 5G and digital twin technologies. *Process Integration and Optimization for Sustainability*, *9*(2), 431–470. DOI: 10.1007/s41660-024-00478-4

Mansour, D. E. A., Numair, M., Zalhaf, A. S., Ramadan, R., Darwish, M. M., Huang, Q., Hussien, M. G., & Abdel-Rahim, O. (2023). Applications of IoT and digital twin in electrical power systems: A comprehensive survey. *IET Generation, Transmission & Distribution*, *17*(20), 4457–4479. DOI: 10.1049/gtd2.12940

Narayanan, V. L. (2024). Reinforcement learning in wind energy—A review. *International Journal of Green Energy*, *21*(9), 1945–1968. DOI: 10.1080/15435075.2023.2281329

Nguyen, V. H., Tran, Q. T., Besanger, Y., Jung, M., & Nguyen, T. L. (2022). Digital twin integrated power-hardware-in-the-loop for the assessment of distributed renewable energy resources. *Electrical Engineering*, *104*(2), 377–388. DOI: 10.1007/s00202-021-01246-0

Ohalete, N. C., Aderibigbe, A. O., Ani, E. C., Ohenhen, P. E., Daraojimba, D. O., & Odulaja, B. A.Nzubechukwu Chukwudum OhaleteAdebayo Olusegun AderibigbeEmmanuel Chigozie AniPeter Efosa OhenhenDonald Obinna DaraojimbaBukola A. Odulaja. (2023). AI-driven solutions in renewable energy: A review of data science applications in solar and wind energy optimization. *World Journal of Advanced Research and Reviews*, *20*(3), 401–417. DOI: 10.30574/wjarr.2023.20.3.2433

Pujana, A., Esteras, M., Perea, E., Maqueda, E., & Calvez, P. (2023). Hybrid-model-based digital twin of the drivetrain of a wind turbine and its application for failure synthetic data generation. *Energies*, *16*(2), 861. DOI: 10.3390/en16020861

Qays, M. O., Ahmed, M. M., Parvez Mahmud, M. A., Abu-Siada, A., Muyeen, S. M., Hossain, M. L., Yasmin, F., & Rahman, M. M. (2022). Monitoring of renewable energy systems by IoT-aided SCADA system. *Energy Science & Engineering*, *10*(6), 1874–1885. DOI: 10.1002/ese3.1130

Rababaah, A. R. (2023). Comparative study of deep learning models versus machine learning models for wind turbine intelligent health diagnosis systems. *Arabian Journal for Science and Engineering*, *48*(8), 10875–10899. DOI: 10.1007/s13369-023-07810-z

Rahmani-Sane, G., Azad, S., Ameli, M. T., & Haghani, S. (2025). The applications of artificial intelligence and digital twin in power systems: An in-depth review. *IEEE Access : Practical Innovations, Open Solutions*, *13*, 108573–108608. DOI: 10.1109/ACCESS.2025.3580340

Ranawaka, A., Alahakoon, D., Sun, Y., & Hewapathirana, K. (2024). Leveraging the synergy of digital twins and artificial intelligence for sustainable power grids: A scoping review. *Energies*, *17*(21), 5342. DOI: 10.3390/en17215342

Rekeraho, A., Cotfas, D. T., Cotfas, P. A., Bălan, T. C., Tuyishime, E., & Acheampong, R. (2024). Cybersecurity challenges in IoT-based smart renewable energy. *International Journal of Information Security*, *23*(1), 101–117. DOI: 10.1007/s10207-023-00732-9

Rodriguez, E. (2025). Digital twin dynamics in smart energy grids: Advanced integration and practical implementations. *Digital Twin Dynamics*, *1*(1), 33–46.

Rosca, C. M., & Stancu, A. (2025). A comprehensive review of machine learning models for optimizing wind power processes. *Applied Sciences (Basel, Switzerland)*, *15*(7), 3758. DOI: 10.3390/app15073758

Sadiq, M., Ali, S. W., Terriche, Y., Mutarraf, M. U., Hassan, M. A., Hamid, K., Ali, Z., Sze, J. Y., Su, C.-L., & Guerrero, J. M. (2021). Future greener seaports: A review of new infrastructure, challenges, and energy efficiency measures. *IEEE Access : Practical Innovations, Open Solutions*, *9*, 75568–75587. DOI: 10.1109/ACCESS.2021.3081430

Safari, A., Daneshvar, M., & Anvari-Moghaddam, A. (2024). Energy intelligence: A systematic review of artificial intelligence for energy management. *Applied Sciences (Basel, Switzerland)*, *14*(23), 11112. DOI: 10.3390/app142311112

Shafik, W. (2024). Introduction to renewable energy technologies. In *Soft computing in renewable energy technologies* (pp. 1–31). CRC Press. DOI: 10.1201/9781003462460-1

Sheng, S., & O'Connor, R. (2023). Reliability of wind turbines. In *Wind energy engineering* (pp. 195–211). Academic Press. DOI: 10.1016/B978-0-323-99353-1.00016-5

Shittu, H., & Shittu, M. (2024). AI powered digital twins for predictive maintenance and operational optimization of renewable energy systems. *International Journal of Science, Architecture. Technology and Environment*, *1*, 151–163.

Smart, E. E., Olanrewaju, L. O., Usman, J., Otaru, K., & Umar, D. (2025). Artificial intelligence (AI) in renewable energy forecasting and optimization. *Renewable Energy*, *10*, 11.

Song, Z., Hackl, C. M., Anand, A., Thommessen, A., Petzschmann, J., Kamel, O., Braunbehrens, R., Kaifel, A., Roos, C., & Hauptmann, S. (2023). Digital twins for the future power system: An overview and a future perspective. *Sustainability (Basel)*, *15*(6), 5259. DOI: 10.3390/su15065259

Stadtmann, F., Rasheed, A., Kvamsdal, T., Johannessen, K. A., San, O., Kölle, K., Tande, J. O., Barstad, I., Benhamou, A., Brathaug, T., Christiansen, T., Firle, A.-L., Fjeldly, A., Frøyd, L., Gleim, A., Høiberget, A., Meissner, C., Nygård, G., Olsen, J., & Skogås, J. O. (2023). Digital twins in wind energy: Emerging technologies and industry-informed future directions. *IEEE Access : Practical Innovations, Open Solutions*, *11*, 110762–110795. DOI: 10.1109/ACCESS.2023.3321320

Strasser, T. I., Widl, E., Kuchenbuch, R. A., Lázaro-Elorriaga, L., Laraudogoitia, B. T., Ginocchi, M., . . . Algaba, E. R. (2024, November). Towards interoperability testing of smart energy systems–An overview and discussion of possibilities. In *IET Conference Proceedings CP904* (Vol. 2024, No. 29, pp. 263–268). The Institution of Engineering and Technology.

Udo, W. S., Kwakye, J. M., Ekechukwu, D. E., & Ogundipe, O. B. (2024). Optimizing wind energy systems using machine learning for predictive maintenance and efficiency enhancement. *Journal of Renewable Energy Technology*, *28*(3), 312–330.

Ukoba, K., Olatunji, K. O., Adeoye, E., Jen, T. C., & Madyira, D. M. (2024). Optimizing renewable energy systems through artificial intelligence: Review and future prospects. *Energy & Environment*, *35*(7), 3833–3879. DOI: 10.1177/0958305X241256293

Wang, N., Chen, Q., Zhu, L., & Sun, H. (2022). Integration of data-driven and physics-based modeling of wind waves in a shallow estuary. *Ocean Modelling (Oxford)*, *172*, 101978. DOI: 10.1016/j.ocemod.2022.101978

Yan, J., Nuertayi, A., Yan, Y., Liu, S., & Liu, Y. (2023). Hybrid physical and data-driven modeling for dynamic operation characteristic simulation of wind turbine. *Renewable Energy*, *215*, 118958. DOI: 10.1016/j.renene.2023.118958

Yuan, X., Wang, S., & Zhao, X. (2025, March). Dynamic quality management in wind power construction: Integrating digital twins with BIM and IoT. In *2025 IEEE International Conference on Electronics, Energy Systems and Power Engineering (EESPE)* (pp. 916–919). IEEE. DOI: 10.1109/EESPE63401.2025.10986901

Zhang, X., Tao, J., & Noshadravan, A. (2026). Probabilistic digital twin for reliability-based maintenance optimization of offshore wind turbines. *Renewable Energy*, *256*, 123777. DOI: 10.1016/j.renene.2025.123777

Zhong, D., Xia, Z., Zhu, Y., & Duan, J. (2023). Overview of predictive maintenance based on digital twin technology. *Heliyon*, *9*(4), e14534. DOI: 10.1016/j.heliyon.2023.e14534

Chapter 6
Cybersecurity and Data Privacy in Wind Power Systems

Ibrahim Mwammenywa
http://orcid.org/0000-0003-2735-643X
University of Dar es Salaam, Tanzania

ABSTRACT

Rapid global expansion of wind energy both onshore and offshore has transformed wind farms into highly digitalized, data-driven infrastructures. While digitalization enhances efficiency, predictive maintenance, and market integration, it also amplifies exposure to cyber threats. This chapter provides a structured overview of cybersecurity landscape specific to wind power systems, highlighting vulnerabilities at the levels of turbine control, supervisory and communication networks, cloud-based analytics, and supply-chain dependencies. It situates these risks within the evolving regulatory and standards framework like IEC 62443, EU NIS2 Directive, and General Data Protection Regulation (GDPR), and explores how compliance requirements intersect with operational realities. It also reviews recent advances in anomaly detection, fault prediction, and privacy-preserving analytics. Finally, it proposes a practical defense-in-depth roadmap integrating technical safeguards, organizational measures, and policy considerations to strengthen the resilience and trustworthiness of wind power systems.

INTRODUCTION

Wind energy has emerged as a cornerstone of the global transition toward low-carbon electricity. Unlike fossil fuel–based systems, modern wind power systems

DOI: 10.4018/979-8-2600-0106-6.ch006

are cyber-physical systems. For efficiency and reliability, they heavily depend on digital infrastructure. They vastly integrate sensors, industrial control systems (ICS), supervisory control and data acquisition (SCADA) platforms, and increasingly edge-to-cloud analytics, as shown in Figure 1. These technologies enable condition monitoring, predictive maintenance, improved forecasting, and smoother integration into complex electricity markets.

Figure 1. Wind Turbine Digital Control and Access Architecture

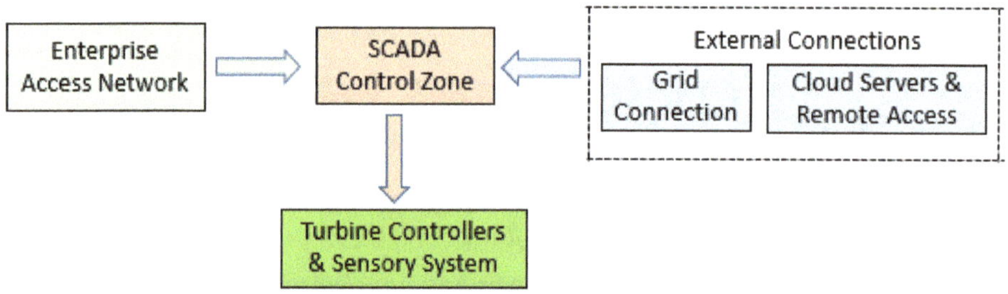

Over the past decade, digitalization has accelerated the convergence of operational technology (OT) and information technology (IT) in the wind sector. This integration allows continuous real-time data exchange between turbine controllers, substations, and enterprise systems (Mazare & Ramezani, 2024). Advanced digital twins and AI-based analytics are now standard in monitoring turbine performance and predicting component failure (Beretta et al., 2021). Moreover, cloud-based energy management platforms facilitate remote operation and grid-balancing coordination, which is crucial as distributed renewable generation increases globally (World Economic Forum, 2024).

However, the same digitalization that drives efficiency also creates new vulnerabilities. Every layer of connectivity—from turbine controllers and programmable logic controllers (PLCs) to remote gateways and cloud applications—represents a potential attack path, as shown in Figure 2. Recent analyses have shown that OT systems in wind farms are particularly exposed to threats such as ransomware, supply-chain compromise, and unauthorized remote access (IEC, 2024; DNV, 2023). Cyber adversaries who gain access to a turbine's control logic could degrade generation, push the machinery into unsafe operational states, or trigger premature component failures. Meanwhile, unauthorized access to operational datasets can expose sensitive information, such as generation profiles or curtailment patterns. However, compromised access control logs, biometric data, or geolocation records can reveal personally identifiable information (PII) about people who gain unauthorized access (Egan, 2022; ISA, n.d.).

Figure 2. Cyber-vulnerable areas within wind power system (Knack, Kam, & Tam, 2024)

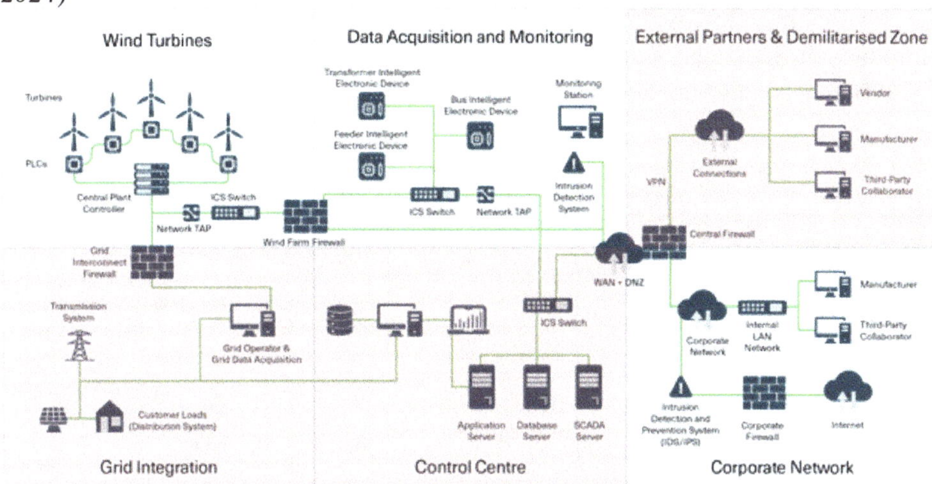

These risks underscore the necessity of treating wind-energy systems as a special case in cyber-physical security domain. Unlike conventional wind power systems, modern wind power systems operate with ICS, SCADA platforms, and cloud-based data pipelines. Using the distributed sensor system, real-time monitoring, and predictive analytics improves the operational capabilities. However, these systems become the potential cyber-attack, thus, demanding specialized strategies that differ from traditional power generation systems or IT-only networks (Mazare & Ramezani, 2024; Beretta et al., 2021).

Recent technical reports and academic studies emphasize that ensuring the secure operation of wind power plants extends beyond technical reliability. It includes safeguarding economic value, protecting human safety, and maintaining the credibility of renewable energy as a central pillar of the global energy transition (World Economic Forum, 2024; DNV, 2023). Consequently, the stakes associated with cyber incidents—operational, financial, and reputational—make robust cybersecurity measures essential. Cyber-protection measures are essential not only to maintain operational continuity and data integrity but also to ensure investor confidence and compliance with international standards such as IEC 62443 and ISO 27019.

Research Gap and Sectoral Distinctions

The cybersecurity has been extensively studied in traditional ICS. However, the unique cyber-physical interdependencies, cloud-connected architectures, and regulatory exposure of modern wind power systems remain underexplored. Majority of existing studies often treat wind power systems as generic industrial systems, overlooking cyber-physical aspects. The cyber-physical architecture such as turbine-level autonomy, fleet-wide data aggregation, and offshore operational constraints introduce distinct cyber-attack surfaces and recovery challenges. Table 1 explains the difference in cybersecurity issues compared to other energy sectors.

Table 1. How Wind-Energy Cybersecurity Differs from Other Sectors

Sector	Key Characteristics	Cybersecurity Focus	Why Wind Differs
Conventional ICS (e.g., oil, gas, manufacturing)	Centralized plants, stable environments	Network segmentation, patch management	Wind farms are geographically distributed, with intermittent connectivity and vendor-managed assets.
Smart Grid / DER Systems	High data exchange with grid operators	Data integrity, authentication	Wind involves turbine-level control loops and aeroelastic feedback, where cyber events can affect physical stability.
IT Cloud Infrastructure	Homogeneous virtual systems	Identity, access, encryption	Wind integrates real-time OT control with cloud analytics, bridging incompatible trust domains.
Solar PV	Static assets, predictable generation	Inverter firmware, SCADA	Wind involves dynamic mechanical stresses and high control bandwidth, making anomaly detection and fault isolation more complex.

There is a pressing need for integrated, field-validated cybersecurity frameworks tailored to the distributed, data-driven, and safety-critical nature of wind-energy systems—an area still insufficiently addressed compared to other energy sectors.

Unique Cyber Risk Profile of Wind Power Systems

Wind power systems exhibit a cyber risk profile that is different from other industrial infrastructures due to their distributed design, technological heterogeneity, and reliance on remote operations. The modern wind farms are made of multiple dispersed turbines and auxiliary systems that must operate in a coordinated manner while interfacing with cyber networks. These cyber networks are for remote operational, economic, and maintenance purposes. Reliance on the cyber networks introduces a complex interplay of conventional IT threats alongside highly specific OT risks, necessitating tailored cybersecurity strategies.

Utility-scale onshore and offshore wind farms often comprise dozens to hundreds of turbines that spread across a large farm/area. Each turbine typically contains a combination of proprietary PLCs, sensors, and embedded communication modules. These components are often supplied by different manufacturers who may have varying hardware and software standards. This heterogeneity complicates routine cybersecurity operations such as patch management, secure configuration, and vulnerability remediation. For example. updates and security measures applicable to one model may not be compatible with others, increasing the likelihood of unpatched vulnerabilities (NREL, 2019). Furthermore, the integration of third-party software and vendor-supplied control modules adds additional layers of complexity. This heterogeneity nature of the system brings a challenging of maintaining a unified cyber-security posture across all assets (Montoya et al., 2024).

Geographical dispersion of wind power systems' components, further amplifies the cyber risk landscape. For instance, offshore systems, the harsh marine environment constrains the frequency of on-site maintenance and software updates, thereby increasing dependence on secure remote access and updates (Van Schendel & Varenhorst, 2025; Taher et al., 2022). The remoteness of these facilities also introduces latency and reliability challenges in monitoring and control, which can exacerbate the impact of cyber incidents when they occur. Both offshore and remote onshore wind platforms rely on long communication chains such as satellite or cellular links (4G/5G). Within the long telecommunication links, each communication hop represents a cyber-attack or interception point.

The convergence of OT and IT networks is a defining feature of modern wind power systems, but it introduces additional risks. Operational data streams that were historically confined within isolated SCADA environments. However, in the modern wind power systems, data are now routinely exported to enterprise IT and cloud platforms for advanced analytics, including forecasting, predictive maintenance, and energy market optimization (Mazare & Ramezani, 2024). While this integration significantly improves operational efficiency and economic performance, it simultaneously broadens the blast radius of potential attacks. Any compromise in the corporate IT environment can propagate into safety-critical OT systems, and potentially affecting the whole wind power system (Morrison, Liu, & Lin, 2022). Therefore, cyber defense strategies must account for threats that span traditional IT boundaries into mission-critical OT domains.

Case Studies

Case Study 1: Ørsted IT Network Intrusion (2020).

In 2020, Ørsted experienced a cyber intrusion in its corporate IT infrastructure. The company temporarily isolated its SCADA systems to prevent lateral movement between IT and OT layers; thus, no turbines were compromised. This precautionary isolation resulted into several hours of limited operations and an estimated financial loss of €2–3 million due to curtailed maintenance and dispatch (Danish Energy Agency, 2021).

Case Study 2: German Wind Fleet Ransomware Incident (2022).

In 2022, a coordinated ransomware disrupted remote monitoring for more than 5,800 wind turbines—roughly 11 gigawatts of capacity, across Germany. Although turbines continued autonomous generation, some operations such as adjusting setpoints or conducting predictive maintenance were disturbed for several days. This incident resulted in a decline of an overall efficiency by 2–3% and temporarily halted condition-based maintenance routines (Tagesschau, 2022; WindEurope, 2023).

Case Study 3: Firmware Supply Chain Breach (Asia-Pacific, 2023).

In 2023, a leading wind operator in Asia-Pacific discovered a malicious firmware injected into turbine controllers. This happened through a third-party vendor's update system. The breach was contained before any safety compromise occurred; however, it resulted in unplanned restarts and a 6% rise in unscheduled maintenance visits within two months (DNV, 2023). This case highlights the urgent need for secure firmware update mechanisms.

These cases collectively highlight that cyber incidents in wind power systems are not theoretical. They have measurable technical and financial consequences that are beyond technical IT and OT domains.

The consequences of cyber compromise in wind power systems are disproportionately severe. Malicious attacks on turbine setpoints or SCADA commands can induce curtailment, accelerate wear and tear, or push mechanical components beyond safe operational limits. The attacks have a potential to cascade the effects across the farm or grid. This may result into significant revenue losses, unplanned maintenance costs, and reputational damage to operators (Egan, 2022). Beyond economic impacts, there are direct safety implications such as unauthorized manipulation of turbines, cranes, or other mechanical systems can endanger onsite personnel and local communities (Idaho National Laboratory, 2024).

Several specific vulnerabilities are inherent to wind-farm design and operations. At the turbine-controller level, many legacy communication protocols lack robust authentication and encryption system. These protocols are susceptible to spoofing, command injection, or replay attacks. Moreover, SCADA and RTU systems are often exposed to other networks through inadequate segmentation or insecure third-party remote-service portals (DNV, 2023). Furthermore, edge gateways may transmit telemetry without encryption or rely on outdated Virtual Private Network (VPN) configurations, while cloud platforms can centralize extensive datasets under over-privileged accounts, creating significant attack surfaces (NREL, 2019). Additionally, human and organizational factors further compound the risk profile. For instance, usage of default passwords, misconfigured accounts, incomplete incident-response plans, and limited cybersecurity training remain among the most common causes of real-world compromise (Egan, 2022). These vulnerabilities demonstrate that cyber risks in wind power systems are not purely technical. They are also socio-technical in nature, requiring coordinated organizational policies and workforce awareness.

Additionally, wind power systems often rely on third-party vendors for software and patch updates, remote monitoring, and turbine management. While these arrangements improve operational efficiency, they introduce supply-chain risks. A malicious actors can exploit weak vendor security practices to compromise multiple assets in the wind power system simultaneously (World Economic Forum, 2024). Furthermore, adoption of AI-enabled predictive maintenance and real-time performance optimization, comes with new attack vectors. Attackers can manipulate sensor readings to generate false predictive maintenance alerts or mask abnormal operational behaviors, potentially delaying detection and response (Beretta et al., 2021).

Geographical and environmental factors also play a role in shaping cyber risks, especially on offshore farms. Harsh weather conditions and limited accessibility, create a reliance on long-distance communications. The long-distance communication to remote locations may depend on cellular or satellite links for SCADA connectivity, which are inherently more vulnerable to interception, jamming, or spoofing than wired local networks. These environmental constraints necessitate robust and secure communication protocols to ensure operational continuity.

Together, the technological, organizational, and environmental characteristics create a cyber risk landscape that is geographically dispersed and technologically fragmented. This cyber risk landscape is also operationally high-stakes. Recognizing this unique cyber-risk profile is essential for designing effective defense strategies. The effective defensive strategy that ensures both digital resilience and physical safety. It, therefore, must combine both technical measures and organizational processes. The technical measures may include network segmentation, encrypted communications, and continuous monitoring; and organizational processes include incident-response planning, workforce training, and vendor risk management. By understanding the complex interplay of these factors, operators can protect not only the continuity and reliability of wind power generation but also the economic, safety, and environmental value inherent in modern wind power systems.

EVOLVING THREAT LANDSCAPE IN WIND POWER SYSTEMS

The cybersecurity threats facing wind power systems are no longer hypothetical. Over the past decade, documented incidents have demonstrated the existence of both criminal and state-sponsored actors who target wind power infrastructures. The motives vary from financial gain and extortion to espionage and geopolitical signaling; but the methods increasingly exploit the specific digital dependencies of wind power systems.

The most common threat is ransomware, by the financially motivated attackers. Attackers exploit weak remote-access portals or compromised employee credentials to encrypt critical data and disrupt operations. In some cases, attackers exfiltrate sensitive operational datasets and threaten to release them unless a ransom is paid. For wind power system operators, such breaches can halt generation scheduling, damage customer trust, and trigger regulatory scrutiny (World Economic Forum, 2024).

Supply chain compromise is another major risk vector. Turbine OEMs, software vendors, and third-party service providers are some of the connected stakeholders to the operators. All of them sometimes maintain privileged access to deployed wind power systems. A single breach in one supplier's development environment can affect the whole wind power system. The effect can cascade to other wind systems worldwide through malicious updates or insecure default configurations (Wind River, n.d.).

Advanced persistent threats (APTs) which are often linked to state-sponsored attacks. They normally focus on reconnaissance and sabotage. Reports suggest that reconnaissance of European and Asian offshore wind power systems has already occurred, with attackers probing SCADA systems for vulnerabilities (NREL, 2019). In extreme scenarios, attackers may attempt to disable grid-scale renewable assets as part of broader hybrid warfare strategies, leveraging the economic and symbolic value of wind power system.

The growing role of data-driven operations introduces novel threat categories. Wind operators increasingly rely on predictive analytics models, digital twins, and AI-based anomaly detection to optimize performance. These big-data systems are susceptible to cyber-attacks such as data poisoning and model evasion attacks. The attackers subtly manipulate data or models to blind detection systems (Lyu et al., 2023a; Morrison et al., 2022; Culler, 2021). Compromised systems and models could misclassify faults, mask intrusions, or trigger unnecessary maintenance costs.

Moreover, there are privacy-related threats. Attackers can compromise energy profiles generated by SCADA and smart meter data. If such datasets are compromised by the attackers; and not sufficiently aggregated or anonymized, they may violate privacy rights under frameworks like the EU General Data Protection Regulation (GDPR) (Van Schendel & Varenhorst, 2025; Cordova, 2025). Thus, cybersecurity in wind power systems should also safeguard the rights and trust of stakeholders.

Finally, insider threats which are whether malicious or accidental, remain among the most persistent risks. These threats may be due to misconfigured firewalls, failure to safeguard credentials, or negligence in applying patches have been at the root of many real-world incidents (Egan, 2022). Establishment of cyber-security system and deployment of technical controls is essential given the complexity of wind power systems and the reliance on different vendors.

Considering all these evolving threats in wind power systems, these systems are no longer immune from the cyber-attacks. The threats and attacks are increasingly becoming sophisticated as the wind power systems expand and interconnect to other grids and digital systems.

DEFENSE STRATEGIES AND GOVERNANCE FRAMEWORKS

Securing wind power systems requires more than conventional IT and OT defenses. It requires strategies comprising international standards and robust industrial practices and governance. Wind power systems operate differently from conventional IT environments, where patch cycles and centralized monitoring are the norm. They require an effective defense that spans both technical and organizational layers of resilience, covering OT, industrial control, and distributed infrastructure.

International standards and frameworks provide the defense foundation. For example, ISA/IEC 62443 series is widely adopted across critical infrastructure. It defines a layered approach for securing industrial automation and control systems. Its principles including defense-in-depth, least privilege, and secure development lifecycle, which are directly applicable to wind power systems (ISA, n.d.; Wind River, n.d.). Another example, NREL's guidelines which emphasize on securing communications, authenticating control commands, and monitoring system integrity (NREL, 2019). DNV's joint industry projects extend this further. It is extended by tailoring OT cyber frameworks to the realities of offshore platforms, where physical inaccessibility and environmental stresses make traditional patching and monitoring difficult (DNV, 2023).

At the technical control level, defense strategies cover network segmentation, monitoring, and anomaly detection. Network segmentation refers to strictly separating IT corporate systems from OT turbine controllers. It remains a first line of defense. Intrusion detection systems (IDS) normally are adapted for ICS protocols to flag anomalous traffic patterns. Moreover, machine learning and AI–based approaches are increasingly employed to detect deviations in SCADA data (Campoverde-Vilela et al., 2023; Lyu et al., 2023b; Morrison, Liu, & Lin, 2022). However, deploying AI-based detection systems introduces its own risks and costs such as adversarial manipulation, robust model governance and validation requirements (Ennis & Colangelo, 2022).

Incident readiness and resilience engineering are equally vital. Operators must establish playbooks that account for different scenarios ranging from ransomware lockouts to APT-led sabotage. Both regulators and insurers highly recommend drills and red-team exercises to validate readiness. Embedding secure-by-design principles—such as hardened default configurations and cryptographic authentication in turbines—reduces long-term attack surface. Moreover, lifecycle considerations, from procurement through decommissioning must address supply chain security.

This is important in ensuring that vendors provide transparency in software bills of materials (SBOMs) and patching responsibilities.

Furthermore, the policy and governance dimension cannot be overlooked. Legal frameworks such as the EU GDPR impose obligations to safeguard personal and operational data. However, reports such as CERRE's reports on energy data sharing, highlight tensions between open data initiatives and privacy rights (Mazare, Ramezani, & Smith, 2024; Ennis & Colangelo, 2022). Regulators are beginning to treat renewable energy assets as part of critical national infrastructure, mandating stricter reporting of cyber incidents and minimum control baselines. Internationally, forums like the World Economic Forum are promoting cross-border collaboration to secure offshore wind, recognizing that cyber threats rarely respect jurisdictional boundaries (World Economic Forum, 2024).

Defense-in-Depth for Wind Power Systems

In practice, the most resilient organizations combine multiple defensive layers, as shown in Figure 3 below, into a holistic security posture: they align technical controls with recognized standards, bake resilience into organizational processes, and ensure legal/policy compliance.

Figure 3. Defense-in-Depth layered cyber security architecture for the power system (Ryu et al., 2024)

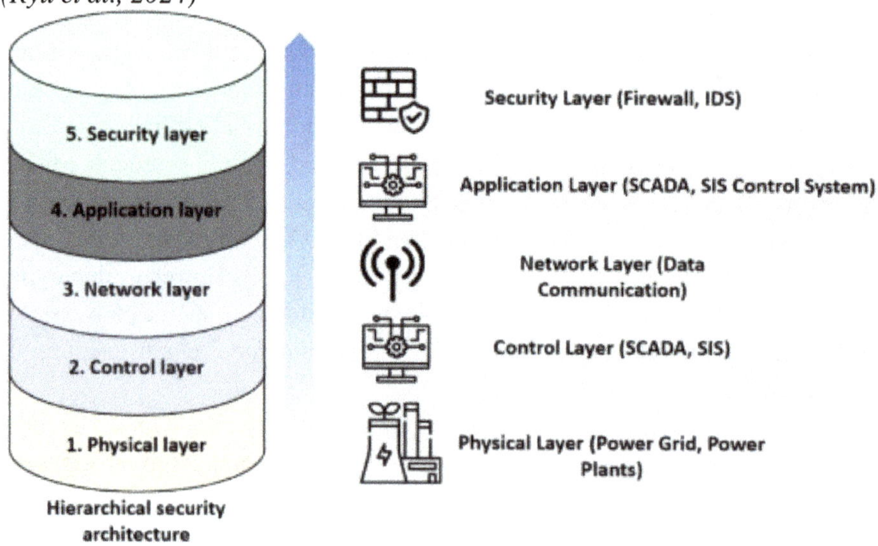

For wind farms this means implementing a defense-in-depth model that describes each layer with concrete best practices and supporting evidence, as follows:

Device & control-level protections (field & PLC/turbine level): Best practice is to harden devices at the source: enforce unique device identities (PKI certificates for controllers), disable unused services, enforce minimal local accounts (no default credentials), and require signed firmware and secure update pipelines from OEMs. Embedding cryptographic authentication and runtime safety interlocks reduces the chance of a remote compromise on the controllers. These fundamentals are emphasized by different standards including the IACS standards (NREL, 2019; ISA, n.d.).

OT network architecture (zoning, segmentation & controlled gateways): It applies the IEC/ISA 62443 zone-and-conduit model. The model groups assets into functionally coherent zones (turbine cell, plant SCADA, DMZ, enterprise IT) and enforces strict controls at conduits (protocol filtering, deep-packet inspection for IEC protocols, and fail-close boundary mechanisms). It employs jump hosts and ephemeral; audited remote access rather than direct vendor SSH into PLCs. Employing the zone segmentation is highly recommended as it both limits lateral movement and enables safe isolation (fail-to-safe) during incidents (NREL, 2019; ISA, n.d.).

IT/Cloud & supply-chain controls: Do not treat cloud and analytics services as merely "convenience"—apply least privilege to APIs, rotate and vault keys, require SBOMs and secure-by-design evidence from suppliers, and mandate cryptographic signing of updates. Contractually require vendors to follow secure development lifecycles, to notify on breaches, and to allow independent verification. Recent DER/wind reviews stress that supply-chain and vendor access controls are among the highest leverage mitigations for fleet-scale risk (Lyu et al., 2023a).

Detection, safety assurance & hybrid monitoring: This combines ML/statistical anomaly detection with physics-aware checks and formal safety analysis. Data-driven detectors such as PCA, LSTM, DSI-Net and other sequence models are powerful for spotting unusual patterns in SCADA streams and early mechanical faults. However, they can be fooled or masked by filtering. Nonetheless, pairing these methods with control-theory checks (safe set enforcement) and reachability analysis produces stronger guarantees. Different scholarly studies show improved fault lead-times for hybrid approaches (PCA/ML) and demonstrate the value of reachability during cyber-induced unsafe states (Idaho National Laboratory, 2024; Campoverde-Vilela et al., 2023; Bouramdane et al., 2023). ML raises alerts, physics checks validate whether commands could drive the system into an unsafe region, and formal reachability helps quantify worst-case outcomes.

Organizational readiness, exercises & privacy-by-design: Technical controls must be backed by organizational measures. Operators must enforce organizational measures such as tested incident playbooks (safety triage, containment, regulator notification), regular tabletop and red-team exercises (including offshore logistics), and forensic readiness (tamper-evident log storage). In parallel, they must implement DPIAs where PII or high-granularity operational data are processed by using time-aggregation, differential privacy or federated learning when sharing datasets for research or fleet analytics to reduce re-identification risk. Recent studies recommend these governance and privacy steps as necessary complements to technical security measures (Saleem et al., 2024; Taher et al., 2022).

LESSONS FROM INCIDENTS AND RESEARCH FRONTIERS

Cybersecurity concerns in wind power systems are not just theoretical. There is a growing set of documented incidents, forensic reviews and academic studies that show how vulnerabilities manifest in practice. They are shaping the next wave of defensive strategies. Proactive defensive measures are born of combining lessons learned from real-world events with the analytical power of academic/scientific models.

Operational Incidents and Case Reviews

Wind farms have not yet suffered globally disruptive cyberattacks on the scale of other energy sectors. However, there are series of incidents highlighting recurring weaknesses. Different reviews and research projects by industry analysts and scholars have pinpointed weaknesses. Egan (2022) shows that misconfigurations, insecure vendor access, and weak segmentation were common root causes across several outages and near misses. In many cases, the initial foothold was gained through corporate IT networks, with attackers laterally moving into SCADA systems.

Other incidents demonstrate the fragility of supply-chain security. Arsal et al., (2025) describes that tampering with firmware or exploiting third-party remote-maintenance platforms can introduce backdoors deep inside turbine controllers. This is common challenge to offshore operators who often rely heavily on vendor remote services, making them particularly exposed to poorly secured gateways. The 2022–2024 retrospective studies confirm that operational resilience depends as much on vendor governance and controls as on technical patching.

Academic Advances in Detection and Safety Analysis

Research efforts mirror the operational challenges in wind power systems. They concentrate on systematic detection and safety assurance. Studies applying principal component analysis (PCA) and other statistical techniques have demonstrated effective early warning for drivetrain and bearing anomalies (Campoverde-Vilela et al., 2023). More recent ML models (including LSTM sequence models and hybrid neural networks) capture temporal and spatial dependencies across turbine fleets, showing measurable improvements in fault lead time (Lyu et al., 2023b).

The formal reachability analysis has proven valuable especially at the intersection of safety and cybersecurity. By modeling turbine dynamics, researchers show that even modest malicious command can push actuators into damaging regimes. This proves the need for safety envelopes and interlocks embedded at the control-software level (Idaho National Laboratory, 2024). These methods also provide operators with mechanisms to test "worst-case" cyber-induced states before deploying real-world mitigations.

Privacy-Preserving Data Analytics

Beyond anomaly detection, a growing research frontier concerns the privacy of operational datasets. Studies warn that high-resolution wind generation and consumption data can reveal commercially sensitive information or indirectly expose patterns of local energy use (Morrison et al., 2022). Scholars such as Culler (2021) and Lyu et al., (2023a) evaluate methods including differential privacy, time aggregation, and synthetic data generation, showing that privacy can be preserved without rendering datasets useless for forecasting or predictive maintenance. Early results are promising but highlight a delicate privacy–utility trade-off, particularly when data are used in adversarial contexts where attackers may attempt re-identification.

Convergence of Lessons

Taken together, operational cases and research findings reinforce several critical themes in wind-energy cybersecurity:

Weak fundamentals remain the most exploited entry points: Despite advances in monitoring, anomaly detection, and regulatory compliance, many real-world incidents continue to exploit basic security lapses. Default credentials on turbine controllers, poor network segmentation between SCADA and enterprise IT, and unsecured vendor access portals remain widespread vulnerabilities (Egan, 2022; NREL, 2019). For example, attackers have repeatedly leveraged factory-set passwords or unencrypted remote connections to gain control over turbine operations, causing curtailment, forced maintenance, and revenue losses. These cases underscore that foundational cybersecurity hygiene is non-negotiable. Activities such as patching, credential management, network segmentation, and vendor access controls form the baseline higher-order defenses. Operational discipline in these activities is particularly important in offshore and remote installations, where physical intervention is costly and time-consuming (Idaho National Laboratory, 2024).

Hybrid defense is essential: Modern wind power systems generate highly complex cyber-physical data streams, including turbine vibration, rotor speed, environmental conditions, and network traffic logs. Purely statistical or ML models for anomaly detection are insufficient if the underlying physics of turbine operation are not considered. Conversely, physics-only approaches may miss subtle such as network-based intrusions or adversarial manipulations (Lyu et al., 2023a; Ennis & Colangelo, 2022). A hybrid defense strategy integrates physics-aware modeling, statistical anomaly detection, and AI-driven pattern recognition to provide robust security in both OT and IT domains. For instance, combining rotor-speed predictive models with ML-based network traffic analysis enables operators to differentiate between legitimate operational variations and potential cyber intrusions. This reduces false positives while improving early detection.

Privacy and security converge in modern wind projects: Wind power systems become more data-driven and the cybersecurity and privacy converge more. While centralized analytic platforms aggregate operational data, community energy schemes and smart-grid integration involve sharing information across operators, regulators, and end-users. Therefore, ensuring compliance with GDPR and similar frameworks must be considered in cybersecurity planning (Mazare, Ramezani, & Smith, 2024). This compliance demands that privacy considerations be embedded in analytic and operation systems design. The design considers making privacy-by-design and security-by-design complementary pillars of wind-farm resilience. Measures such as differential privacy, data aggregation, and federated learning allow operators to leverage analytics without exposing sensitive operational or personnel information.

Formal methods bridge theory and practice: Formal verification and model-checking methods provide structured guarantees about wind power system. It checks the behavior under both normal operations and cyber-attack scenarios. In the wind-energy domain, formal methods can simulate turbine control loops, SCADA communication protocols, and emergency shut-down procedures to prove safety properties even when parts of the system are compromised (Mazare & Ramezani, 2024; Ennis & Colangelo, 2022). The combination of formal methods with empirical testing and adversarial simulations offers stability confidence to operators. Operators can gain confidence that control systems will respond safely to unexpected inputs or malicious commands. This approach is particularly valuable for systems where failures can have cascading financial and environmental consequences.

Integrated learning from incidents and research outputs: Operators can analyze which vulnerabilities were exploited historically. Retrospective incident reviews on attacks, misconfigurations, and near misses provide the sector with diagnostic insights. Moreover, forward-looking academic research provides a prescriptive guidance. Research offers models, frameworks, and methods to anticipate and mitigate future threats. Therefore, operators that adopt research-driven and evidence-based methods to gain a strategic advantage. Daily operations integrating these approaches can safeguard turbines, optimize revenue, and maintain public trust and regulatory compliance. Therefore, cybersecurity and privacy are not optional overheads. They are core competitive differentiators in wind power systems where digital resilience directly correlates with financial performance, operational continuity, and reputational capital.

OPERATOR PLAYBOOK FOR RESILIENCE

Operator Playbook for Resilience

Ensuring cybersecurity in wind power systems is not limited to preventing cyber-attacks. It involves the ability to respond effectively to incidents and mitigate their consequences. It builds organizational capacity to recover quickly, as shown in the Figure 4. A pragmatic operator playbook can be developed based on the operational experience of wind farms, together with guidance from industry standards and regulatory frameworks. This playbook enables operators and vendors to systematically address threats across the entire incident lifecycle, ensuring both operational continuity and personnel safety.

Figure 4. Proposed Operator Playbook for Resilience

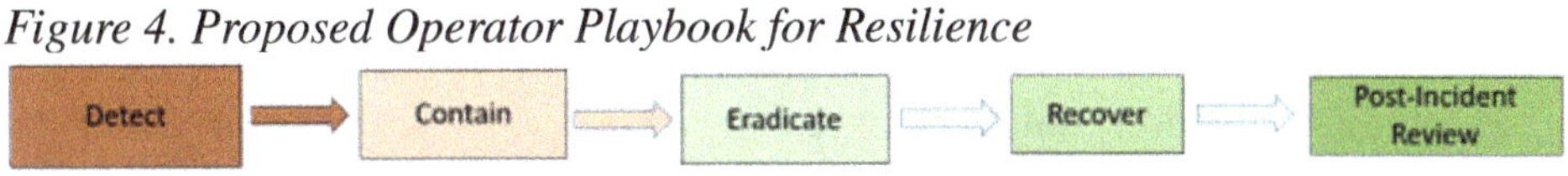

Detection

Effective incident management begins with early detection. Wind farms generate enormous volumes of data, including turbine performance metrics, SCADA logs, network traffic, and environmental sensor data. Leveraging this data for real-time monitoring allows operators to detect anomalies indicative of cyber intrusions, hardware malfunctions, or software failures (Lyu et al., 2023a; Ennis & Colangelo, 2022).

Detection mechanisms can range from signature-based intrusion detection systems (IDS) to physics-informed anomaly detection. Advanced approaches utilize ML or hybrid models to flag subtle deviations that may escape traditional monitoring. Importantly, these systems must balance sensitivity with specificity, minimizing false positives that could trigger unnecessary operational disruptions while ensuring high-confidence alerts are acted upon promptly (Mazare & Ramezani, 2024).

Detection is also supported by forensic readiness, through a continuously collection of SCADA snapshots, controller states, and network logs in tamper-evident storage. This enables operators to reconstruct incidents and identify root causes quickly. This approach is essential for regulatory compliance and insurance purposes (Idaho National Laboratory, 2024; DNV, 2023).

Containment

Rapid containment of a detected incident is critical to limit damage. Wind power systems benefit from segmentation and zoning given by IEC 62443 principles. This allows any compromised part such as SCADA segments or turbine clusters to be isolated automatically without affecting the entire farm (ISA, n.d.). Moreover, remote disconnection routines should be "fail-safe," preserving the safety of personnel and equipment while preventing escalation of malicious activity into other domains (NREL, 2019).

Containment also involves controlling access and communication channels. Vendor portals, cloud connections, and inter-farm networks must be temporarily restricted or placed in a monitored state during containment. Operators' playbook should have a clear predefined escalation protocols, including notifications to grid operators, regulators, and internal security teams to ensure compliance and situational awareness (World Economic Forum, 2024).

Human coordination complements technical measures. Incident response drills, particularly for offshore wind farms with limited onsite staff, prepare operators for rapid manual interventions if automated containment fails. Regular tabletop exercises, including IT, OT, legal, and operations personnel, reinforce roles, decision-making pathways, and communication protocols under realistic scenarios (DNV, 2023).

Eradication

After containment, the focus shifts to eradication. Eradication refers to eliminating the root cause of the incident. This may involve patching vulnerable controllers, removing malware from SCADA networks, or reconfiguring access credentials. In heterogeneous wind systems, eradication is complicated by the diversity of turbine models, proprietary PLCs, and vendor-supplied modules, requiring coordinated action across multiple vendors (Montoya et al., 2024; NREL, 2019).

Forensic data collected during detection is critical during eradication. Logs and snapshots allow operators to determine the attack vector, affected assets, and any potential persistence mechanisms. Moreover, supply-chain considerations are particularly important. Vendors should provide signed firmware updates, breach notifications, and independent security assessments to ensure eradication efforts do not inadvertently reintroduce vulnerabilities (Mazare, Ramezani, & Smith, 2024; Wind River, n.d.).

Recovery

Recovery entails restoring normal operations with minimal disruption while ensuring that the system is secure. For instance, operators must follow procedures for safely bringing turbines, SCADA segments, and edge devices back online, verifying system integrity at each step (Idaho National Laboratory, 2024). Recovery plans should incorporate redundancy and resiliency features. These features include failover control networks, mirrored cloud instances, and contingency power routing to maintain minimum viable output during restoration (Mazare & Ramezani, 2024).

Continuous monitoring remains essential during recovery. Anomalies may persist or recur, especially, if attackers exploited latent vulnerabilities. AI-driven monitoring tools, combined with human oversight, help to confirm that all operational parameters have been restored (Chen et al., 2025). Corrective actions including enhanced patching, configuration hardening, and updating remote-access policies may also be implemented during the recovery phase (Kathirvel et al., 2025).

Post-Incident Review

Finally, post-incident review focusing on learning from the event to strengthen future resilience. Comprehensive reviews should examine technical, organizational, and procedural aspects. These will help in identifying gaps in detection, containment, eradication, or recovery. Review findings should feed back into supplier assessments, contract management, firmware lifecycle policies, and remote-support contingencies (DNV, 2023).

Post-incident analysis also informs regulatory reporting and audit processes. It ensures lessons learned are integrated into risk assessments, incident-response playbooks, and cybersecurity training programs (NREL, 2019). Furthermore, sharing anonymized insights across the industry can improve sector-wide resilience by highlighting emerging threats and effective countermeasures (World Economic Forum, 2024; Ennis & Colangelo, 2022).

Case Study: North Sea Offshore Exercise (2024).

In 2024, Equinor and DNV conducted a joint simulation replicating simultaneous loss of SCADA connectivity and turbine controller compromise. Detection time averaged six minutes, with automatic isolation achieved in twelve minutes and full restoration within one hour and forty minutes. When compared to a baseline exercise without automation, restoration time improved by over 50% and containment efficiency increased by 60%. These findings empirically support the operational benefits of standardized zoning, automated triage, and cross-disciplinary rehearsal in reducing downtime and preserving safety margins.

The operational resilience of wind power systems depends on both human and systemic preparedness. While defense-in-depth architectures such as IEC 62443 provide structural safeguards, operational playbooks ensure speed and coordination during incidents.

Proactive Risk Management and Architecture

Resilience is reinforced by proactive measures that reduce the likelihood and impact of incidents. Key strategies include:

> **Asset inventory and criticality classification**: A comprehensive asset inventory is foundational to effective cybersecurity. Operators must maintain an up-to-date catalog of all hardware, software, and network components, including turbines, SCADA controllers, edge gateways, sensors, and communication links (NREL, 2019). Each asset should be classified according to its operational criticality, including its impact on safety, revenue generation, and grid stability. For example, a main SCADA controller may be critical for real-time turbine coordination, whereas a weather-monitoring sensor may have a less impact. Mapping these assets within the overall network architecture allows operators to prioritize protection and allocate resources efficiently. Some operators implement dynamic dashboards that visualize asset dependencies. These dashboards alert security teams when high-criticality components are exposed to changes, such as unpatched firmware or unusual network traffic.

This approach leads to a faster incident triage, and effective disaster-recovery planning.

Segmentation, zoning and hardening: Segmentation and hardening of wind-farm systems is important for effective cybersecurity. It provides a clear separation between turbine clusters, SCADA zones, and enterprise IT networks (ISA, n.d.). For instance, turbines can be grouped into subzones with independent controllers, limiting the impact of a breach in a single turbine or cluster. Hardened configurations include enforcing secure communication protocols, disabling unused ports, and applying baseline security templates to controllers and gateways. Practical examples include deploying firewalls between OT and IT domains, configuring VLANs for isolated turbine groups, and implementing intrusion detection sensors at zone boundaries. Regular vulnerability assessments and penetration tests are essential. They validate the effectiveness of these security domains, ensuring segmentation controls remain operational under real-world conditions.

Identity and access management (IAM): Strong IAM practices are essential to prevent unauthorized access to critical systems. Operators should implement multifactor authentication (MFA) for remote access for all vendor and technical operator. This should be coupled with Public Key Infrastructure (PKI)–based device identities for SCADA and turbine controllers (NREL, 2019). Examples include using hardware tokens, biometric verification, or one-time passwords for human operators, while device identities ensure only authorized hardware communicates within the OT network. IAM policies also include role-based access control (RBAC), restricting operator permissions to only those functions necessary for their role. Logging and monitoring of access attempts provide traceability and support rapid response to suspicious activity, such as unusual login times or unexpected access from offshore locations.

Supply-chain controls: Normally, wind farms depend on third-party vendors for firmware updates, maintenance, and remote monitoring. This brings potential supply-chain risks. Effective mitigation requires vendors to adopt secure software development lifecycles (SDLCs) and establishment of contractual obligations for incident reporting and independent audits (Wind River, n.d.). Operators may also implement vendor risk assessments, evaluating suppliers for past security incidents, compliance certifications, and patch management practices. For example, an offshore turbine vendor may be required to provide quarterly security audit reports and demonstrate that firmware updates are tested for both functional performance and security integrity before deployment.

Privacy-by-design: Modern wind farms collect data beyond operational metrics, sometimes including personal information from employees, subcontractors, or nearby communities. Implementing privacy-by-design principles ensures that data collection, storage, and sharing comply with regulatory requirements such as GDPR (Mazare, Ramezani, & Smith, 2024). Operators should conduct Data Protection Impact Assessments (DPIAs) to evaluate the risks associated with processing personal data, identifying mitigation strategies such as anonymization, aggregation, or differential privacy when sharing data externally. For example, turbine performance datasets shared with third-party analytics providers may be aggregated over time or location to prevent identification of individual technicians' activities. Embedding privacy into the design of monitoring, analytics, and cloud systems strengthens trust and ensures compliance with evolving legal frameworks.

The integration of detection, containment, eradication, recovery, and post-incident review into a coherent operational playbook ensures that wind operators can maintain both cybersecurity and operational continuity, even under adverse conditions.

Data Privacy, GDPR, and Privacy-Preserving Implementation in Wind Operations

Digitalization in wind-energy systems has transformed how operational and environmental data are collected, analyzed, and monetized. Each turbine, substation, and control platform now generates terabytes of high-frequency telemetry — including power curves, vibration spectra, temperature logs, and even operator credentials. As these data streams are aggregated in cloud-based analytics or shared across multi-tenant platforms, data privacy becomes a critical operational and regulatory concern. Under the European Union's General Data Protection Regulation (GDPR), data derived from identifiable individuals — such as maintenance logs linked to staff, location-tracked vehicles, or residential prosumer data from hybrid farms — must be processed with explicit safeguards (Vianny et al., 2025).

Unlike traditional IT systems, wind-energy cyber-physical systems (CPS) blur the boundaries between personal, operational, and environmental data. Turbine controllers and supervisory control and data acquisition (SCADA) systems collect parameters that can indirectly reveal personal behavior patterns (e.g., shift schedules or maintenance activity). At the same time, high-resolution operational data, such as vibration profiles or output forecasts, can constitute commercially sensitive information. Thus, cybersecurity and data privacy are not separate layers but intertwined defense objectives.

The GDPR Framework in Wind Operations

GDPR establishes a risk-based approach to data governance, demanding that operators demonstrate both technical and organizational compliance. In the wind-energy context, this translates into several concrete obligations:

Lawfulness and Purpose Limitation: Data collected from turbines, drones, or worker devices must have clearly defined operational purposes — e.g., predictive maintenance or grid coordination. Using the same datasets for secondary analytics or AI training requires re-assessment under GDPR Articles 6–7.
Data Minimization and Retention Control: Only essential variables should be stored long-term. For example, instead of retaining high-frequency SCADA traces indefinitely, operators may keep aggregated daily metrics.
Data Protection Impact Assessments (DPIAs): Required under Article 35, DPIAs help identify privacy risks in planned data-processing activities such as cloud telemetry sharing or AI-based anomaly detection.
Cross-Border Data Transfers: Many wind developers operate transnational fleets. Any transfer of operational data outside the EU — e.g., to cloud analytics in the U.S. or Asia — must follow adequacy decisions or Standard Contractual Clauses (SCCs).

Integrating GDPR compliance into daily operations requires collaboration between data protection officers (DPOs), IT/OT engineers, and vendor partners — ensuring that privacy constraints do not undermine situational awareness or incident response efficiency.

Privacy-Preserving Methods for Wind-Data Analytics

To reconcile privacy obligations with data-driven optimization, the sector increasingly adopts privacy-preserving computation. These techniques enable meaningful analytics without exposing sensitive or identifiable data. Key methods include:

Time Aggregation: By aggregating high-frequency turbine telemetry into coarser time intervals (e.g., hourly averages), operators reduce identifiability while preserving statistical trends useful for condition monitoring (van Schendel & Varenhorst, 2025).
Differential Privacy (DP): DP adds mathematically controlled noise to datasets, ensuring that the inclusion or exclusion of any single data record does not affect aggregate results. It is a useful technique particularly for publishing benchmarks or performance data without revealing proprietary information.

Federated Learning (FL): FL allows models to be trained locally at each wind farm or control center, sharing only anonymized model updates. This approach minimizes data exposure to third parties while maintaining predictive accuracy unlike using a centralized raw data system.
Homomorphic Encryption (HE): HE enables computations on encrypted data. Operators can outsource analytics to cloud services without decrypting sensitive data. Although computationally intensive, it is promising for highly regulated environments such as offshore wind control systems.

Adopting these methods demonstrates **"privacy-by-design"**. This architectural philosophy is mandated by GDPR Article 25 to ensure that privacy is embedded at the system design stage rather than retrofitted post-deployment.

Balancing Data Utility and Confidentiality

Another key challenge is balancing data utility (for performance optimization) with confidentiality (for compliance and trust). Over-anonymization may degrade the precision of condition monitoring or predictive maintenance algorithms, whereas under-protection exposes sensitive commercial or personal data. A balanced governance model includes:

Tiered access levels based on user roles (operators, vendors, auditors).
Dynamic data masking for remote diagnostic sessions.
Audit trails ensuring accountability for each data access or modification.

Furthermore, privacy and cybersecurity controls must reinforce one another. Encryption, authentication, and zoning strategies not only prevent unauthorized system access but also protect data integrity and confidentiality.

Example: Privacy-Aware SCADA Data Pipeline

A modern wind farm's data pipeline can be conceptualized as shown in Figure 5 below.

Figure 5. Privacy-Preserving Data Flow in a Wind Farm

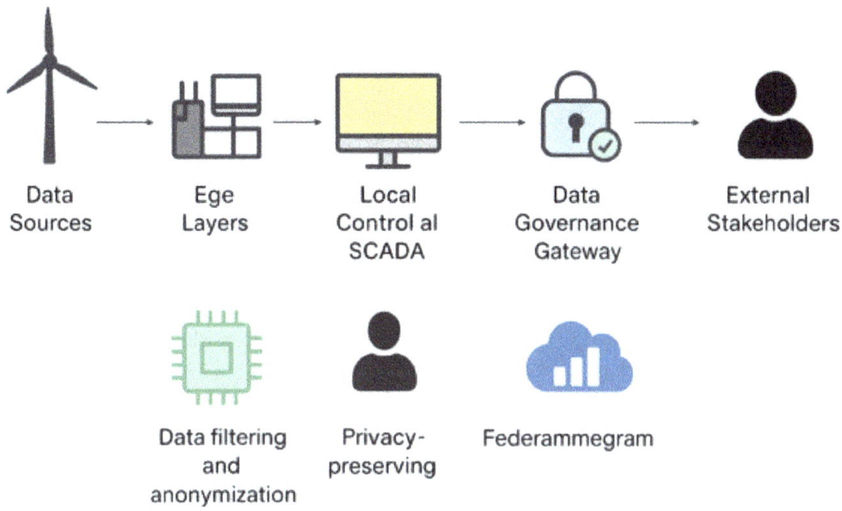

The Main Data Flow:

Data Sources: Turbine sensors, operator terminals, weather stations.
Edge Layer: Data filtering and anonymization (removal of personal identifiers, aggregation).
Local Control System (SCADA): Secure communication with IEC 62443 zoning and encrypted data links.
Data Governance Gateway: Applies DPIA results, enforces access rules, and injects differential privacy noise when exporting datasets.
Cloud Analytics Platform: Runs federated or encrypted learning models without raw data transfer.
External Stakeholders: Regulators, grid operators, and research institutions receive privacy-preserving aggregated outputs.

Privacy Safeguards:

Data Filtering and Anonymization
Applied mainly between the Edge Layer and the Data Governance Gateway.
Removes personally identifiable information (PII) or sensitive operational metadata (e.g., exact timestamps or location data).

Techniques: k-anonymity, pseudonymization, or noise injection before data aggregation.
Privacy-Preserving Processing
Enforced throughout the pipeline (from Edge to Cloud).
Includes cryptographic or algorithmic protections like homomorphic encryption, secure multi-party computation, and differential privacy.
Ensures that even if data must be shared, the underlying sensitive patterns remain private.
Federated Learning / Privacy-Aware Analytics
Operates between the Cloud Analytics Platform and External Stakeholders.
Instead of sending raw data, local models are trained at each site; only encrypted parameters or gradients are shared.
Preserves data sovereignty while still enabling fleet-wide optimization.

Case Study: Fraunhofer IEE Privacy-Preserving Telemetry Aggregation (2025).

A consortium led by Fraunhofer IEE piloted a differential privacy framework for sharing operational telemetry across multi-operator wind farms in Europe. Raw one-second SCADA data were aggregated into five-minute intervals, and differential privacy noise with $\varepsilon = 0.5$ was applied before cloud upload. The resulting dataset preserved 98.7% analytic accuracy for power-curve modeling while meeting GDPR anonymization criteria (van Schendel & Varenhorst, 2025). The study provides practical evidence that privacy-preserving analytics can retain operational utility without compromising compliance.

Quantitative Example: According to the European Union Agency for Cybersecurity (ENISA, 2024), the average downtime cost per wind turbine following a cyber incident range between €1,200 and €2,500 per hour. Simulations by the National Renewable Energy Laboratory (NREL, 2024) found that patch delays exceeding sixty days can raise the likelihood of compromise by 34%. Moreover, the RESCue Pilot (NREL, 2024) demonstrated that hybrid anomaly detection—combining machine learning with physics-based constraints—reduced false alarms by 41% compared to purely statistical models.

Collectively, these quantitative findings emphasize that the economic and privacy stakes in wind cybersecurity are substantial. Therefore, the systematic implementation of both cybersecurity and privacy-by-design measures yields measurable returns in resilience and compliance.

Future Directions

Currently, there are ongoing researches focusing on quantifying privacy loss in industrial telemetry. These studies also focus on automating compliance through machine-readable privacy policies embedded in control software. Moreover, emerging standards are beginning to bridge the gap between cybersecurity and data protection. For example, the ISO/IEC 27557:2022 framework on privacy engineering and the forthcoming European Energy Data Space (EEDS) initiative emphasize interoperability and lawful processing in renewable systems.

In summary, treating privacy as a core component of resilience, should not be considered as a legal afterthought. Operators can enable secure innovation while ensuring alignment with ethical and regulatory expectations. The integration of data privacy, GDPR compliance, and privacy-preserving technologies is essential. It maintains stakeholder trust in digitalized wind energy systems.

CONCLUSION AND RECOMMENDATIONS

Conclusion

Wind power systems have evolved from isolated mechanical assets into highly digitalized cyber-physical infrastructures which are central to the low-carbon transition. Modern wind farms integrate ICS, SCADA platforms, edge and cloud analytics, and dense sensor networks to enable forecasting, predictive maintenance, and efficient grid participation. These digital capabilities improve operational performance but also create a distinct cyber and privacy risk profile (DNV, 2023; NREL, 2019; ISA, n.d.).

A major contributor to this risk is asset heterogeneity. Wind farms may contain hundreds of turbines from multiple vendors, each with proprietary controllers, PLCs, and communication protocols. This diversity complicates patching and secure configuration. Moreover, geographically dispersed sites, especially offshore, depend on long communication chains involving RTUs, relays, and cellular or satellite links (van Schendel & Varenhorst, 2025; Knack et al., 2024). Each link introduces opportunities for interception or failure, and limited onsite access increases reliance on secure remote connectivity (Arsal et al., 2025).

Convergence of OT and IT domains intensifies these challenges. Operational data once confined to SCADA, now flows to enterprise networks and cloud platforms for analytics and forecasting (Lyu et al., 2023a; Ennis & Colangelo, 2022). As a result, breaches in corporate IT can reach safety-critical OT.

The consequences span operational, financial, and reputational domains. Malicious manipulations can curtail generation, accelerate equipment wear, or cause unsafe conditions. Financial losses arise from downtime, maintenance, and contractual penalties. Moreover, reputational harm can undermine investor confidence. Increasing volumes of high-resolution telemetry data introduce privacy risks governed by regulations such as GDPR, driving the need for aggregation, anonymization, and differential privacy (Mazare, Ramezani, & Smith, 2024).

Mitigation strategies exist. Standards such as IEC/ISA 62443 provide frameworks for network segmentation, device hardening, and secure development. Regulations like NIS2 impose requirements for incident response, continuity planning, and reporting (World Economic Forum, 2024). Furthermore, research advances, ranging from anomaly detection to privacy-preserving analytics, offer further tools for proactive threat identification and intrusion prevention (Lyu et al., 2023a; Mazare & Ramezani, 2024).

Continuous monitoring and adaptive learning are essential as threats, regulations, and technologies evolve. AI-enabled hybrid detection systems support real-time anomaly identification, while forensic readiness ensures accurate reconstruction and remediation (Ennis & Colangelo, 2022). Organizational preparedness—drills, tabletop exercises, and iterative improvements—further enhances resilience (DNV, 2023). However, implementing these measures across diverse, distributed fleets remains challenging. Operators must align technical controls including MFA, supply-chain security, firmware signing, and secure remote access, with organizational processes such as incident-response playbooks, training, and post-incident reviews (Arsal et al., 2025; NREL, 2019; Wind River, n.d.). Effective resilience depends on integrating these technical and procedural safeguards.

Given the expanding role of wind energy in global electricity supply, wind farms must therefore achieve both digital and operational robustness. Cybersecurity and privacy now directly influence grid stability, energy security, and economic resilience especially for wind farms. In summary, wind systems exemplify the opportunities and challenges of modern cyber-physical infrastructure. It is highly connected, geographically dispersed, and heterogeneous, yet increasingly supported by standards, technologies, and governance practices. Sustainable progress will depend on integrating technical defenses, proactive risk management, and continuous organizational learning to ensure safe, efficient, and resilient wind-energy operations.

Recommendations

To transition from fragmented cybersecurity efforts toward systemic resilience in wind-energy systems, several priorities emerge. These recommendations address both technical and organizational dimensions, ensuring that wind farms remain operationally reliable, secure, and privacy-compliant in the face of evolving cyber threats.

Harden the Basics: The first priority is to establish foundational cybersecurity hygiene across all wind-farm assets. Eliminating default credentials, unprotected vendor access, and unsegmented networks is essential to reduce the attack surface. For instance, many legacy turbines are shipped with factory-set passwords or unsecured remote-access portals, which can be exploited by attackers to gain control over SCADA or PLC systems (Egan, 2022). Operators should implement network segmentation following IEC 62443 principles, separating turbine clusters, SCADA zones, and enterprise IT into discrete security domains, and enforce device hardening practices such as disabling unused services, applying security patches promptly, and configuring firewalls and access controls (NREL, 2019; ISA, n.d.). Regular vulnerability assessments and penetration testing should be conducted to verify that these basic controls are effective, particularly for remote offshore installations where physical access is limited.

Strengthen Identity and Supply Chains: Robust identity and access management (IAM) is crucial to prevent unauthorized intrusions. Operators should enforce multifactor authentication (MFA) for all human and vendor access and implement PKI-based device identities to authenticate turbines, controllers, and edge devices (NREL, 2019). IAM policies should include role-based access control (RBAC) to ensure personnel only access systems relevant to their duties. Supply-chain security is equally critical. Vendors must be required to implement secure development lifecycles (SDLCs), digitally sign firmware updates, and provide contractual clauses for breach notifications and independent security audits (Wind River, n.d.). These measures reduce the risk of compromised third-party components propagating into operational networks, a concern highlighted by documented supply-chain attacks in renewable energy (Bouramdane et al., 2023). Coordinated vendor assessments, including cybersecurity audits and firmware integrity verification, further reinforce resilience.

Advance Detection Capabilities: Proactive anomaly detection is central to early warning and incident mitigation. Operators should deploy hybrid detection frameworks that integrate physics-aware models, machine learning, and reachability analysis to identify abnormal turbine behavior or network activity (Lyu et al., 2023a; Ennis & Colangelo, 2022). Physics-aware models ensure that alerts correspond to physically plausible events, reducing false positives, while machine-learning components detect subtle deviations in operational patterns or network traffic indicative of novel attacks. Detection systems should be validated against adversarial and adaptive scenarios, simulating sophisticated attack strategies, including command injection, data spoofing, and stealthy lateral movement across the network. This testing ensures that detectors remain robust in real-world operational environments and can provide actionable intelligence without overwhelming operators with spurious alerts.

Integrate Privacy into Operations: Wind farms increasingly collect and process personal and operational data, raising privacy concerns under frameworks such as GDPR. Operators should conduct Data Protection Impact Assessments

(DPIAs) for all datasets containing personally identifiable information (PII) or operationally sensitive metrics, identifying risks and mitigation strategies (Cordova, 2025). Technical measures for privacy include time aggregation, which reduces granularity to prevent tracking of individual personnel or turbine operators; differential privacy, which adds controlled noise to datasets to preserve statistical integrity while obscuring sensitive details; and federated learning, which enables collaborative machine-learning model training across distributed turbines without centralized data transfer. These strategies balance data utility with confidentiality, allowing operators and researchers to analyze system performance while maintaining compliance and stakeholder trust.

Institutionalize Incident Response and Exercises: Maintaining tested incident-response playbooks is essential for operational resilience. Playbooks should define clear procedures for triage, containment, isolation, and communication, addressing both cyber and physical risks (Arsal et al., 2025; NREL, 2019). For example, automated routines for turbine disconnection or SCADA isolation can reduce response time while safeguarding personnel and equipment. Regular cross-disciplinary exercises, including IT, OT, operations, legal, and vendor teams, reinforce preparedness and clarify roles. Offshore wind farms, with complex logistics and limited on-site personnel, benefit particularly from drills simulating pump-ashore communication chains, remote-access contingencies, and multi-vendor coordination. Post-exercise evaluations should feed back into playbook updates, strengthening future response capabilities.

Foster Research and Collaboration: Finally, joint research and collaboration accelerate innovation and improve sector-wide resilience. Academic–industry partnerships can focus on adversarial testing of anomaly-detection systems, secure-by-design turbine controllers, and large-scale evaluation of privacy-preserving analytics (Lyu et al., 2023a; Ennis & Colangelo, 2022). Shared data repositories, simulation environments, and testbeds allow researchers and operators to evaluate new security solutions under realistic operational constraints, bridging the gap between theoretical research and field deployment. Collaborative frameworks also support knowledge exchange across vendors and operators, enabling rapid dissemination of threat intelligence, vulnerability disclosures, and best practices. Standardizing these practices across fleets and regions contributes to systemic resilience, reducing the probability that isolated incidents escalate into widespread operational disruptions.

Outlook

Cybersecurity and privacy in wind energy are no longer peripheral considerations—they have become foundational to the reliability and credibility of the renewable energy transition. As wind power assumes a growing share of the global electricity mix, operators, regulators, and investors alike require assurance that these highly digitalized systems are resilient against evolving cyber threats. A single breach can propagate across interconnected control and cloud systems, causing operational downtime, financial loss, and erosion of public trust in renewables as a dependable decarbonization pathway (Arsal et al., 2025; NREL, 2019).

A defense-in-depth approach remains the most effective means to manage this risk. By layering technical, organizational, and procedural safeguards across IT, OT, and cloud environments, operators can ensure that a compromise in one domain does not cascade across the entire system. This entails continuous network segmentation, multi-factor authentication, encrypted telemetry, and automated incident-response playbooks that emphasize both containment and recovery (Chen et al., 2025; DNV, 2023). In practical terms, segmentation under IEC 62443 and the adoption of automated isolation routines can prevent intrusions from spreading between turbines, substations, and corporate systems.

Equally critical is the operationalization of international standards. Frameworks such as IEC 62443, ISO 27019, and the EU's NIS2 Directive now form the backbone of regulatory compliance in the energy sector. When embedded into everyday practices—from turbine-level firmware updates to supply-chain contracts—these standards enable structured governance, real-time threat monitoring, and coordinated incident reporting. Rather than being mere compliance tools, such standards operationalize cyber resilience as a measurable organizational capability (ISA, n.d.; World Economic Forum, 2024).

Parallel to these defensive measures, privacy-preserving innovation has emerged as a vital enabler of trust and data-driven optimization. High-resolution operational data, if improperly handled, can expose sensitive details about infrastructure, personnel, or local communities. Integrating privacy by design—through techniques such as differential privacy, federated learning, and secure multi-party computation—allows analytics and predictive maintenance to proceed without compromising confidentiality (Cordova, 2025; van Schendel & Varenhorst, 2025). Empirical pilots have shown that federated learning models retain over 95% accuracy compared with centralized training while maintaining full data sovereignty at the site level.

Looking forward, cybersecurity and privacy will increasingly converge into a unified discipline of resilient governance. Operators must transition from reactive compliance to proactive assurance—embedding cyber risk management, privacy impact assessment, and adversarial testing into procurement, design, and maintenance lifecycles. The sector's maturity will depend not only on technological robustness but also on the transparency of its practices and the trust it commands from stakeholders.

Summary: Linking Threats to Mitigation Strategies

The synthesis of operational evidence and research findings reinforces several key relationships between threats and mitigations in wind energy systems:

Network Intrusions often exploit weak authentication and unsegmented networks; mitigations include IEC 62443 zoning, least-privilege principles, and continuous monitoring.
Ransomware and Supply-Chain Compromises exploit remote-access dependencies; mitigations include firmware signing, vendor audits, and secure onboarding policies.
Data Leakage and Privacy Breaches arise from ungoverned data sharing and over-privileged cloud access; mitigations include data minimization, encryption, and privacy-preserving analytics.

Together, these mappings highlight how technical, operational, and governance strategies must act in concert to maintain resilience across the full lifecycle of wind power systems as shown in Figure 6.

Figure 6. Threat–Mitigation Matrix of Wind Energy Cybersecurity

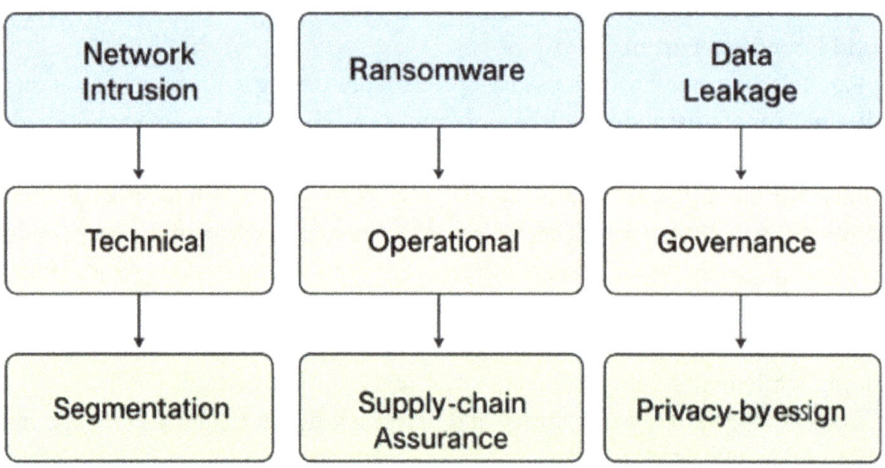

REFERENCES

Arsal, M., Asad, H., Kamel, T., & Khan, A. (2025). *Cyber-safety assessment of wind turbines: A reachability analysis approach against cyber-attacks*. City, University of London. https://openaccess.city.ac.uk/id/eprint/35350/

Beretta, M., Julian, A., Sepulveda, J., Cusidó, J., & Porro, O. (2021). An ensemble learning solution for predictive maintenance of wind turbines main bearing. *Sensors (Basel)*, *21*(4), 1512. https://doi.org/10.3390/s21041512

Bouramdane, A. A.. (2023). *Cyberattacks in smart grids: Challenges and solving the multi-criteria decision-making for cybersecurity options. Cybersecurity*. MDPI., https://doi.org/10.3390/jcp3040031

Campoverde-Vilela, L., Feijóo, M. D. C., Vidal, Y., Sampietro, J., & Tutivén, C. (2023). Anomaly-based fault detection in wind turbine main bearings. *Wind Energy Science*, *8*, 557–574. https://doi.org/10.5194/wes-8-557-2023

Chen, J., Yan, J., Kemmeugne, A., Kassouf, M., & Debbabi, M. (2025). Cybersecurity of distributed energy resource systems in the smart grid: A survey. *Applied Energy*, *383*, 125364. https://doi.org/10.1016/j.apenergy.2025.125364

Cordova, J. G. L. (2025). The right to energy and the data protection and privacy rights: Towards a coherent enforcement under EU law within the energy transition. *European Journal of Risk Regulation*. Advance online publication. https://doi.org/10.1017/err.2025.30

Culler, M. J. (2021). *Cybersecurity guide for distributed wind* (Technical Report INL/EXT-21-64999). Idaho National Laboratory. https://www.osti.gov/biblio/1826578

Danish Energy Agency. (2021). *Cybersecurity annual review 2021*. Danish Ministry of Climate, Energy and Utilities. https://www.scirp.org/reference/referencespapers?referenceid=3884055

DNV. (2023). *OT cyber security for offshore wind: Joint industry project (JIP)*. DNV. https://www.dnv.com/group/joint-industry-projects/ot-cyber-security-for-offshore-wind/

Egan, M. (2022). *A retrospective on 2022 cyber incidents in the wind energy sector* (Graduate project). Boise State University. https://scholarworks.boisestate.edu/cyber_gradproj/3/

Ennis, S., & Colangelo, G. (2022). *Energy data sharing: The case of EV smart charging*. Centre on Regulation in Europe (CERRE). https://cerre.eu/publications/energy-data-sharing-and-the-case-of-ev-smart-charging/

European Union Agency for Cybersecurity (ENISA). (2024). *Threat landscape for the energy sector 2024*. https://www.enisa.europa.eu/publications/enisa-threat-landscape-2024

Idaho National Laboratory. (2024). *Attack surface of wind energy technologies in the United States* (INL Threat Assessment Report). https://inl.gov/content/uploads/2024/02/INL-Wind-Threat-Assessment-v5.0.pdf

International Electrotechnical Commission (IEC). (2024). *IEC 62443 series – Industrial communication networks – Network and system security*. IEC. https://www.isa.org/standards-and-publications/isa-standards/isa-iec-62443-series-of-standards

International Society of Automation (ISA). (n.d.). *ISA/IEC 62443 series of standards — Standards overview*. https://www.isa.org/standards-and-publications/isa-standards/isa-iec-62443-series-of-standards

Kathirvel, A., Gobalakrinshan, N., & Naren, A. K. (2025). Cybersecurity in industrial automation using AI. In *AI-enhanced cybersecurity for industrial automation*. IGI Global., https://doi.org/10.4018/979-8-3373-3241-3.ch024

Knack, A., Kam, Y. H. S., & Tam, K. (2024). *Enhancing the cyber resilience of offshore wind* (CETaS Research Report). The Alan Turing Institute. https://cetas.turing.ac.uk/publications/enhancing-cyber-resilience-offshore-wind

Lyu, H., Zhang, Q., & Chen, Y. (2023a). Anomaly detection of wind turbine driveline based on sequence decomposition interactive network (DSI-Net). *Sensors (Basel)*, *23*, 8964. https://doi.org/10.3390/s23218964

Lyu, Q., He, Y., Wu, S., Li, D., & Wang, X. (2023b). Anomaly detection of wind turbine driveline based on sequence decomposition interactive network (DSI-Net). *Sensors (Basel)*, *23*(21), 8964. https://doi.org/10.3390/s23218964

Mazare, M., Ramezani, H., & Smith, A. (2024). Cyber-physical threat mitigation in wind energy systems. *Energy Informatics*, *7*(1), 49. https://doi.org/10.1186/s42162-024-00449-6

Mazare, M., & Ramezani, H. (2024). Enhancing cybersecurity in wind turbines: A resilient reinforcement learning-based optimal control for mitigating FDI attacks. *Applied Energy*, *373*, 123939. https://doi.org/10.1016/j.apenergy.2024.123939

Montoya, M. C., Rubio-Medrano, C. E., & Kareem, A. (2024). On the cybersecurity of smart structures under wind. *Journal of Wind Engineering and Industrial Aerodynamics*, *251*, 105777. https://doi.org/10.1016/j.jweia.2024.105777

Morrison, R., Liu, X., & Lin, Z. (2022). Anomaly detection in wind turbine SCADA data for power-curve cleaning. *Renewable Energy*, *184*, 473–486. https://doi.org/10.1016/j.renene.2021.11.118

National Renewable Energy Laboratory (NREL). (2019). *Key cyber security requirements for wind farms (Systems Engineering Workshop)*. https://www.nrel.gov/wind/systems-engineering-workshop-2019

National Renewable Energy Laboratory (NREL). (2024). *Renewable energy and storage cybersecurity research (RESCue) pilot final report* (NREL/TP-5T00-89921). https://www.nrel.gov/docs/fy24osti/89921.pdf

Ryu, D., Lee, S., Yang, S., Jeong, J., Lee, Y., & Shin, D. (2024). Enhancing cybersecurity in energy IT infrastructure through a layered defense approach to major malware threats. *Applied Sciences (Basel, Switzerland)*, *14*(22), 10342. https://doi.org/10.3390/app142210342

Saleem, D., ... (2024). *Renewable energy and storage cybersecurity research (RESCue) pilot final report* (NREL/TP-5T00-89921). National Renewable Energy Laboratory. https://www.nrel.gov/docs/fy24osti/89921.pdf

Tagesschau. (2022, March 2). *Cyberattack affects wind turbines in Germany*. https://www.tagesschau.de

Taher, F., & Liu, J., AbdulAzeem, M., & Zhan, Z. (2022). Cybersecurity challenges in offshore renewable energy systems: A survey. *Energies*, *15*(14), 5012. https://doi.org/10.3390/en15145012

Van Schendel, S., & Varenhorst, I. A. M. (2025). A time-aggregation approach for reducing identifiability in household energy data. *Utilities Policy*, *94*, 101900. https://doi.org/10.1016/j.jup.2025.101900

Vianny, M., Kathirvel, A., Rithik, G., Naren, K., & Anthony Raj, A. (2025). Private AI—An overview. In *AI-driven breakthroughs in antimicrobial resistance*. IGI Global., https://doi.org/10.4018/979-8-3693-7550-1

WindEurope. (2023). *Wind energy cyber resilience report 2023*. WindEurope Association. https://eepublicdownloads.entsoe.eu/clean-documents/Network%20codes%20documents/NC%20CS/WindEurope-position-paper-cybersecurity-framework-fit-for-wind-energy.pdf

Wind River. (n.d.). *Wind River—Secure development lifecycle / IEC 62443 guidance and resources.* https://www.windriver.com/resource/wind-river-security-IEC62443

World Economic Forum. (2024, November 15). *Cybersecurity guidance for the global offshore wind industry.* https://www.weforum.org/stories/2024/11/offshore-wind-industry-cybersecurity/

Chapter 7

The Development of Wind Farm Projects in India to Promote Renewable Energy Infrastructure in the Country

Aditya Singh
http://orcid.org/0000-0001-9347-5627
Amrita Vishwa Vidyapeetham, India

ABSTRACT

Wind energy is a natural source of energy which is renewable in nature and eco-friendly in nature if used properly. However, in the recent centuries, humans used non-renewable energies for their development and support urbanization, which was a harmful decision in the long run. In the recent decade, efforts on using wind energy have been put to slowly and steadily replace the non-renewable energy up to certain extent in order to save environment from further harm. In this chapter, a number of major Wind Farm Projects will be discussed and challenges or problems faced in their construction. Then, gaps in the current research in regard to wind farm projects are will also be covered in the chapter. Additionally, data from numerous sources will be collected to perform graphical analyses to find out the benefits of constructing Wind Farm projects and how it will be helpful in India's development in the future.

DOI: 10.4018/979-8-2600-0106-6.ch007

INTRODUCTION

Wind Farms are often known by different names including wind power plants, wind parks or wind parks, which normally consist of wind turbines at the same location which are utilized to generate electricity. Wind Farms differ in size which range from small scale ones having small number of wind turbines to large scale ones having hundreds of wind turbines which covers widespread areas. The electric grid is freely connected to every wind farm and these farms require small portions of land in ratio to its capacity of renewable energy generation. For example, as per a study in 2020 in the country Italy, the 2030's objectives of National Energy Plan in regard to the total wind power, only needed an area equivalent the size of its smallest province named Prato. The wind generators which are used in groups, have the starting range from six hundred kilowatt till five megawatts spread across a given piece of land to get the most gain of the flowing wind in that area. On carefully observing wind turbines, it can be found that they comprise of nacelle, tower as well as rotor present on their upper part along with numerous blades, which points in the wind's direction. Then, propellers are said to turn around a given horizontal axle which could act on an electricity generator. This leads to the generation of renewable energy by every wind turbine present in the area in order to transmit the generated energy to the ground level with the assistance of underground electrical cables which could further transport it to wind power plant's transformer. Then, it is responsible to redistribute the energy to source the electric grid as well as meet the requirements of energy of not only companies but also houses. As per the recorded history, wind was utilized for over thousand of years to push the vessels to sail the seas or move the grain mills. Nonetheless, the nineteenth century saw a turning point where wind turbine was utilized to produce electricity from the wind was created. Even though, wind power generation took place in UK and USA in the 19th century, but the credit for the first modern wind power generation is given to Denmark. With technological advancements over the recent centuries, it led to the current development, where wind energy is expected to be one of the major renewable sources that can help in accomplishing climate – neutral economy in European continent by mid-21st century. The length of the blades, size of the turbines as well as location of wind farms decide the energy amount, they could generate. The nonstop advancement in the design and manufacturing of the wind turbines along with the enhancement in the infrastructure, were able to decrease the wind energy costs as well as led it to the position of main drivers in the transition of energy. As per the report of WindEurope in the year 2021, wind farms in Europe covered around fifteen percent of electricity demand, but in a number of countries in Europe, it exceeded the twenty percent electricity demand which includes Spain, Germany, in addition to Portugal.

Figure 1. Skidby Windmill in East Yorkshire, England, UK (a) during 1950 and (b) during 2003, originally built in 1821.

Source: (a) East Riding Archives, Wikimedia, (b) Charles Rispin, Wikimedia

Figure 2. Wind Farm in Copenhagen, Denmark, showing the example of modern wind farms

Source: CGP Grey, Wikimedia

Figure 3. Wind Farm and transmission lines in East Sussex, England, UK

Source: Diliff, Wikimedia

Figure 4. Wind Farm on top of the mountain in Biscay, Spain

Source: Mariordo (Mario Roberto Duran Ortiz), Wikimedia

Figure 5. Wind Farm in Molsheim, Germany

Source: Karsten Würth, Wikimedia

Figure 6. Wind Farm located in Portugal

Source: Paulo Valdivieso, Wikimedia

Types of Wind Farms

There are three types of wind farms, which are as follows:

- Onshore Wind Farms
- Nearshore Wind Farms
- Offshore Wind Farms

Onshore Wind Farms

These Wind Farms are the most popular ones, and they are present on land at least 3 km far from the coast. They use the terrestrial air currents to produce electricity. Normally, they are present in the areas which have low habitat or conservation value. Due to their proximity to the electric grid and easy availability, they are preferred ones.

Figure 7. Onshore Wind Farm located in Monte de a Runa in Spain

Source: Mavarelacruz, Wikimedia

Nearshore Wind Farms

These farms are present on land, but they are less than 3 km far from the coast, making them closer to the coast compared to the Onshore Wind Farms. The Nearshore Wind Farms has the ability to harness not only sea winds but also terrestrial winds to generate electricity. However, they are the least known to the public.

Figure 8. Nearshore Wind Farm located in Gothenburg in Sweden

Source: Jopparn, Wikimedia

Offshore Wind Farms

These Wind Farms are constructed in the open sea numerous miles from the coast, and they are expected to harness sea winds to produce electricity. Generally, the open water area where they are present, corresponds to the oceans where wind force at lower altitude is more regular and higher in comparison to the terrestrial winds flowing on the land, but these farms can also be present in the freshwater. It is preferred that they are present fifty feet below the water surface and at least two hundred nautical miles away from the coast, in order to not disturb the essential sailing routes or already existing businesses near the coast.

Figure 9. Middelgrunden Wind Farm, an Offshore Wind Farm in Denmark

Source: Photo by Kim Hansen, Postprocessing by Richard Bartz & Kim Hansen, Wikimedia

Indian Scenario

In the recent years, the capacity of India's wind power production has greatly increased. By the end of the month of March 2024, the country's total installed wind power capacity exceeded 45.8 GW, which also makes India to have 4th largest wind power installed capacity across the globe. The western, southern in addition to northwestern states are the parts of the country to have the accumulation of wind power capacity. It was observed in the past that the cost of wind power was drastically decreasing, which could be observed in the month of December 2017, where it reached Rs 2.43/kWh without the application of any indirect or direct subsidy on it. Though, its levelized tariff increased and reached Rs 3.17/kWh by the month of May in the year 2023. To decrease the developer's risk as well as increase the clarity in the auctions of wind power which are tariff-based ones, the Union Government made an announcement to apply certain guidelines by the last month of the year 2017. The installation of wind power just requires two percent of the area of the Wind Farms, where rest of the area could be utilized for plantations, agriculture, and so on. Further, during the increase of falling grid frequency, they are able to give fast frequency response in the former. As per the records, in the year 2021, India was the 5th country in the world in the production of wind power. The start of the wind power development took place after the eminent power engineer Maneklal Sankalchand Thacker began a project with CSIR to find the potential of harnessing the power of wind in the country by the last month of the year 1952. This was also supported by a Symposium organized by CSIR along with UNESCO, held in the latter half of 1954 in New Delhi, which further convinced the British power engineer & authority on the generation of wind power, E. W. Golding, to recommend nonstop as well as widespread surveys in India's various regions. With research and technological development over the decades, India slowly reached to its current position in the generation of wind power, where significant milestones started from the year 2011 in this direction. By the year 2024, India became the 4th country in the world to have the most installed wind power capacity.

Figure 10. Wind Farm in Tirunelveli in Tamil Nadu, India

Source: Shankaran Murugan, Wikimedia

Figure 11. Wind Farm in Jaisalmer in Rajasthan, India

Source: Nagarjun Kandukuru, Wikimedia

In India, seventy percent of the generation of annual wind power is from the month of May till the month of September which overlaps with the duration of the Southwest monsoon in the country and three-fifth of the generation of wind power takes place in the nighttime, which makes them complementary with the Solar Farms in India. The top Indian states with the maximum installed wind power capacity are Gujrat, Tamil Nadu, Maharashtra, Karnataka and Rajasthan, but states like Odisha, Madhya Pradesh and union territory like Ladakh also have good scope in the wind energy sector. Installed capacity of 175 GW from renewable sources was fixed as a grand aim in the first quarter of 2021, which comprised of sixty gigawatt of wind power capacity as well as one hundred gigawatt of solar power capacity by the year 2022, but by the end of the year 2022, only slightly over two-thirds of its grand aim was achieved. Nevertheless, by the first half of this year 2024, still wind and solar goals are somewhat left to achieve of the stated grand aim. Though, the growth of total renewable power installed capacity was good as during FY 2017 – 2018, it was just seventy gigawatts. As per the report of National Institute for Wind Energy located in Chennai, the western states have high potential in terms of fast, steady, in addition to stable flow of the wind, which is followed by the southern states in India. Further to repower the wind powered projects, the policy released by the Central Government (Union Government) of the country was that the repowering potential is over twenty-five thousand megawatts. The installation of additional wind turbines was part of this policy, having minimum three megawatts capacity along with height of the hubs above one hundred twenty metres which will positioned in between the already present turbines without affecting the performance of one another. Then, the coast of Gujrat and Tamil Nadu are highly favourable to have offshore wind energy potential in the country, which can also be followed by the coast of Maharashtra and Kerala. In the month of September 2015, the Cabinet of India permitted the National Offshore Wind Energy Policy after taking the initial prefeasibility examination reports on the development of offshore wind farms in the states of Gujrat as well as Tamil Nadu published by the consortium earlier around mid-June that year into consideration. This authorized the Ministry of New & Renewable Energy to serve as the Nodal Ministry intended for the utilization of offshore areas which comes under the Exclusive Economic Zone of India. The Expression of Interest was called by the Nodal Ministry as well as Nodal Agency in order to invite bidders to develop the 1st one thousand megawatts offshore wind farm of commercial scale in the country, near Gujrat's coast. This Expression of Interest was published in the 1st half of the year 2018, which stated the detection of the proposed area under the EU funded study by FOWIND and FOWPI, which specified the location of the proposed wind farm to be twenty-three to forty kilometres far from the port of Pipavav present in the Gulf of Khambhat and this area might cover around four hundred square kilometres. Until last year i.e. 2023, there were no active offshore wind farms in India, but it is anticipated that significant milestones will be achieved by India by the year 2030 in this regard. India's current standing in the world in regard to installed wind power generation is due to the development of Onshore Wind Farms, but at the same time Offshore Wind Farms are also important given the vast coastal areas present in the country. In the beginning, India had the National Wind Power Policy, but later on the development of policies for Offshore and Onshore Wind Farms took place. Then, two important initiatives taken by Government of India in the development of Wind Farms were:

- National Offshore Wind Energy Policy
- National Wind – Solar Hybrid Policy

National Offshore Wind Energy Policy

The notification of this policy came into existence in the month of October 2015, with the aim to progress with offshore wind energy development in the Exclusive Economic Zone of India which surrounds the over seven thousand five hundred long coastlines of the country.

National Wind – Solar Hybrid Policy

This policy came into existence in May 2018, with the main aim to create a framework which could be used to promote large scale grid linked wind – solar photovoltaic hybrid systems in order to not only effectively but also optimally use the solar as well as wind resources, land, in addition to transmission infrastructure.

Then, the Government of India had been endorsing wind farm projects with the assistance of investment by the private sector, through several financial as well as fiscal inducements like Concessional Custom Duty Exemption on particular parts of the generators used for wind power, and Accelerated Depreciation Benefit. Also, the projects which were commissioned before the 1st quarter of the year 2017, were having support of Generation Based Incentive Scheme in the country. Then, other inducements to increase the installation capacity of the wind energy in India, comprise of technical assistance like detection of sites having potential for wind power as well as examination of wind resources in the country via National Institute of Wind Energy which is situated in the Tamil Nadu state. Then, Inter State Transmission System fee was waived for interstate sale of wind as well as solar power aimed at projects which will be commissioned by the end of June of the next year i.e. 2025. Also, for the Tariff Based Competitive Bidding Process, the government stated guidelines which could be used to procure power from the grids which are linked to the projects relying on wind power with the aim to give an outline intended for wind power procurance via a bidding process which could be transparent in nature. This comprise of making the whole process standardized for everyone as well as explaining the numerous involved stakeholder's responsibilities in addition to their roles. The distribution licensees are made sure by such guidelines to allow the wind power procurement at competitive pricing in a cost-efficient approach. Furthermore, the Wind Renewable Purchase Obligation's trajectory to be announced till the year 2030 in the country.

What are the Essential Factors That Decide the Wind Farm Projects?

The essential factors that are considered in the case of wind farm projects are:

- Wind conditions
- Siting considerations
- Electricity grid considerations
- Physical access

Wind Conditions: It is important to understand the wind power density in different regions in a country and to observe the wind speed in those regions. Since, to produce more amount of electricity through wind turbines, it is required for wind speed on average to be faster in that place in comparison to other places. Therefore, faster winds in a place makes it highly preferable to develop wind farms economically.

Siting Considerations: For wind farm projects, location of the place becomes considerably important which couldn't be ignored. If the location of the place is not preferrable for the wind farm projects, then it will be unwise to go with their development.

Electricity Grid Considerations: For the development of wind farm projects on a large scale, the most important step is to search for areas having sufficient Available Transfer Capacity, as it is a way to measure the transmission system's remaining capacity which will be available for additional integration of 2 interconnected areas which doesn't require any major upgrade in the available substations as well as in transmission lines.

Physical Access: It is important to determine whether the place is even physically accessible or not. In case the place is having favourable wind conditions, but it is inaccessible then, the development of wind farm project in that place will become very costly, requiring longer duration of time to complete the project but with uncertainty.

Objectives

There are several objectives of this chapter, which are as follows:

- To understand wind farms and their types in general and to understand the gaps in the current studies.
- To understand the importance of wind farms in promoting sustainable development and to understand some of the recent wind farm projects in India.
- To understand the stand of India and its policies in association with wind farms
- To understand the current trend and scope of wind farms in the future, particularly in India.

MOTIVATION

The author performed this study on wind farms because in the past decades the country India had been focusing on sustainable development and especially in the past decade, the Government of India had been highly active in promoting projects which could assist in the sustainable development of the country. Therefore, wind farms are also one of the major projects which are going on in the country to fulfil its sustainability goals. Through this book chapter, the author tried to cover some of novel wind farm projects which either got completed recently or still undergoing in India. Further, how such projects could really be beneficial for the country and their possible future in the country, was covered by this chapter.

How Does Wind Farm Projects Promote Renewable Energy Infrastructure and What is the Need of Developing Wind Farms in India?

India is one of the major emitters of CO2 in the world and with the increase of global warming, climate change is taking place at a drastic rate. Biswas et al. (2024) discussed about uses of green energy which is derived from renewable energy sources, which could lead to development of humans in a sustainable manner. Due to which the Government of India is focusing on sustainable development of the country which includes public transport projects discussed by Singh (2023) to decrease the carbon footprint of the country, dam projects to generate hydroelectricity to decrease the consumption of fossil fuels to produce electricity to meet the energy demands as mentioned by Singh (2023). Moreover, the country is building a number of infrastructure projects which includes not only dams to produce hydroelectricity but also projects running as well as operating on renewable energy which was conversed by Singh (2023). Further, India's sustainable planning and development were explained by Singh (2023). However, to meet the Net Zero Target by 2070, India not only needs sustainable construction practices but also needs to focus on harnessing its wind power potential. India is blessed by wind energy and fortunately in the recent decades, India was able to utilize its wind potential up to some extent. Still, that is not sufficient for the country to achieve its lofty target in the future. Further, wind energy is another type of renewable energy for which India is having enormous unutilized potential. So, through the development of Wind Farms, it is possible to decrease the air pollution considerably by replacing the fossil fuels usage to produce electricity, which will decrease the carbon footprint and can control the rate of climate change up to some extent. Through the development of wind farm projects, it will help incorporating renewable energy infrastructure which will be beneficial for the country's sustainable development in the long run.

What is the Need of AI in Wind Energy Systems and How AI Could be Used in Wind Energy Systems Which are Important for Development of Wind Farms in the Modern World?

Even though policy frameworks, site selection, market conditions, as well as infrastructure planning fundamentally lead the development of projects like Wind Farms, but the latest developments in AI have increasingly been used as assistance tools in order to improve the decision-making across the various stages of the lifecycle of wind energy, making AI as an enabling technology in the wind energy system. Then, AI is essential in Wind Forecasting and Resource Assessment, as in wind farm projects, the economic as well as technical viability is important, which could be done by assessment of wind resources with high accuracy. Techniques which are on the basis of Artificial Intelligence, like ML assisted data driven Pattern Recognition in addition to forecasting, have been increasingly implemented to enhance long-term and short-term wind power as well as speed output predictions. The mentioned tools assist in capacity planning, improved turbine placement, in addition to integration of grids, especially in regions having high temporal variability in the patterns of winds like semi-arid as well as coastal zones in the country India. Notably, AI is able to improve statistical and conventional meteorological methods instead of switching them, which enables enhanced reliability in operational planning and feasibility assessments. Then, AI is also needed to optimize maintenance and condition monitoring. In the recent years, AI assisted system of condition monitoring are increasingly utilized to analyse sensor data from blades, gearboxes, and turbines, in order to detect early signs of failure, fatigue, or wear. Then, AI based models for predictive maintenance allows operational costs to be optimized, asset life to be enhanced, in addition to decreased downtime, which all are crucial for the deployment of wind farms on large scale. For instance, in India conditions, these tools are especially applicable for wind farms which are located in harsh or remote environs where manual reactive maintenance as well inspection could be inefficient and expensive. Also, AI assist in Grid Integration as well as Market Decisions, which could be seen at the system level, which Artificial Intelligence is implemented to assist such decisions which gird based and market based, like integration with hybrid renewable systems, demand response alignment, and dispatch optimization. These applications support not only policymakers but also operators in enhancing the economic competitiveness as well as managing intermittency of wind energy inside the electricity markets which keeps on evolving with time. Thus, AI doesn't act as a detached solution, but it functions as a decision support layer which could complement market mechanisms, infrastructure investments, in addition to regulatory frameworks, which all governs the deployment of wind energy. In general, it could be said that Artificial Intelligence helps in the deployment of wind energy by improving market responsiveness, operational efficiency, in addition to forecasting accuracy, whereas the main success of projects like wind farms still depend on infrastructure readiness, policy alignment, and sound engineering design. Artificial Intelligence positioning as an enabling tool makes sure that the systems of wind energy keep their main role in execution and planning of renewable energy.

LITERATURE REVIEW

The section of Literature Review is divided into nine different parts which grouped similar types of literatures under each theme, which focuses on decision support frameworks, spatial optimization, & site suitability; national positioning, deployment at macro level, energy transition context, & market dominance; marine restrictions, offshore potential, colocations, & floating wind; forecasting accuracy, spatial temporal examination, & estimation of wind availability; energy storage, compatibility of grids, HVDC, & voltage stability; market competitiveness, cost benefit, investment risk, & profitability; enabling tools for monitoring, forecasting, & control; public perception, environmental licensing, & project risk; optimization of algorithms, placement of turbine, & farm layout; which are as shown in the subsequent subsections.

Wind Energy Market Trends and Landscape at National & Global Levels

Siram et al. (2022) studied wind energy, and they worked on altering the Indian landscape of renewable energy through the influence of their studied area. They found out that the country was generating roughly one-fifth of its energy from renewable sources at that time and roughly three-fifth of its energy was generated by the country's western as well as northern parts. Further, they covered the challenges and critical aspects in wind energy's development in the country. Next, von Krauland and Jacobson (2024) studied onshore wind energy in India, and they worked on co-locating not only solar but also wind resources in the country along with manifold wind speed limits. They stated that the around one and a half percent of the total land area of the country will be sufficient to satisfy the demand of electricity of the country for the year 2019. Also, Yadav et al. (2024) considered the Biratnagar present in Nepal to be their case, and they took wind as well as solar farms into account to decarbonize the airport. They discussed several aspects of energy production which included environmental, economic in addition to energy. They performed a sensitivity analysis of wind speed, wind type as well as interest, and they compared both types of farms in the airport, through their study. Then, Biswas et al. (2024) studied sustainable energy, and they worked on the ways to explore their studied area so that the natural power could be released to develop future.

Wind Potential Mapping, Forecasting, & Resource Assessment

Islek and Yuksel (2022) studied wind farms, and they worked on finding their possible stable locations as well as they examined the future potential of wind power in order to understand the forecasted changes which could occur in the Black Sea through their research. Further, they developed plans for such projects in their studied area. Later on, Basak et al. (2023) studied the offshore based wind energy, and they worked on predicting its spatiotemporal potential in the case of India by taking the assistance of CMIP6 datasets to create a combination of differential weight. They utilized 9 CMIP6 global change models in order to develop dissimilar ensemble variants to perform the sensitivity of their studied area to the changes in the climate. Similarly, Sakuru and Ramana (2023) studied wind power, and they used ERA5 reanalysis to find out its potential in the case of India. They considered the data starting from 1979 until the year 2018 at one hundred metre to perform the above reanalysis of the wind. They further used the three parameter Weibull function's mixture to represent the distribution of the speed of wind. Then, Lopez et al. (2023) studied wind farms, and they utilized computational fluid dynamics to understand not only how atmospheric stability affects the design of their studied area but also the examination of wind resources especially in area of Andes Mountains, present in Ecuador. They utilized the values of wind speed to classify the potential of wind power and designed a hypothetical model to obtain suitable outcomes. Also, Prasad et al. (2023) studied wind projects which are offshore ones, and they used reanalysis datasets with the assistance of LIDAR validation to improve such resources in the Indian state Gujrat. They took the reanalysis datasets of over forty years in order to determine the best dependable reanalysis datasets to examine the mentioned resource. Next, Roga et al. (2024) considered the case of central part of India, and they worked on examining the potential of wind energy in their considered case. They stated that to protect environment as well as to develop economically, sustainable energy is essential. To perform a thorough analysis of the potential of wind energy, they incorporated the software windographer in a particular place called as Nagpur to obtain favourable outcome. On the other hand, Mahata et al. (2024) studied the production of wind power in the Indian state Gujrat, and they worked on comparing a number of time-series prediction models for their studied area. They stated that in their comparative study they included RNN, ARIMA, LSTM, ANN as well as GRU, and to verify their models, they gathered data from metrology and Central Electrical Authority in India. While, Huang et al. (2024) studied wind farm projects, and they worked on applying model selection prediction system intended for theoretical power production as well as wind speed on the basis of the conversion in addition to categorization of wind. They stated that their system to predict wind comprises not only a prediction model, but also categorization and clustering. Moreover, Pessanha et al. (2024) studied the wind farms which were offshore ones, and they worked on understanding their capacity factors particularly on the coastal areas of Brazil, with the assistance of Bayesian method.

Spatial Planning, Site Selection of Wind Farms, & GIS-MCDM Methods

Saraswat et al. (2021) studied wind in addition to the solar farms, and they worked on modelling techniques which were based on GIS as well as Multi – Criteria Decision Making to examine their locations in the country India. They found out that the Indian state Rajasthan is the most appropriate land not only for wind but also for solar farms. Later on, Nagababu et al. (2022) worked on 2 stage algorithm which was based not only on GIS but also on Multi – criteria – techniques, in order to detect the potential regions for the installation of wind farms at the micro level in the Indian conditions. With the assistance from their proposed algorithm, they stated that over 1.8 million sq. km. area is appropriate in the country to develop wind farms. Then, Daoudi et al. (2022) studied the Morocco - UK green energy export project and they worked on analysing the installation of the 1st onshore wind farm-based project. They considered the case of Tiznit province along with the data of twenty-four years wind speed to obtain suitable outcome. Afterwards, Wimhurst et al. (2023) utilized the model which was a logistic regression one and with its assistance, they worked on forecasting the appropriate site for commercial purpose wind farms in USA. They chose the datasets of aggregate predictor to the size of twenty nationwide hexagonal shaped grid cells as well as to validate the performance of their model, they permitted the locations of the current commercialized wind farms. While Daoudi et al. (2023) studied an analysis approach which was data envelopment one and they used that method to rank the sites of the wind farms. They stated that the offshore wind farm's effectiveness was supported by their obtained static as well as dynamic results. Also, Wimhurst et al. (2023) studied wind farms, and they worked on regulating the utilized factors in the appropriateness models of wind farms by reviewing the past published studies. They not only performed a systematic review of the location factors in such studies, but they selected as well as incorporated those factors by condensing them across 5 themes to accomplish their study. Followed by, Yildiz (2024) considered the place Balikesir present in Turkey, and they worked on selecting site for wind farms through the use of spatial MCDM method. He claimed that it was the 1st time that to select site of onshore wind turbine, seaport criterion was utilized. He also found out that 2.34% of the selected city is within the utmost suitable class intended for the installation of wind farms and that whole area is slightly over three hundred thirty-four square kilometres. Similarly, Yousefi et al. (2024) used GIS, and they combined it with analytical hierarchy process in order to develop the latter along with MCDM to select appropriate site for wind farms. They claimed to perform a tyring review of applicable literature to detect the criteria of other researchers and their novel AHP development would be useful in measuring the potential of the wind energy. While Raju M et al. (2024) studied wind farm, and they worked on the way to optimally placing the constant hub height wind turbines in it with the assistance of twin archive leading decomposition on the basis of an algorithm which was multi – objective – evolutionary one. They used adaptive offspring strategy to advance the algorithm named TAG – DMOEA in order to obtain results of 6 cases which they further compared with other ten modern algorithms.

Offshore Wind Energy Systems and Marine Considerations

de Vasconcelos et al. (2022) studied offshore wind farms, and they worked on their licensing in regard to the environment, where they covered the consequences of policies as well as guidelines particularly for the new markets. They stated that to avoid terrible negative effects on the marine environs, studies corresponding to the strategic environmental examination is necessary and spatial planning of maritime can speed up the one stop shop i.e. environmental licensing of such projects. Then, Bi and Law (2023) studied not only floating solar but also offshore wind farms, and they worked on co-locating them along with the coverage of performance of solar power due to the effects of wave as well as high wind conditions. For a hybrid farm based on their studied area, they validated the co-locating synergy to obtain the desired results. Similarly, Kolli et al. (2023) studied offshore wind-based farms, and they worked on considering them as a source for supplementary coolant power in order to improve the nuclear power plant's seismic resilience. They stated that the performance of cooling systems is greatly linked with the safety of nuclear plants and through their case study of wind farms, they were able to successfully propose a strategy on seismic resilience. Next, Hong et al. (2024) performed a thorough survey in order to understand the opportunities as well as challenges behind the installation of wind farms which were floating offshore ones. They stated that for the environmental in addition to weather examinations, the models which were installation cost prediction ones must be improved. Likewise, Watson et al. (2024) studied the offshore based wind farms, and they worked on finding out how they affect the ecosystem on a global scale. They then consolidated this sector's world-wide proof on the services associated with ecosystems and stated that wind farms frequently affect the former along with biodiversity in a negative manner. Also, Edwards-Jones et al. (2024) studied offshore wind farms, and they worked on their delivery as well as planning while considering the comprehensions of stakeholders in inserting the marine net gain for them. They stated that the sustainable delivery of the earlier mentioned farms has the potential to deal with both the challenges of not only loss of biodiversity but also the climate change. Then, Jalili et al. (2024) studied the decommissioning of offshore wind farm-based projects and then they performed not only environmental but economic assessments to assist in the process of making such decisions in regard to the earlier stated decommissions. On the basis of the field experience as well as data available, they developed their proposed method, and for various parameters, they further performed a decommissioning cost sensitivity analysis to accomplish their research goals. Furthermore, Bonthu et al. (2024) studied the coast of India, and they worked on taking ecological protections into account while finding out the potential of the offshore wind energy to produce desired outcomes.

Financial Viability, Investment, and Techno-Economic Analysis

Bhattacharjee et al. (2022) studied an Indian offshore wind farm, and they used Bio motivated meta - heuristic plan to increase its annual profit. The location of their study was the Bay of Khambhat and for optimization purpose, they used a genetic algorithm which was a modified one. Then, Ziemba (2022) studied the investments in the wind farm-based projects which are offshore ones, and they worked on their uncertain multi criteria analysis to understand the Polish Economic Zone in the Baltic Sea region. They stated that they combined methodological, stochastic as well as fuzzy uncertainties to create an approach to give suitable outcome of their research. Next, Asamoah et al. (2023) studied wind farm at 4 different locations which included Denu, Atiteti, Anloga as well as Sege, in order to examine the fifty-megawatt utility scale wind farm to understand its financial, technical in addition to environmental aspects. They found out that such a wind farm is feasible in all the studied locations. Later on, Yadav et al. (2024) took wind as well as solar farms into account to decarbonize the Biratnagar airport.

System Stability, Power Quality, & Grid Integration

Amuthan et al. (2023) studied doubly fed induction generator wind farm, and they worked on an adjustment mechanism on the basis of internet of things especially for a direct reference model adaptive IMC to assist their studied area's voltage sag. They claimed that their proposed neural network MIT obtained one ms response time as well as 0.15% torque ripple superior to the present Fuzzy based MIT approach. While earlier, Magesh et al. (2022) studied wind farms which were grid connected ones and they worked on simulating as well as measuring the problems related to the quality of their power. They claimed that their study was essential to not only understand the root causes of the mentioned problems but also to enhance the performance of the wind farms. Afterwards, Ngoenmeesri et al. (2023) studied wind farms, and they worked on improving the quality of power of such farm present in Thailand with the assistance of an energy storage system of five Megawatt-hour Li – ion battery. They stated that to decrease the number of feeder trips, they developed a battery energy storage system to obtain favourable results. Next, Vijay et al. (2024) studied offshore based wind farms which utilizes voltage source converters along with high voltage DC transmission technologies, in order to not only mitigate but also analyse the sub synchronous resonance to obtain suitable results. Moreover, Hussain et al. (2024) studied wind as well as solar photovoltaic system and they worked on finding viable synergy between their hybrid, which could be utilized in a green building's power supply in the place Kota in the Indian state Rajasthan. They optimized the mentioned hybrid system with the assistance of iHOGA to accomplish their study.

Performance Improvement, Layout Design, & Optimization

Gyatso et al. (2023) considered the region of high plateau, and they worked on examining the performance of wind power in those areas, where they included the field test of a wind farm situated on the plateau of Qinghai – Tibet as their study case. They stated that they were the 1st such farms in the world present above five thousand metres, where they were able to systematically perform the field study, and they found that over eight thousand four hundred houses could get electricity through the twenty-two-megawatt wind farm. While earlier, Bhattacharjee et al. (2022) worked on Bio motivated meta - heuristic plan to increase the annual profit od offshore wind farm in India at Bay of Khambhat. Next, Raju M et al. (2024) worked on multi-objective-evolutionary based algorithm to optimally placing the constant hub height wind turbines in it with the assistance of twin archive leading decomposition, to generate desired results. Furthermore, El Jaadi et al. (2024) utilized a novel method which was based on particle swarm optimization as well as on artificial neural networks, to improve the extraction of electricity in a way to optimize the layout of wind farms. They claimed that the hybrid technique proposed by them could assist in improving the output of the energy along with the consideration of topographical limits in addition to the patterns of the wind specific to the sites.

Social Acceptance, Environmental Impact, & Risk Assessment

Somi et al. (2021) studied onshore wind farms and with the assistance of fuzzy case-based reasoning, they created a risk breakdown matrix for such projects. They stated that their used FCBR is appropriate for identifying the risk in construction and the info-based approaches are perfect for identifying risk in the construction projects which are new ones. They further tried to identify one hundred and sixty-nine work package level risks for the above stated projects to give new insights for the safe construction of future projects. Later on, Mohamed et al. (2023) studied wind farm-based projects, particularly the onshore ones and they worked on identifying the risk involved in them, on the basis of context driven ontologies, by following domain centric method. To improve the sharing, storage, as well as reuse of knowledge with the aim to identify risk, they completed not only automating but also mapping the representation of risk info in addition to the considering the context of the projects. While le Maitre et al. (2024) worked on determining the residents react during the development of various faces of such projects, to get fruitful outcomes. Similarly, Watson et al. (2024) worked on determining the ways ecosystem is affected by wind farms which are offshore ones at the global scale. On the other hand, Edwards-Jones et al. (2024) worked on the delivery and planning of wind farms which are offshore ones, to accomplish their study.

Intelligent Support Systems and Digital Tools

Amuthan et al. (2023) worked on adjustment mechanisms to proceed with doubly fed induction generator wind farm, on the basis of IoT for a direct reference model adaptive IMC, to achieve their desired study goals. Later on, Huang et al. (2024) worked on implementing model selection prediction system intended for theoretical power production, to generate desired results. Then, Mahata et al. (2024) worked on comparing a number of time-series prediction models to proceed with the production of wind power in Gujrat, to accomplish their study.

RESEARCH GAPS

After going through a number of published papers in journals, proceedings and books, only the ones from the past few years were selected for further consideration. The author found that the researchers not only in India but in other countries have actively worked on wind farms, but he couldn't find much regarding specific ongoing wind farm projects in India or the ones which just recently got completed in India.

MAIN FOCUS OF THE CHAPTER

This book chapter focusses on the classification of wind farms and explains briefly about wind farms. It further talks about the evolution of wind farms in brief in the world, and then it covers the Indian scenario, including how the wind power projects got started in the country, the policies and initiatives of the Government of India towards the development of wind farm projects, as well as the future plans and goals of the government in this matter. The chapter also covers the problems and issues related to the development of wind farm projects in the country. Moreover, the gaps in the existing literature were covered and some novel wind farm projects which are either ongoing ones or just got completed recently were discussed in this chapter. It also focussed on analysing the available information from different sources to understand the current trend and future of wind farms in India particularly, to understand how they can help the country in achieving its future sustainable development goals.

ISSUES, CONTROVERSIES, PROBLEMS

Wind farm projects are a good option towards sustainable development as they can serve as a source of green energy and has the ability to meet our energy demands up to certain extent. However, some issues and problems are also created with their development like how they affect the visual impact of that place and how they affect the landscape of the place where they are built. Since, they need more space to function properly, so rural as well as wild areas are preferred for their construction. However, this brings up the concern of countryside's industrialization along with possible occurrence of habitat loss. This might decrease the arrival of tourists in that place, if the place was popular for its habitat and natural scenery in the past which would have been negatively affected after their construction. There are chances of them interfering with the nearby radars. Moreover, there are challenges with the grid connection which wind farm projects face, as many places need to upgrade their existing grids for them. Then, acquiring land and obtaining various permits from the concerned authorities is also time consuming as well as hard to complete. Plus, wind farms being an intermittent source of energy also makes power supply coming based on fossil fuels to be utilized at other times, but the good thing is that India found the answer to this problem by combining them with solar farms. Then, India still doesn't have sufficient technological advancement for Offshore wind farms, as they are still just in the planning phase, which might need years to start implementing them. Also, the Offshore wind farms will require even more hefty investments in comparison to the Onshore wind farms. Even though, wind farms are excellent option not only as per the present situation but also in the long term. Still the development process of wind farm projects is also important, in order to not harm the environment and cause disappearance of the existing habitat of the place, where they are being built.

SOME MAJOR RECENT WIND FARM PROJECTS IN INDIA

In this section, some of the current wind farm projects which are either completed in the past few years or are still under construction are discussed:

- Sidhpur Wind Farm
- Hatalageri Onshore Wind Farm
- Mulanur Wind Power Project
- Gujrat Offshore Wind Project
- Tamil Nadu Offshore Wind Project
- Bhuj Wind Project
- AGEL Khayda Wind Farm-I
- Greenko Jaisalmer Wind Farm
- JSW Tamil Nadu Wind Farm
- JSW Wind Farm Projects, Karnataka

Sidhpur Wind Farm

This wind farm project is located in the Indian state Gujrat's Sidhpur, with a nameplate capacity of 251 megawatt. It is an Onshore Wind Farm as well as its ownership divided equally by CLP India PVT LTD (Apparva) and CDPQ. It was expected to go into operation by the year 2022, but according to the 1st quarter of this year 2024, 25.2-megawatt interstate transmission system linked project was commissioned by the company KP Energy Limited as part of the 6th phase of the project. This completed the commissioned net cumulative capacity as 163.8 megawatt. However, by the last week of July this year, the total commissioned net cumulative capacity of the project reached 245.7 megawatts after the successful commission of 23.1 megawatts by K.P. Energy took place as part of the 10th phase of the project, which made the project almost near to its awarded capacity.

Figure 12. Satellite photo of Siddhpur in Gujrat, India

Source: Google Earth

Hatalageri Onshore Wind Farm

This project is an onshore wind farm, which is being developed in the place Hatalageri, present in the Indian state Karnataka. The Indian developer Viviid Renewables joined hands with Copenhagen Infrastructure Partners to build this project, where the former will provide balance of plant services in the project's development. The wind farm is expected to have three hundred megawatts capacity, and it is anticipated that the project will be finished by the end of the next year i.e. 2025.

Figure 13. Satellite photo Hatalageri in Karnataka, India

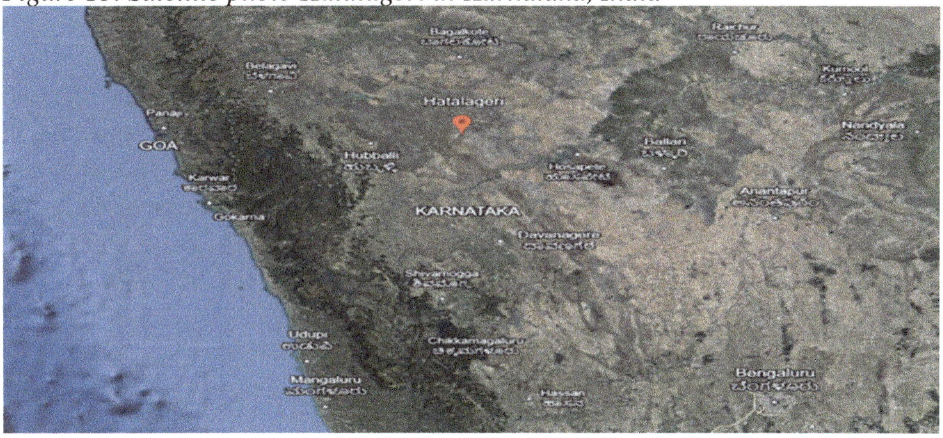

Source: Google Earth

Mulanur Wind Power Project

This is an active wind power project, which was commissioned in the year 2021 in the Mulanur town present in the Indian state Tamil Nadu. The wind farm is an onshore one and has a capacity of three hundred megawatts. Sprng Energy developed this Mulanur Wind Farm and at present its stakes are completely owned by Shell. Further, one hundred twenty-meter-high concrete towers were installed at the site for this project, and the development of this project happened in many phases. Then, the German company Nordex was chosen to supply the wind turbines in the project.

Figure 14. Satellite photo of Mulanur in Tamil Nadu, India

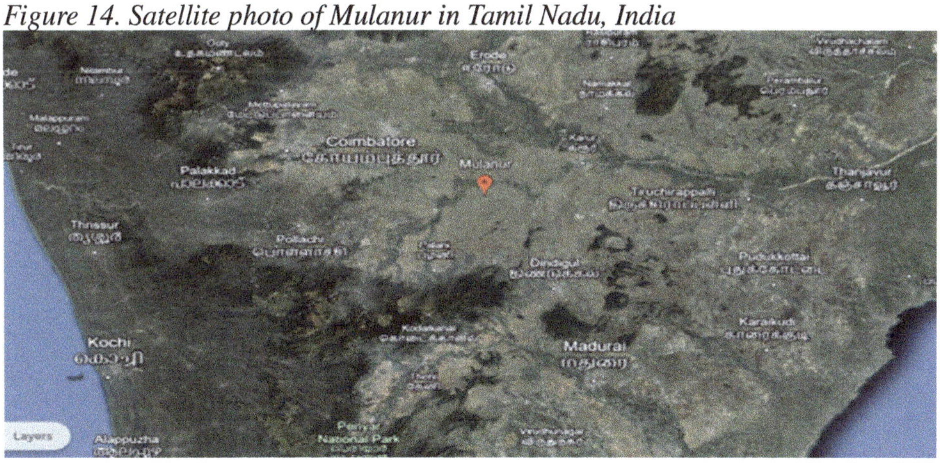

Source: Google Earth

Gujrat Offshore Wind Project

This project is still in the very initial stages and the Union Cabinet approved a scheme named Viability Gap Funding to support this ambitious project. It will be an Offshore Wind Farm with a capacity of five hundred megawatts. This project will be situated around the coast of the Indian state Gujrat and the Union Government has also planned to upgrade the nearby seaport to allow smooth transportation and have seamless logistic supply. As per the plan, the location of the project will in the Gulf of Khambhat and it will be near to the port area of Pipavav in Gujrat.

Figure 15. Satellite photo of Pipavav seaport in Gujrat

Source: Google Earth

Figure 16. Map of Gujrat, India along with Gulf of Khambhat

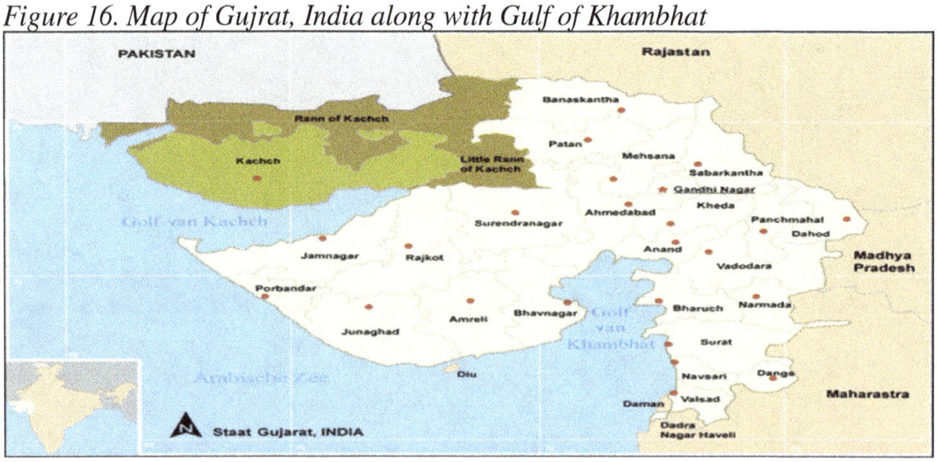

Source: Miljoshi & Lexw, Wikimedia

Tamil Nadu Offshore Wind Project

Just like the earlier project, this project is also in the very initial stages and with the approval of Viability Gap Funding scheme in the month of June this year 2024, the project can proceed further. Its capacity will be five hundred megawatts, and it will around the coast of the Indian state Tamil Nadu. With the earlier offshore project, these two are expected to generate renewable electricity of around 3.72 billion units which could help in decreasing the CO2 emission amounting to 2.98 million tonnes in the span of a quarter of a century's time. The location of this offshore wind farm will be in the Gulf of Mannar, which will be in between the India's southeastern coast and Sri Lanka's western coast.

Figure 17. Satellite photo of India's southeastern coast and Gulf of Mannar

Source: Google Earth

Bhuj Wind Power Project

The Saudi Arabia company Alfanar Energy Private Ltd. owns two three hundred megawatts wind farm projects in the Kutch district of the Indian state Gujrat. In the year 2018, the company was a successful bidder for Bhuj 1 Project after Government of India Enterprise named Solar Corporation of India Ltd. invited proposals on Build Own Operate basis for the aggregate capacity of two thousand megawatts. The project's location was around the village areas near Nakhatrana city and Lakhpat town, which lies in the Kutch district having Bhuj as the headquarters of the district. The Bhuj 1 Project was commissioned in the year 2021, and the towers are 120 m tall. However, for both the projects, the project's total installed capacity is 600 megawatts or 602.8 megawatts to be precise, and they produced electricity will be sold to Solar Energy Corporation of India.

Figure 18. Satellite photo of Lakhpat and Nakhatrana in Gujrat, India

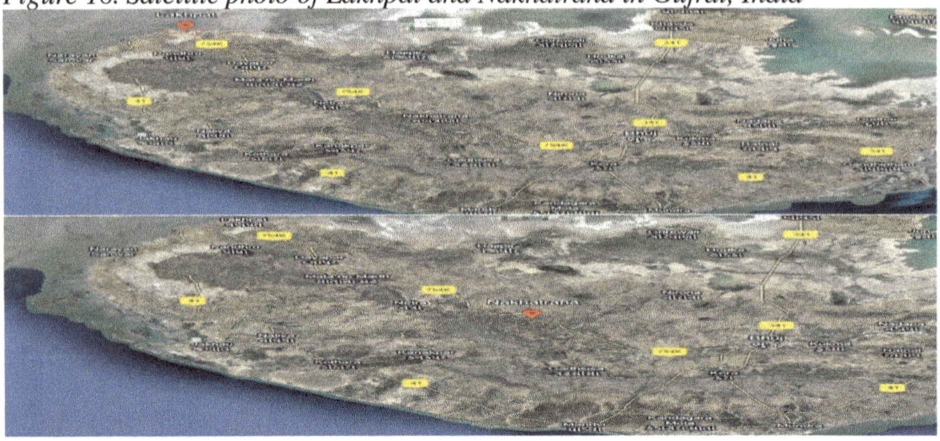

Source: Google Earth

AGEL Khavda Wind Farm-I

This is an Onshore Wind Farm project which is being part of the world's largest Renewable Energy Park situated in Khavda which also has installed solar farms. By the middle of this year 2024, it was stated that the Adani Green Energy Limited has operationalize the two hundred fifty megawatts wind power, at the end, this massive renewable energy project is supposed to have thirty gigawatts of installed capacity in the next five years.

Figure 19. Satellite photo of Khavda in Gujrat, India

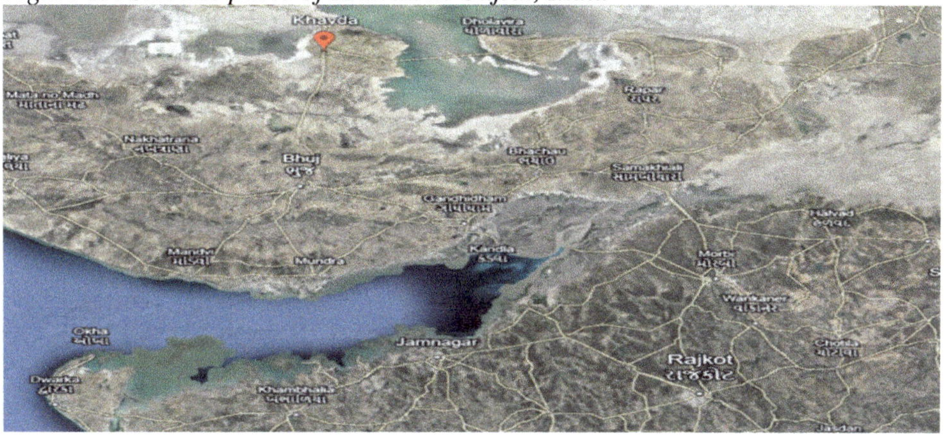

Source: Google Earth

Greenko Jaisalmer Wind Farm

This is an Onshore Wind Farm project to be developed by Greenko Energies. The project is supposed to have a designed capacity of nine hundred megawatts and as the project is still in the permitting stage. This project will be built in Jaisalmer in the Indian state Rajasthan.

Figure 20. Location of Greenko Jaisalmer Wind Farm Project

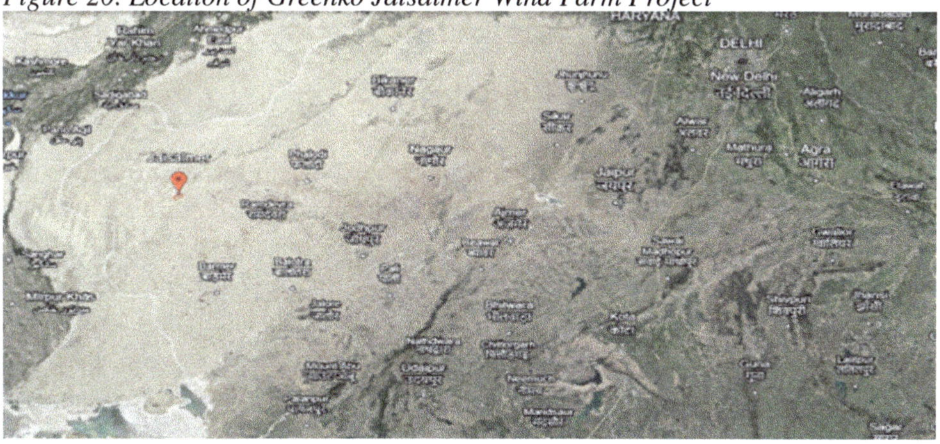

Source: Google Earth

JSW Tamil Nadu Wind Farm

This project is an onshore wind farm which is set to have phased commissioning and development in the Indian state Tamil Nadu. The total design installed capacity of this project is eight hundred and ten megawatts, but as per the reports, last year December (2023), the concerned company commissioned the fifty-one megawatts of wind power, which is supposed to be the phase 1. This project is being undertaken by JSW Renew Energy Limited. The company has a purchasing power agreement with Solar Energy Corporation of India for twenty-five years, in regard to this project.

Figure 21. JSW Tamil Nadu Wind Farm Project location

Source: Google Earth

JSW Wind Farm Projects, Karnataka

JSW Energy Ltd. in the later half of the year 2020 planned to invest in two on-shore wind farm projects in the Indian state Karnataka. The 1st project is situated across Ballari as well as Davangere districts, with the area of one thousand three hundred fifty acres and designed installed capacity up to six hundred megawatts. Whereas the 2nd project's location across districts like Dharwad, Ballari, Davangere as well as Gadag in the Indian state Karnataka, over an area of one thousand eight hundred acres and designed installed capacity of up to eight hundred megawatts. The 1st project was expected to be commissioned by this year 2024, but so far it as per the latest updates it was found to be in the permitting stage. The produced electricity will be sold to JSW Steel for a period of twenty-five years based on the purchasing power agreement. Furthermore, the company has identified 1800 acres of private land for the 2nd project. As per the State Government Proceedings passed in the first month of this year 2024, the company had identified the land in several parts and is expected to provide employment to sixty people through the development of this project.

Figure 22. Location of JSW Wind Farm Projects, Karnataka

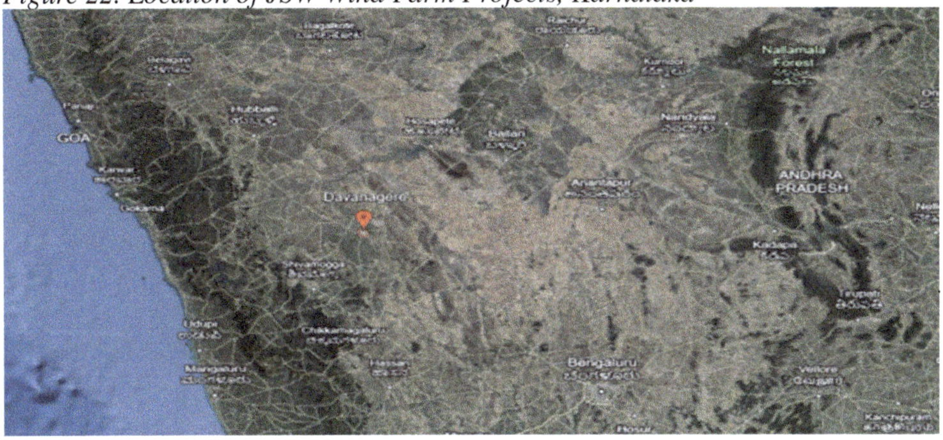

Source: Google Earth

METHODOLGY

The method to perform this scientific study to create this chapter initially started after going through numerous published studies on wind farms, to further filter the scientific and research publications which occurred in the latest years to understand the development of wind farms and wind energy systems not only in India but also in the world, to understand the gaps in the current literature. The chapter then incorporated a conceptual analytical research design on wind farms which was assisted through the synthesis of comparative case studies in order to investigate the wind farm projects development in the country India when considering the renewable energy context at the global level. This chapter is then structured to analyse the current wind energy systems, infrastructure practices, market dynamics, as well as policies in a systematic manner which aligns with the research book chapter's objectives on the strategic insights along with infrastructure development. This methodological method incorporated interpretive examination challenges associated with infrastructure, market, and policy, qualitative synthesis of national and global literatures, in addition to comparative and descriptive analysis of the trajectories of the development of wind farms. This method is suitable for this scientific study as this research book chapter emphasized on wind energy systems strategic deployment, policy evolution, and infrastructure development. The analysis was done based on multiple secondary data sources which included reports from national and international agencies like Ministry of New and Renewable Energy, International Energy Agency, Global Wind Energy Council, and other organizations which are policy oriented; peer reviewed publications on renewable infrastructure and wind energy systems; operational projects on wind farms which are documented case studies; and industry reports, documentation of projects as well as datasets which are publicly available associated with market growth, deployment trends, and capacity of wind farms. These sources collectively allow the interpretive as well as competitive analysis of wind farms and wind energy systems to have robust contextual as well as empirical foundation.

The chapter comprise of four major structured stages as part of its methodological process, which are:

- **Global Contextual Analysis:** This stage consists of synthesis and review of development patterns, infrastructure strategies, as well as technological typologies of wind farms at the global level, which were performed for best practices and international benchmarks to be established.
- **National Level Assessment:** This stage consists of analysis of the deployment trajectory of wind energy at the national level in the country India, in association with market evolution, infrastructure readiness, geographic potential, in addition to policy frameworks, to enable India's positioning in the wind energy landscape at the global level.
- **Case Study Synthesis:** This stage consists of examination of selected major projects on wind farms, in order to identify recurring themes associated with performance outcomes, deployment, planning, and operational challenges. The multiple case studies in this chapter are illustratively utilized to show real world execution.
- **Comparative & Interpretive Examination:** This stage consists of a qualitative comparison between wind farm initiatives in India and global leaders, to point out the strategic lessons, opportunities, structural gaps, which are pertinent to development of sustainable renewable energy, especially considering wind energy systems.

The methodology allowed a holistic examination of the development of wind farms in the country India, by integrating national assessment, global synthesis, as well as interpretation of case studies, to position the country inside the broader renewable energy transitions. This method assists future directions, opportunities, in addition to discussion of challenges based on evidence, to contribute to the literature on the development of sustainable energy infrastructure particularly considering the wind energy systems. Thus, making the study as hybrid multi-case conceptual analytical analysis with market readiness and governance structure.

Further in this section, the challenges observed in the course of the above-mentioned wind farm projects, will be discussed:

- In the case of Sidhpur Wind Farm Project, the pandemic affected the project initially and up to some extent delayed it. However, securing mandatory permissions as well as approvals for each phase is also a time-consuming process.
- On the other hand, Hatalageri Wind Farm Project, hasn't particularly faced any major problem so far, but integration of grid will be a challenge.
- Whereas, the Mulanur Wind Farm Project, faced transportation and operational challenges. The large turbines were difficult to transport and install at the site, as well as the produced power to integrate with the local grid was another hurdle on the path of this project.
- In the case of Gujrat Offshore Wind Project, it is expected to face technological challenges as India has yet to have any offshore wind farm. The grid connectivity, and the construction of the wind farm will itself be a big challenge.
- On the other hand, the Tamil Nadu Wind Project still being in the initial stages like the earlier one is anticipated to face similar challenges. Further, transportation and logistics supply will also be a hurdle for this project, that needs to be overcome.
- The Bhuj Wind Farm Project faced challenges because of Covid-19, which hindered and slowed down its progress and land acquisition for it was also a challenge.
- Whereas the Khavda Wind Farm Project being part of the massive hybrid renewable energy project not only faced land acquisition problem, but the main problem was the geotechnical as well as terrain challenges. Here, the ground water level could be found at shallow depths, presence of soft clays could also be found, along with the soil having high amount of salinity. Further, this place, also lies in the Zone V, which is most prone to the earthquakes. Cengrs worked on finding solutions to such problems in this project.
- In the case of Greenko Jaisalmer Wind Power project, the project is still in the permitting stage, but regulatory and environmental challenges need to be crossed for the project to proceed further.
- On the other hand, JSW Tamil Nadu Wind Farm project faced challenges like charges associated to resources, connectivity problems with the grid, high cost of the infrastructure, possible delays, because of time needed to get tariff approvals, and so on.

- Then, the JSW Wind Farm projects in Karnataka are still in the early development phase. The main problems which were observed so far, was finding the suitable land availability, along with environment and regulatory challenges.

RESULTS AND DISCUSSION

In this section data from different sources are collected to perform graphical analysis in order to understand wind farm's trends and future in India

Table 1. Installed wind energy capacity of India

India's Installed Wind Energy	In MW
Year 2023	43,773
Year 2024	46,656.37
Year 2030 (Planned)	140,000

Source: Press Information Bureau, SolarQuarter, and IEA Wind TCP

Table 2. Renewable Energy aim of India

India's Aim for Renewable Energy	In MW
Total Renewable Energy by 2030	500,000
Wind Energy by 2030	140,000

Source: IEA Wind TCP

Table 3. Offshore Wind Energy Target of India

Offshore Wind Energy Target of India	In MW
Installed Capacity of Offshore Wind Projects to be Issued by 2030	37,000
Total Installed Capacity of all the Planned Offshore Wind Projects in the Future	70,000

Source: EQ Magazine and ET Energy World

Table 4. Employment generation by type of renewable energy in India

Type of Renewable Energy	Employment during FY 2021-22
Wind Energy	25,500
Solar Energy	85,900

Source: CEEW-NRDC Analysis 2022, Council on Energy, Environment and Water

Table 5. Installed capacity of wind energy of India

Year	Expected Addition of Wind Energy Installed Capacity (MW)
2022	3,200
2024	4,600
2026	3,500

Source: Press Information Bureau

Table 6. Installed capacity of Onshore wind energy globally

Global Onshore Wind Installed Capacity	In MW
Year 2021	774,000
Year 2030 (Expected)	1,787,000

Source: Global Wind Energy Council, Global Wind Report 2023, Iberdola, and Renewable Capacity Statistics (IRENA, 2019d) IRENA analysis (IRENA 2019a), IRENA

Table 7. New Wind Energy Installed Capacity of Major Countries in 2023

New Wind Energy Installed Capacity of Major Countries in 2023	In (%)
China	65
USA	5.51
India	2.67

Source: Wood Mackenzie, and World Wind Energy Association

Table 8. Global novel installation of wind energy

New Installations Globally in 2023	In MW
Offshore Wind Energy	10,800
Onshore Wind Energy	106,000

Source: Mercom Clean Energy Insights

Table 9. Share of Indian states in the cumulative installation of wind power

Share of Indian States in the Cumulative Wind Power Installation	Year 2022 (%)	Year 2024 (%)
Tamil Nadu	23.7	23.7
Gujrat	23.5	26.2
Rest of India	52.8	51.1

Source: Mercom Clean Energy Insights, and Ministry of New and Renewable Energy

Table 10. Onshore wind energy capacity by country

Onshore Wind Energy Capacity by Country in 2023	In GW
China	404
USA	147.979
India	44.7

Source: Statista, Mordor Intelligence, Mercom Clean Energy Insights, and Swiss Re Institute

Table 11. Maximum operational Offshore wind farms by countries

Countries	Number of Operational Offshore Wind Farms in 2024
China	129
UK	39
Germany	30

Source: Statista

Table 12. Potential of India's renewable energy

Potential of Renewable Energy of India in 2024	In (%)
Wind Power	55.17
Others	44.83

Source: Energy Statistics India 2025 report – Ministry of Statistics and Programme Implementation, ET Energy World

Table 13. State wise potential of wind energy in India at 120m above ground level

State Wise Wind Energy Potential in India in 2023	In GW
Gujrat	142.56
Rajasthan	127.756
Karnataka	124.155
India	695.509

Source: Press Information Bureau, Ministry of New and Renewable Energy, and India Climate & Energy Dashboard, NITI Aayog

Table 14. Global electricity capacity of wind energy

Electricity Capacity Addition of Wind Energy	Onshore (GW)	Offshore (GW)
Year 2019	594.426	28.305
Year 2023	944.205	73.185

Source: International Renewable Energy Agency

Table 15. Growth of new installed capacity of wind energy globally

New Installed Capacity of Wind Energy	Onshore (MW)	Offshore (MW)
Year 2023	106,000	11,000
Year 2028 (Expected)	145,000	37,000

Source: Statista

Table 16. New installed capacity of wind energy by region

New Installed Capacity of Wind Energy by Region in 2023	In (%)
Asia – Pacific Region	71
Europe	15.77
Rest of the World	13.23

Source: Global Wind Energy Council, Asian Power, WindEurope, and World Wind Energy Association

Figure 23. Installed wind energy capacity of India

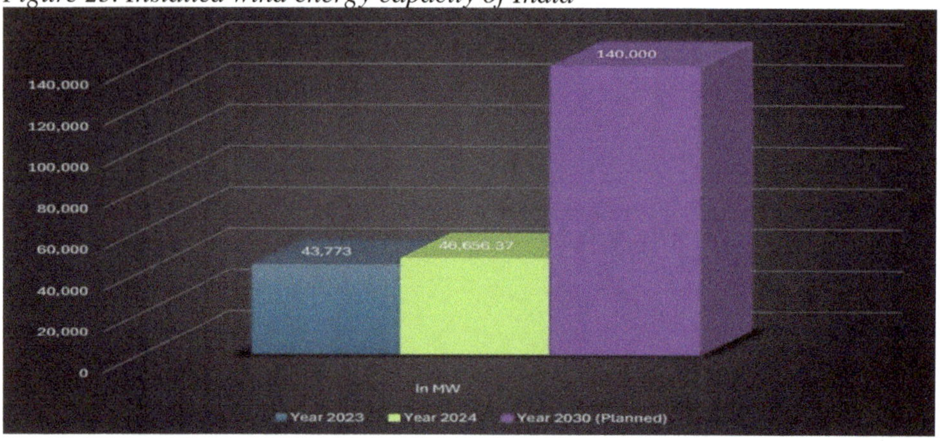

Source: Press Information Bureau, SolarQuarter, and IEA Wind TCP

Figure 24. Renewable Energy aim of India

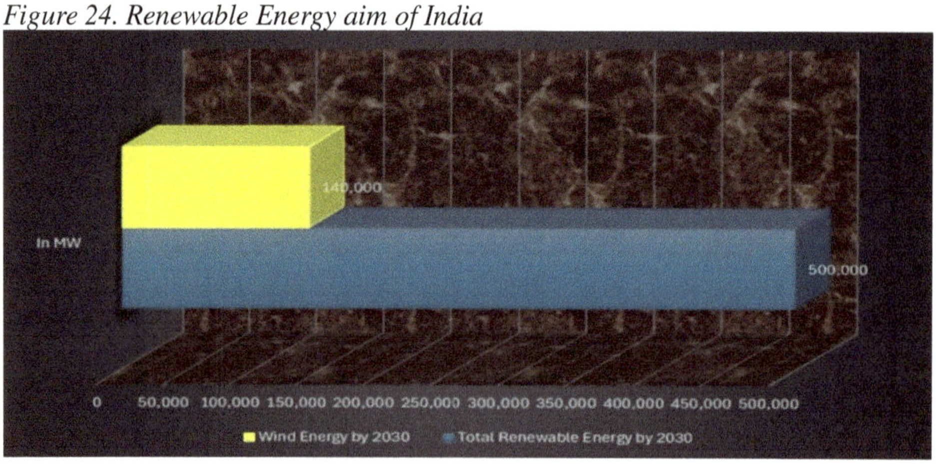

Source: IEA Wind TCP

Figure 25. Offshore Wind Energy Target of India

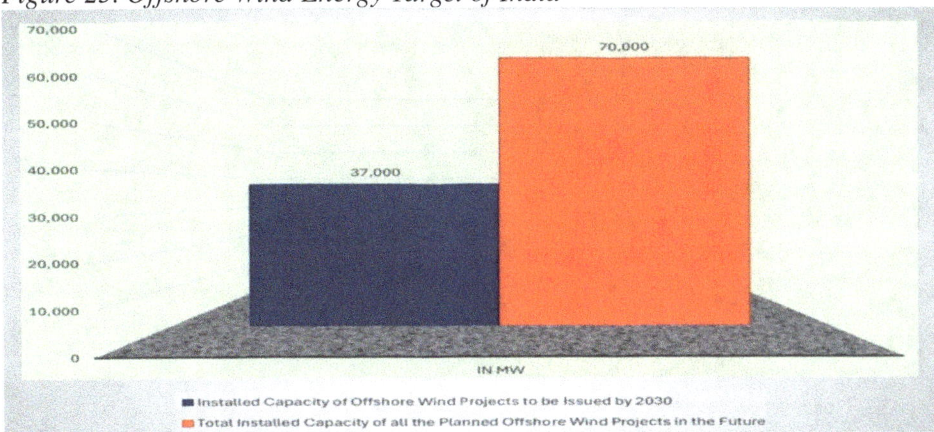

Source: EQ Magazine and ET Energy World

Figure 26. Employment generation by type of renewable energy in India

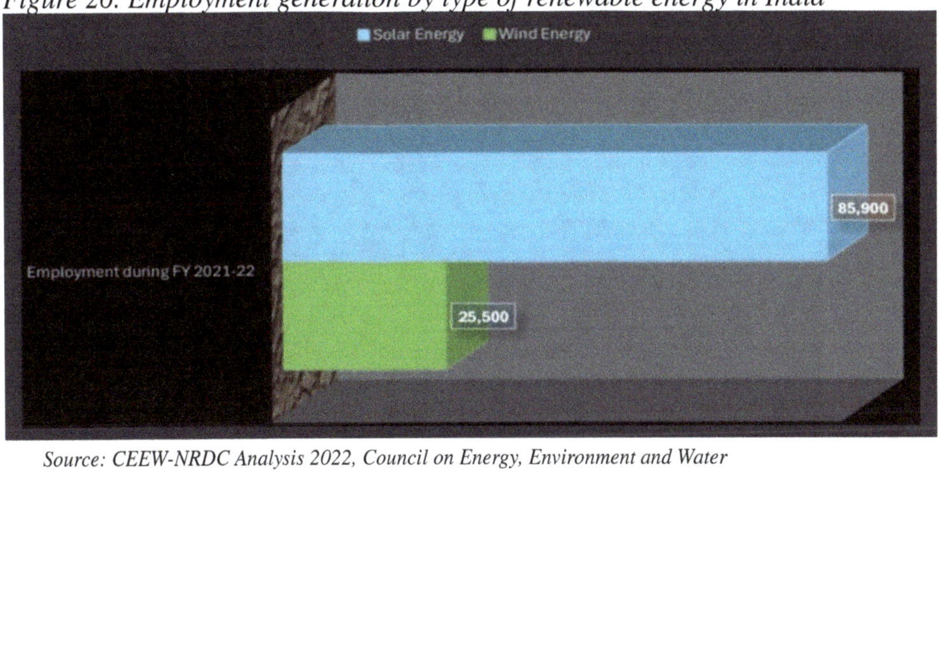

Source: CEEW-NRDC Analysis 2022, Council on Energy, Environment and Water

Figure 27. Installed capacity of wind energy of India

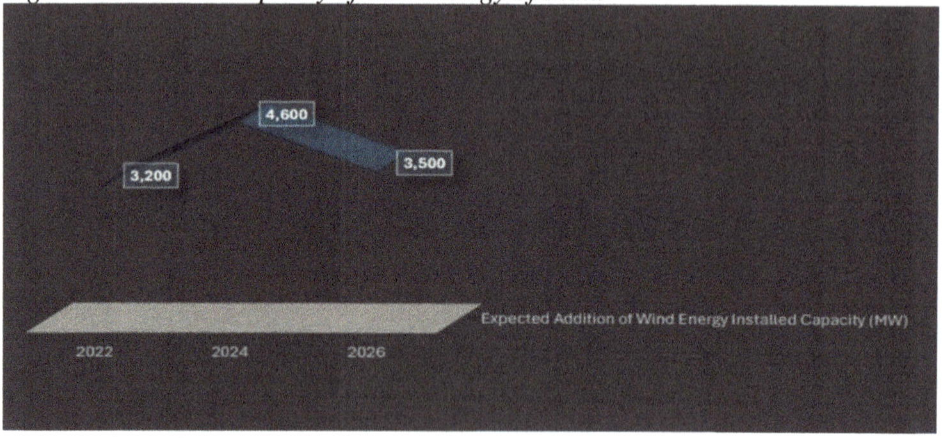

Source: Press Information Bureau

Figure 28. Installed capacity of Onshore wind energy globally

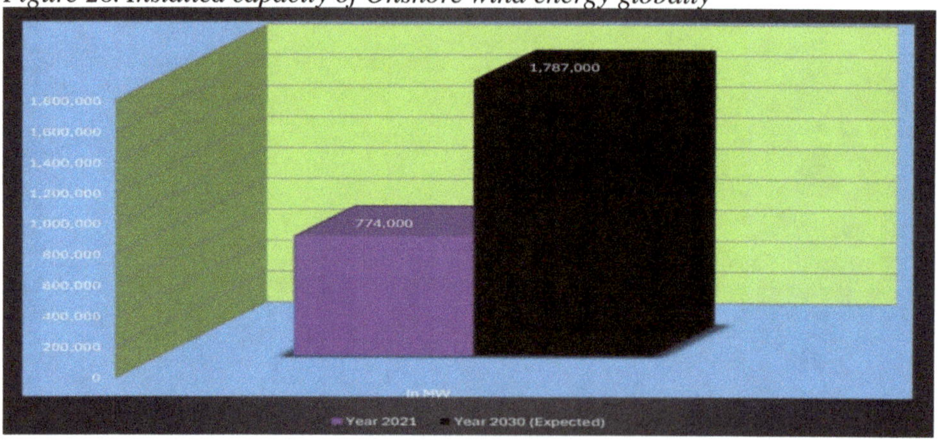

Source: Global Wind Energy Council, Global Wind Report 2023, Iberdola, and Renewable Capacity Statistics (IRENA, 2019d) IRENA analysis (IRENA 2019a), IRENA

Figure 29. New Wind Energy Installed Capacity of Major Countries in 2023

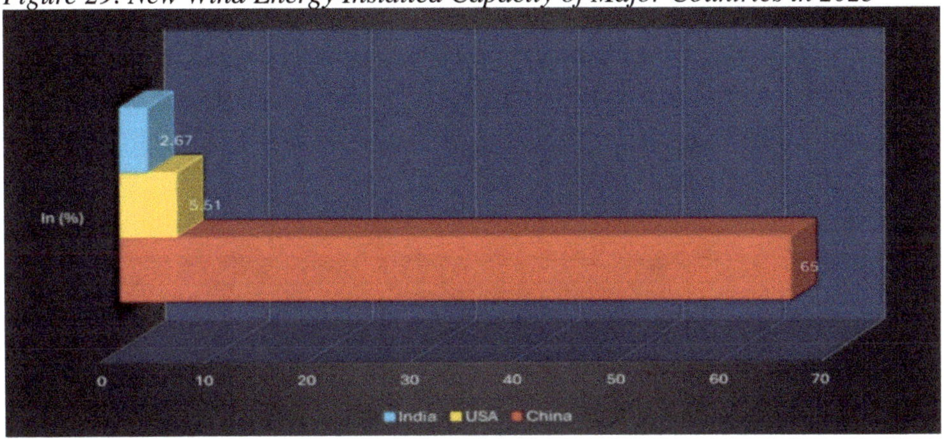

Source: Wood Mackenzie, and World Wind Energy Association

Figure 30. Global novel installation of wind energy

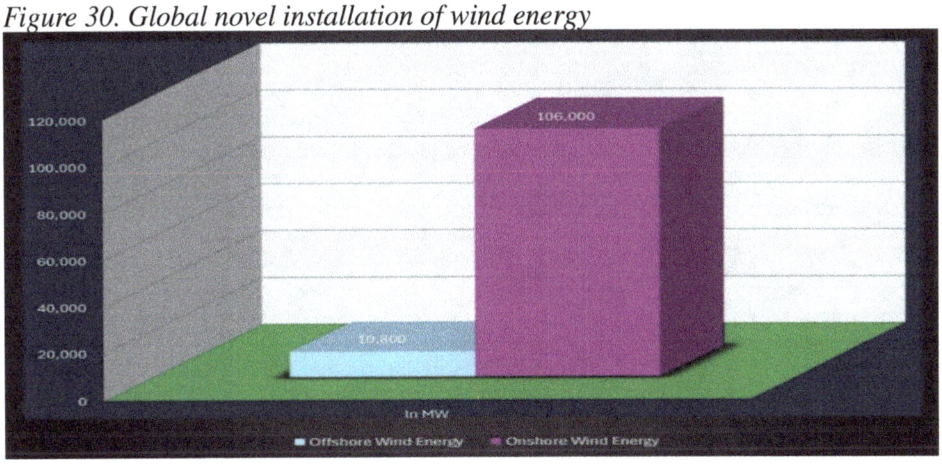

Source: Mercom Clean Energy Insights

Figure 31. Share of Indian states in the cumulative installation of wind power

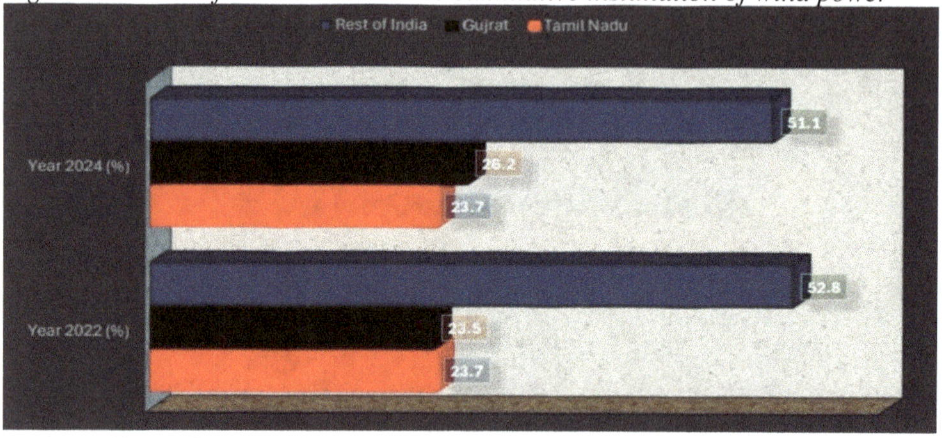

Source: Mercom Clean Energy Insights, and Ministry of New and Renewable Energy

Figure 32. Onshore wind energy capacity by country

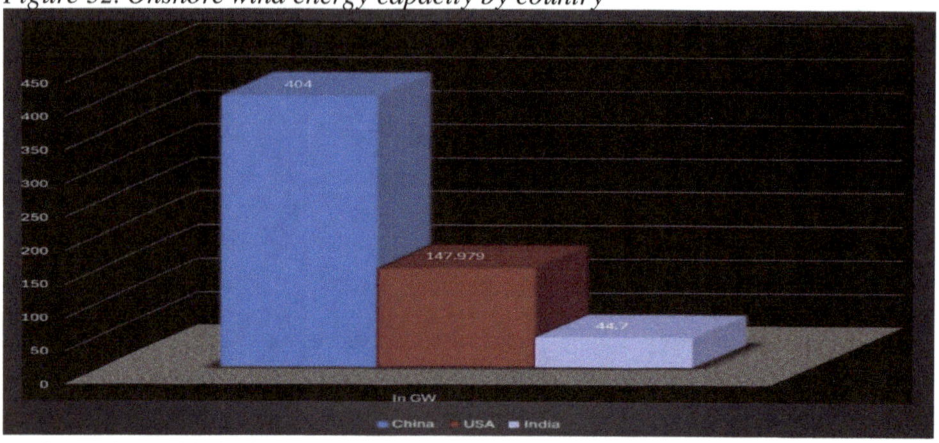

Source: Statista, Mordor Intelligence, Mercom Clean Energy Insights, and Swiss Re Institute

Figure 33. Maximum operational Offshore wind farms by countries

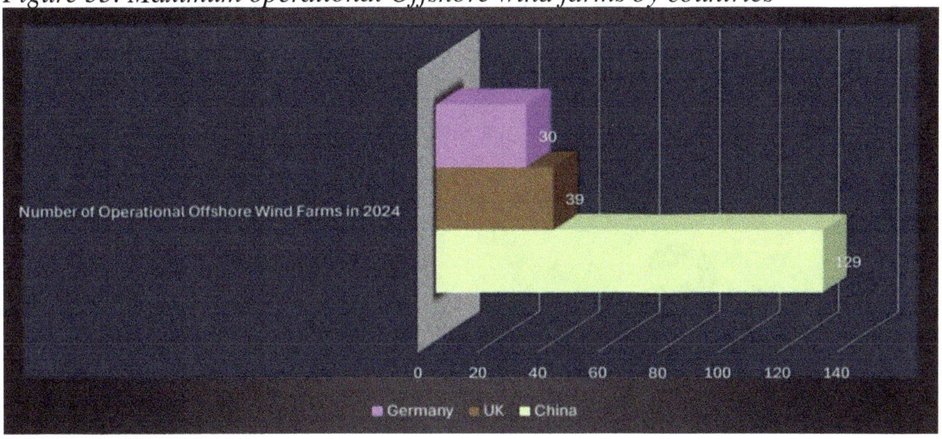

Source: Statista

Figure 34. Potential of India's renewable energy

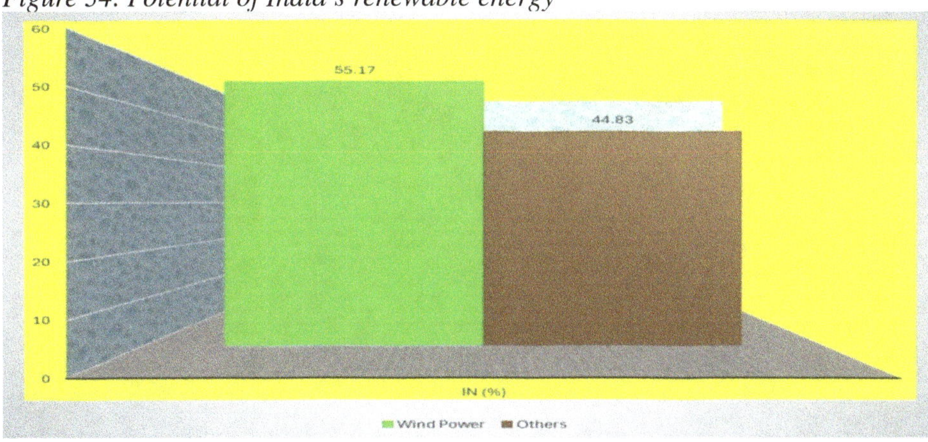

Source: Energy Statistics India 2025 report – Ministry of Statistics and Programme Implementation, ET Energy World

Figure 35. State wise potential of wind energy in India

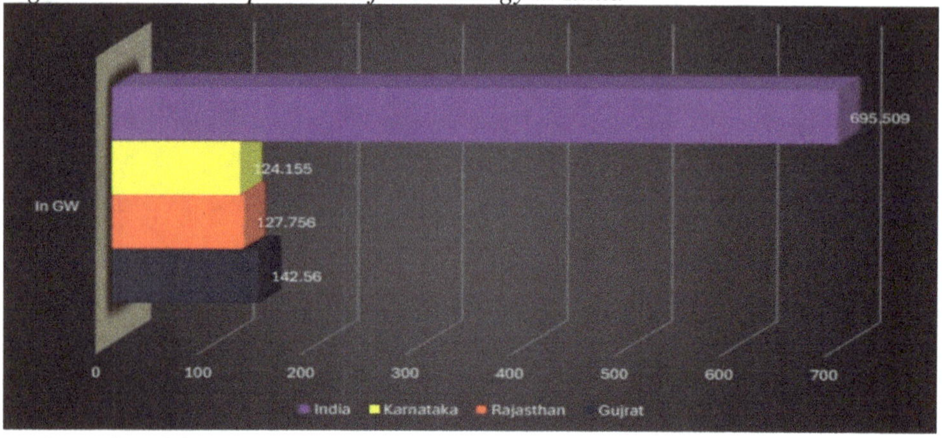

Source: Press Information Bureau, Ministry of New and Renewable Energy, and India Climate & Energy Dashboard, NITI Aayog

Figure 36. Global electricity capacity of wind energy

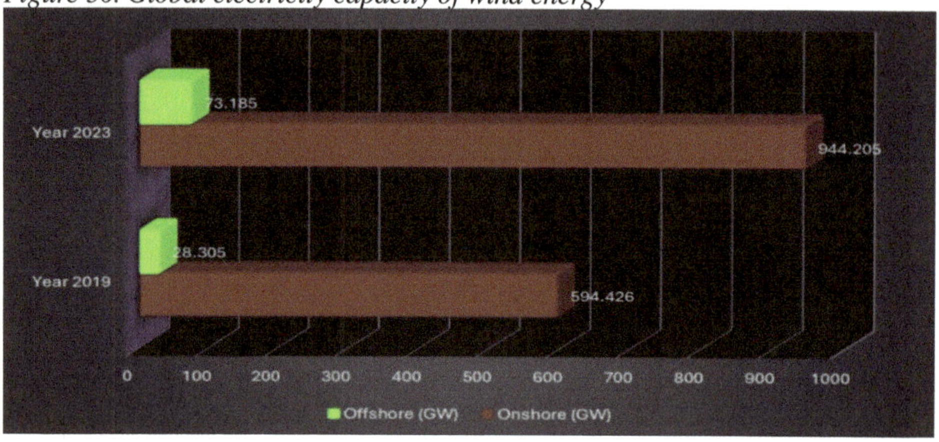

Source: International Renewable Energy Agency

Figure 37. Global new installed capacity of wind energy

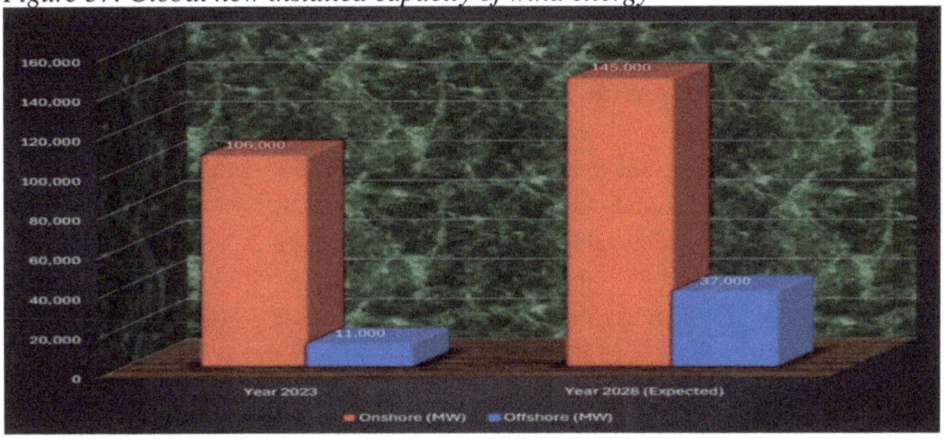

Source: Statista

Figure 38. New installed capacity of wind energy by region

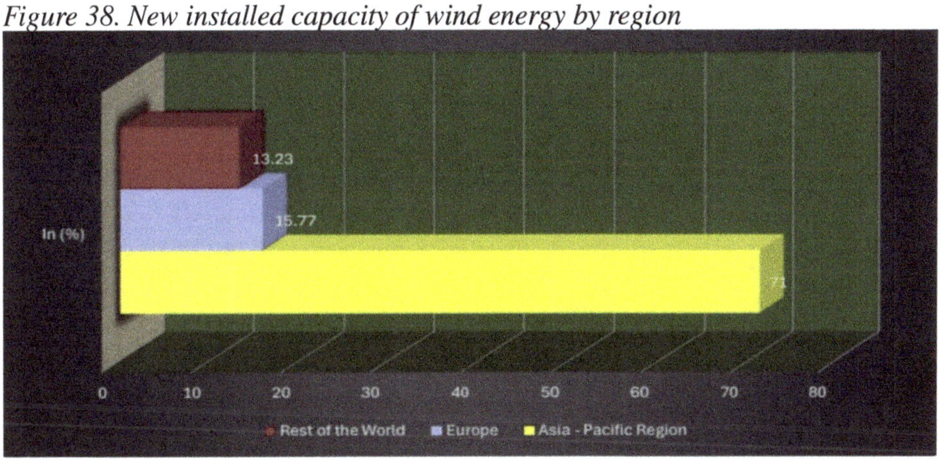

Source: Global Wind Energy Council, Asian Power, WindEurope, and World Wind Energy Association

Major Advantages and Challenges of Wind Farm Projects

In this subsection, based on the above graphs, the challenges in addition to advantages of wind farms is being analysed:

- According to Fig. 23, there was around 6.59% increase in the total installed capacity of wind power of India, but the expected target for 2030 is slightly more than thrice the total installed capacity achieved till 2024 by India, means that the country needs to increase its rate of installation capacity at a faster rate to achieve its 2030 target. Plus, this shows positive trend and India's inclination towards wind energy as well as development of wind farm projects.

- Based on Fig. 24, India's target for the year 2030, 28% of the total renewable energy will corresponds to the wind energy, meaning the wind farm projects will play an important role in achieving the lofty target.
- According to Fig. 25, India is planning to issue 52.86% of the planned Offshore wind farm installed capacity by the year 2030. This shows that in the coming 6 years, a number of big Offshore projects will be planned and focused.
- Based on Fig. 26, the employment offered by the wind farms is 3.37 times less than solar projects in the financial year 2021-22, which is a big challenge for their development, as it is unable to offer much employment to the people.
- According to Fig. 27, the installed wind capacity every year will show maximum increase till 2024 and then the rate of increase will decrease afterwards, means their growth will be somewhat slow.
- Based on Fig. 28, the global installed capacity of Onshore wind energy is expected to increase by 2.309 times in the span of 9 years, which also shows their scope and demand in the world.
- According to Fig. 29, the new addition in the installed capacity of wind energy in 2023 was just 2.67% in comparison to China and USA having 65% as well as 5.51% respectively, meaning China's growth last year was at a different level and of course Germany's and Brazil's progress last year was also better than India.
- Based on Fig. 30, the new installed capacity of Onshore projects globally was 9.81 times more than Offshore projects, might be due to the technological and monetary hurdles on the path of Offshore projects.
- According to Fig. 31, Tamil Nadu and Gujrat were the leading wind energy states in India in 2022 but the latter surpassed the former with a decent margin by 2024, whereas other states are far behind them.
- According to Fig. 32, China is leading the world of installed capacity of Onshore projects with a big margin, followed by USA, India has performed well, but it has a long way to go compared to the world's top 2 countries.
- Based on Fig. 33, China is also the world's leader in the list of most operational Offshore projects, followed by UK, whereas India has yet to debut.
- According to Fig. 34, the potential of wind power in India in 2024 was measured as 55.17% of the total renewable energy which is a major portion that can't be ignored to harness the full potential of its renewable energy.

- Based on Fig. 35, the western states of India like Gujrat and Rajasthan are having the highest potential of wind power as per the las year's assessment (2023), which is followed by the southern state Karnataka.
- According to Fig. 36, the addition of electricity capacity of wind energy globally in the year 2019 was 594.426 GW for Onshore and 28.305 GW for Offshore Wind Farm projects, which is expected to increase by 1.589 times for the Onshore Wind Farm projects as well as around 2.586 times for the Offshore Wind Farm projects in the year 2023, in comparison to 2019. This shows significant rise in the growth of electricity capacity of wind energy worldwide, which will be helpful in the direction of sustainable development.
- Based on Fig. 37, there is an expected increase in the installed capacity of both Onshore and Offshore wind energy over the coming years, which shows their increasing demand in the future in the world, but the latter is lagging behind too much compared to the former maybe mostly because of technological and financial obstacles, despite having a good scope.
- As per Fig. 38, Asia – Pacific Region last year dominated the world one-sidedly in the installation capacity of wind power, mostly due to China and other major countries including India, which was followed by Europe. This shows the increasing demand of wind power in Asia Pacific region and a positive trend for the future.

LIMITATIONS

This book chapter covered a number of novel wind farm projects which either recently got completed or still under construction in India, but apart from these mentioned projects, there are many more wind farms as well as associated projects in India which were neither mentioned nor discussed. Then, there are wind farm projects in other countries in the world, which were also ignored in the chapter. Additionally, this chapter didn't particularly focus one of the mentioned wind farm projects in much detail but instead it covered a number of them in brief.

SOLUTIONS AND RECOMMENDATIONS

The wind farm projects come with a number of problems and challenges. Of course, technical challenges can be solved the technological advancement in the country. However, investors are more inclined towards solar farms in comparison to the wind farms due to favourable policies and monetary benefits. The Government of India already made the supporting framework and policies to support wind farms but some additional incentives or some modification in existing policies to attract more investors towards wind farm projects is essential. Then, wind farms could only serve as intermittent source of energy, which needs to be complemented by the solar farms which India had been already implementing in some of the places in the country to make it complete green energy source in those places, but in many places in the country that is not the case yet. Therefore, more such hybrid projects are required in the country in the future. Furthermore, the country has yet to achieve an offshore wind farm, which is having technical constraint and to achieve the dream of using its long coastal area for the development of offshore wind farms in the future, financial constraints also need to be solved. Then, collaboration with advanced countries who are leading the world in offshore wind energy is really essential to progress with their development at a faster rate in India.

FUTURE RESEARCH DIRECTIONS

The book chapter discussed wind farms and talked about a number of new wind farm projects which are either ongoing or just recently got completed. However, in the future the research could be done in the finding out the financial viability of the discussed wind farm projects in detail. Then, research could be done on other wind farm projects which were not covered in the chapter. Similarly, wind farm projects in other countries in the recent years could be researched in detail in the future. Further, research needs to be done for Offshore wind farms in India and to advance the current technology in India, in order to successfully build offshore wind farms in India.

CONCLUSION

The chapter explained wind farms in general and discussed the evolution of wind farms over the century especially over the decades. It also talked about the various types of possible wind farms and some of the leading countries in the European continent as well as the condition in India. Moreover, how India felt the need to utilize wind energy and how over the decades the country reached its current development in the wind power sector in the world was covered in this chapter. Then, the policies in addition to the framework which the Government of India built to progress in the wind power sector in the world and what are the countries current as well as future plans in this sector was mentioned. The book chapter also covered the gaps in the current literature and discussed a number of novel wind farm projects along with the challenges faced in their development. Additionally, this book chapter graphically analysed the information from different sources which are available to understand the market of wind farms and to understand the future of wind farm projects particularly in the sustainable development of the country India.

ACKNOWLEDGEMENT

This research received no specific grant from any funding agency in the public, commercial, or not-for-profit sectors.

REFERENCES

Amuthan, N., M, M. B., Velrajkumar, P., Sivakumar, N., & Jarin, T. (2023). IOT based adjustment mechanism for direct reference model adaptive IMC to support voltage sag in DFIG wind farm. *Measurement. Sensors*, *27*, 100809. DOI: 10.1016/j.measen.2023.100809

Asamoah, S. S., Parbey, J., Yankey, I. K., & Awuah, A. (2023). Techno-economic assessment of a central grid-connected wind farm in Ghana using RETScreen® Expert. *Heliyon*, *9*(1), e12902. DOI: 10.1016/j.heliyon.2023.e12902 PMID: 36685476

Asian Power, https://asian-power.com/videos/asia-pacific-dominates-global-wind-market-71-share

Basak, D., Nagababu, G., Puppala, H., Patel, J., & Kumar, S. V. V. A. K. (2023). Foreseeing the spatio-temporal offshore wind energy potential of India using a differential weighted ensemble created using CMIP6 datasets. *Regional Studies in Marine Science*, *65*, 103066. DOI: 10.1016/j.rsma.2023.103066

Bhattacharjee, P., Jana, R. K., & Bhattacharya, S. (2022). Boosting the annual profit of an offshore wind farm in India with a bio-inspired meta-heuristic scheme. *Environmental Challenges*, *9*, 100642. DOI: 10.1016/j.envc.2022.100642

Bhuj 1 Project India. (n.d.). White-Paper-Poject-Wind-India_2.pdf

Bi, C., & Law, A. W.-K. (2023). Co-locating offshore wind and floating solar farms – Effect of high wind and wave conditions on solar power performance. *Energy*, *266*, 126437. DOI: 10.1016/j.energy.2022.126437

Biswas, H. S., Kundu, A. K., & Poddar, S. (2024). Releasing the Power of Nature: Exploration of Sustainable Energy for a Flourishing Future. In Yıldırım, S., Yıldırım, D., Demirtaş, I., & Kandpal, V. (Eds.), *Green Transition Impacts on the Economy, Society, and Environment* (pp. 108–124). IGI Global Scientific Publishing., DOI: 10.4018/979-8-3693-3985-5.ch007

Bonthu, S., Purvaja, R., Singh, K. S., Ganguly, D., Muruganandam, R., Paul, T., & Ramesh, R. (2024). Offshore wind energy potential along the Indian Coast considering ecological safeguards. *Ocean and Coastal Management*, *249*, 107017. DOI: 10.1016/j.ocecoaman.2024.107017

CEEW. (2022). https://www.ceew.in/publications/assessing-offshore-wind-power-development-for-indias-power-system-in-2030

Council on Energy, Environment and Water. (2022). https://www.ceew.in/publications/indias-expanding-clean-energy-workforce

Daoudi, M., Mou, A. A. S., Idrissi, A., Ihoume, I., Arbaoui, N., & Benchrifa, M. (2023). New approach to prioritize wind farm sites by data envelopment analysis method: A case study. *Ocean Engineering*, *271*, 113820. DOI: 10.1016/j.oceaneng.2023.113820

Daoudi, M., Mou, A. A. S., & Naceur, L. A. (2022). Analysis of the first onshore wind farm installation near the Morocco-United Kingdom green energy export project. *Scientific African*, *17*, e01388. DOI: 10.1016/j.sciaf.2022.e01388

de Vasconcelos, R. M., Silva, L. L. C., Gonzalez, M. O. A., Santiso, A. M., & de Melo, D. C. (2022). Environmental licensing for offshore wind farms: Guidelines and policy implications for new markets. *Energy Policy*, *171*, 113248. DOI: 10.1016/j.enpol.2022.113248

Edwards-Jones, A., Watson, S. C. L., Szostek, C. L., & Beaumont, N. J. (2024). Stakeholder insights into embedding marine net gain for offshore wind farm planning and delivery. *Environmental Challenges*, *14*, 100814. DOI: 10.1016/j.envc.2023.100814

El Jaadi, M., Haidi, T., Belfqih, A., Farah, M., & Dailmy, A. (2024). Optimizing wind farm layout for enhanced electricity extraction using a new hybrid PSO-ANN method. *Global Energy Interconnection*, *7*(3), 254–269. DOI: 10.1016/j.gloei.2024.06.006

EQ Magazine. (n.d.) https://www.eqmagpro.com/india-plans-37gw-offshore-wind-energy-power-by-2030-to-issue-bids-eq-mag/#:~:text=Their%20estimate%2C%20made%20at%20a%20Group%20of,and%20Gujarat%20have%20agreed%20to%20purchase%20power

ET Energy World. (n.d.) https://energy.economictimes.indiatimes.com/news/renewable/india-identifies-offshore-wind-energy-potential-of-70000-mw-along-gujarat-tn-coasts/73072580

ET Energy World. (n.d.) https://energy.economictimes.indiatimes.com/news/renewable/indias-renewable-power-potential-crosses-2-1-million-mw-wind-energy-leads-with-55-share/119837495

Gyatso, N., Li, Y., Gao, Z., Wang, Q., Li, S., Yin, Q., Chen, J., Jin, P., Liu, Z., Ma, Z., Chen, X., Feng, J., & Dorje, . (2023). Wind power performance assessment at high plateau region: A case study of the wind farm field test on the Qinghai-Tibet plateau. *Applied Energy*, *336*, 120789. DOI: 10.1016/j.apenergy.2023.120789

Hong, S., McMorland, J., Zhang, H., Collu, M., & Halse, K. H. (2024). Floating offshore wind farm installation, challenges and opportunities: A comprehensive survey. *Ocean Engineering*, *304*, 117793. DOI: 10.1016/j.oceaneng.2024.117793

Huang, X., Wang, C., & Zhang, S. (2024). Research and application of a Model selection forecasting system for wind speed and theoretical power generation in wind farms based on classification and wind conversion. *Energy*, *293*, 130606. DOI: 10.1016/j.energy.2024.130606

Hussain, S., Sharma, S. K., & Lal, S. (2024). Feasible synergy between hybrid solar PV and wind system for energy supply of a green building in Kota (India): A case study using iHOGA. *Energy Conversion and Management*, *315*, 118783. DOI: 10.1016/j.enconman.2024.118783

Iberdola. (2023). https://www.iberdrola.com/sustainability/renewables-energy-wind-power

IEA Wind TCP. (n.d.). https://iea-wind.org/about-iea-wind-tcp/members/india/

India Climate & Energy Dashboard, NITI Aayog. (n.d.) https://iced.niti.gov.in/energy/fuel-sources/wind/potential

International Renewable Energy Agency. (n.d.) https://www.irena.org/Energy-Transition/Technology/Wind-energy

IRENA. (2019). https://www.irena.org/-/media/Files/IRENA/Agency/Publication/2019/Oct/IRENA_Future_of_wind_2019.pdf#:~:text=This%20entails%20increasing%20the%20global%20cumulative%20installed,to%20installed%20capacity%20in%202018%20(542%20GW)

Islek, F., & Yuksel, Y. (2022). Evaluation of future wind power potential and their projected changes in the Black Sea and possible stable locations for wind farms. *Ocean Engineering*, *266*(Part 2), 112832. DOI: 10.1016/j.oceaneng.2022.112832

Jalili, S., Maheri, A., Ivanovic, A., Neilson, R., Bentin, M., Kotzur, S., May, R., & Sunner, I. (2024). Economic and environmental assessments to support the decision-making process in the offshore wind farm decommissioning projects. *Renewable and Sustainable Energy Reviews*, Volume 190. *Renewable & Sustainable Energy Reviews*, *114080*, 114080. Advance online publication. DOI: 10.1016/j.rser.2023.114080

JSW Energy Limited. (n.d.). LettertoBSEandNSE51MWRenewEnergysigned-261223.pdf

Kolli, S., Dammala, P. K., Bhattacharya, S., Fan, C., Wang, T., & Cui, L. (2023). Offshore wind farms as additional coolant power sources to enhance seismic resilience of nuclear power plants– A case study. *Nuclear Engineering and Design*, *405*, 112204. DOI: 10.1016/j.nucengdes.2023.112204

le Maitre, J., Ryan, G., & Power, B. (2024). Do concerns about wind farms blow over with time? Residents' acceptance over phases of project development and proximity. *Renewable and Sustainable Energy Reviews*, Volume 189. *Renewable & Sustainable Energy Reviews*, *113839*, 113839. Advance online publication. DOI: 10.1016/j.rser.2023.113839

Lopez, G., Arboleya, P., Nunez, D., Freire, A., & Lopez, D. (2023). Wind resource assessment and influence of atmospheric stability on wind farm design using Computational Fluid Dynamics in the Andes Mountains, Ecuador. *Energy Conversion and Management*, *284*, 116972. DOI: 10.1016/j.enconman.2023.116972

Magesh, T., Devi, G., & Lakshmanan, T. (2022). Measurement and simulation of power quality issues in grid connected wind farms. *Electric Power Systems Research*, *210*, 108142. DOI: 10.1016/j.epsr.2022.108142

Mahata, S., Harsh, P., & Shekher, V. (2024). Comparative study of time-series forecasting models for wind power generation in Gujarat, India. *e-Prime - Advances in Electrical Engineering. Electronics and Energy*, *8*, 100511. DOI: 10.1016/j.prime.2024.100511

Mercom Clean Energy Insights. (n.d.). https://www.mercomindia.com/global-offshore-wind-capacity-rises

Mercom Clean Energy Insights. (n.d.). https://www.mercomindia.com/global-wind-installations-2023

Mercom Clean Energy Insights. (n.d). https://www.mercomindia.com/wind-installations-india-grew-129-with-878-mw-added-in-q3

Mercom Clean Energy Insights. (n.d.). https://www.mercomindia.com/india-growth-wind-installations-2023

Mercom Clean Energy Insights. (n.d.) https://www.mercomindia.com/india-added-wind-capacity-yoy-increase

Mohamed, E., Seresht, N. G., & Rizk, S. A. (2023). Context-driven ontology-based risk identification for onshore wind farm projects: A domain-specific approach. *Advanced Engineering Informatics*, *56*, 101962. DOI: 10.1016/j.aei.2023.101962

Mordor Intelligence, https://www.mordorintelligence.com/industry-reports/north-america-wind-power-market

Nagababu, G., Puppala, H., Pritam, K., & Kantipudi, M. V. V. P. (2022). Two-stage GIS-MCDM based algorithm to identify plausible regions at micro level to install wind farms: A case study of India. *Energy*, *248*, 123594. DOI: 10.1016/j.energy.2022.123594

Ngoenmeesri, R., Chidaruksa, S., Wangkeeree, R., & Sirisamphanwong, C. (2023). Power quality enhancement for Thailand's wind farm using 5 MWh Li-ion battery energy storage system. *Heliyon*, *9*(11), e22029. DOI: 10.1016/j.heliyon.2023.e22029 PMID: 38034671

Pessanha, J. F. M., Almeida, V. A., Melo, A. C. G., Maceira, M. E. P., Mendonca, M. J. C., & Medrano, L. A. (2024). A Bayesian approach to estimate the capacity factor of offshore wind farms on the Brazilian coast. *Electric Power Systems Research*, *234*, 110772. DOI: 10.1016/j.epsr.2024.110772

Prasad, K. M. V. V., Nagababu, G., & Jani, H. K. (2023). Enhancing offshore wind resource assessment with LIDAR-validated reanalysis datasets: A case study in Gujarat, India. *International Journal of Thermofluids*, *18*, 100320. DOI: 10.1016/j.ijft.2023.100320

Press Information Bureau. (n.d.) https://pib.gov.in/PressReleaseIframePage.aspx?PRID=1944759

Press Information Bureau. (n.d.)https://pib.gov.in/PressReleaseIframePage.aspx?PRID=1947133

Proceedings of the Government of Karnataka. (n.d.). 0_0_51113123412151GO600M-WJSWRenewableEnergy(Vijayanagar)Ltd.pdf

Proceedings of the Government of Karnataka. (n.d.) JSW Renewable Energy ltd.pdf

Proceedings of the Government of Karnataka. (n.d.) JSWRenewEnergyFourLtd_2.pdf

Proceedings of the Government of Karnataka. (n.d.) JSWRenewEnergyFourLtd.pdf

Raju, M. S. S., Mohapatra, P., Dutta, S., Mallipeddi, R., & Das, K. N. (2024). Optimal placement of fixed hub height wind turbines in a wind farm using twin archive guided decomposition based multi-objective evolutionary algorithm. *Engineering Applications of Artificial Intelligence*, *130*, 107735. DOI: 10.1016/j.engappai.2023.107735

Renewable Energy World, https://www.renewableenergyworld.com/wind-power/2023-was-a-record-year-for-wind-installations-as-world-ramps-up-clean-energy-report-says/

Roga, S., Dahiwale, H., Bardhan, S., & Sinha, S. (2024). Wind energy potential assessment: a case study in Central India. *Proceedings of the Institution of Civil Engineers – Energy*, Volume 177, Issue 3, Pages 130-148. DOI: 10.1680/jener.22.00016

Sakuru, S. K. V. S., & Ramana, M. V. (2023). Wind power potential over India using the ERA5 reanalysis. *Sustainable Energy Technologies and Assessments*, *56*, 103038. DOI: 10.1016/j.seta.2023.103038

Saraswat, S. K., Digalwar, A. K., Yadav, S. S., & Kumar, G. (2021). MCDM and GIS based modelling technique for assessment of solar and wind farm locations in India. *Renewable Energy*, *169*, 865–884. DOI: 10.1016/j.renene.2021.01.056

Singh, A. (2023). The Current Scenario of Sustainable Planning and Development of the Indian Built Environment. In Cobîrzan, N., Muntean, R., & Felseghi, R. (Eds.), *Circular Economy Implementation for Sustainability in the Built Environment* (pp. 106–138). IGI Global., DOI: 10.4018/978-1-6684-8238-4.ch005

Singh, A. (2023). The Current Trends and Need of Infrastructure Projects in the Country India. In Popescu, C., Yu, P., & Wei, Y. (Eds.), *Achieving the Sustainable Development Goals Through Infrastructure Development* (pp. 235–261). IGI Global., DOI: 10.4018/979-8-3693-0794-6.ch010

Singh, A. (2023). The Development and Significance of Bengaluru Suburban Rail Project. In Yilmaz, Z., Golem, S., & Costescu, D. (Eds.), *Handbook of Research on Promoting Sustainable Public Transportation Strategies in Urban Environments* (pp. 267–292). IGI Global., DOI: 10.4018/978-1-6684-5996-6.ch013

Singh, A. (2023). The Need and Development of Dams in India as Well as in the World for a Better Future. In Lousada, S. (Ed.), *Geoinformatics in Support of Urban Politics and the Development of Civil Engineering* (pp. 156–180). IGI Global., DOI: 10.4018/978-1-6684-6449-6.ch007

Singh, A. (2023). The Present and Future Role of Artificial Intelligence in the Promotion of Sustainable Practices in the Construction Sector. In Kahraman, C., Sari, I. U., Oztaysi, B., Cebi, S., Cevik Onar, S., & Tolga, A. Ç. (Eds.), *Intelligent and Fuzzy Systems. INFUS 2023. Lecture Notes in Networks and Systems* (Vol. 759). Springer., DOI: 10.1007/978-3-031-39777-6_2

Siram, O., Sahoo, N., & Saha, U. K. (2022). Changing landscape of India's renewable energy and the contribution of wind energy. *Cleaner Engineering and Technology, 8*, 100506. DOI: 10.1016/j.clet.2022.100506

SolarQuarter. (n.d.). https://solarquarter.com/2024/07/11/indias-solar-power-revolution-leading-with-85474-31-mw-and-57-72-of-renewable-energy-by-june-2024/#:~:text=Wind%20power%20has%20also%20seen%20significant%20growth,embrace%20the%20vast%20potential%20of%20solar%20energy

Somi, S., Seresht, N. G., & Fayek, A. R. (2021). Developing a risk breakdown matrix for onshore wind farm projects using fuzzy case-based reasoning. *Journal of Cleaner Production, 311*, 127572. DOI: 10.1016/j.jclepro.2021.127572

Statista. (n.d.). https://www.statista.com/statistics/963139/china-capacity-of-onshore-wind-energy/

Statista. (n.d.). https://www.statista.com/statistics/865745/india-wind-energy-capacity/#:~:text=The%20wind%20energy%20capacity%20in%20the%20south,India%20from%202009%20to%202023%20(in%20megawatts)&text=Figures%20are%20compiled%20from%20several%20editions%20of%20this%20report

Statista. (n.d.). https://www.statista.com/statistics/264257/number-of-offshore-wind-farms-worldwide-by-country/

Statista. (n.d.). https://www.statista.com/statistics/1368205/global-wind-capacity-additions-forecast-by-type/

Swiss Re Institute, https://www.swissre.com/institute/research/topics-and-risk-dialogues/climate-and-natural-catastrophe-risk/climate-change-wind-power.html

Vijay, S., Swaminathan, G. V., & Periasamy, S. (2024). Analysis and mitigation of subsynchronous resonance in VSC-HVDC-based offshore wind farms. In Chenniappan, S., Padmanaban, S., & Palanisamy, S. (Eds.), *Power Systems Operation with 100% Renewable Energy Sources* (pp. 149–164)., DOI: 10.1016/B978-0-443-15578-9.00010-8

von Krauland, A.-K., & Jacobson, M. Z. (2024). India onshore wind energy atlas accounting for altitude and land use restrictions and co-located solar. *Cell Reports Sustainability, 1*(5), 100083. DOI: 10.1016/j.crsus.2024.100083

Watson, S. C. L., Somerfield, P. J., Lemasson, A. J., Knights, A. M., Edwards-Jones, A., Nunes, J., Pascoe, C., McNeill, C. L., Schratzberger, M., Thompson, M. S. A., Couce, E., Szostek, C. L., Baxter, H., & Beaumont, N. J. (2024). The global impact of offshore wind farms on ecosystem services. *Ocean and Coastal Management, 249*, 107023. DOI: 10.1016/j.ocecoaman.2024.107023

Wikimedia. (n.d.). https://commons.wikimedia.org/wiki/File:Wind_farm_(1828952868).jpg

Wikimedia. (n.d). https://commons.wikimedia.org/wiki/File:Wind_Turbines_and_Power_Lines,_East_Sussex,_England_-_April_2009.jpg

Wikimedia. (n.d.). https://commons.wikimedia.org/wiki/File:Skidby_Mill_in_Winter_-_geograph.org.uk_-_233142.jpg

Wikimedia (n.d.). https://commons.wikimedia.org/wiki/File:DK-84_-_K%C3%B8benhavn_-_Energy_-_Copenhagen_-_Denmark_-_Wind_Farm_(4890897324).jpg

Wikimedia. (n.d.).https://commons.wikimedia.org/wiki/File:Biscay_Spain_wind_farm_05_2012_1756.jpg

Wikimedia, (n.d.). https://commons.wikimedia.org/wiki/File:Middelgrunden_wind_farm_2009-07-01_edit_filtered.jpg

Wikimedia. (n.d.).https://commons.wikimedia.org/wiki/File:Wind_energy_turbines_in_M%C3%B6lsheim,_Germany.jpg

Wikimedia. (n.d.).https://commons.wikimedia.org/wiki/File:Monte_de_A_Ru%C3%B1a._Foto_-_Alexandre_Varela.jpg

Wikimedia. (n.d.).https://commons.wikimedia.org/wiki/File:Skidby_Mill_1950_(archive_ref_DDX272-11)_(27968260583).jpg

Wikimedial (n.d.). https://commons.wikimedia.org/wiki/File:Wind_power_in_south_india_-_panoramio.jpg

Wikimedia. (n.d.).https://commons.wikimedia.org/wiki/File:Wind_power_in_Gothenburg_3.JPG

Wikimedia. (n.d.). https://commons.wikimedia.org/wiki/File:A_Wind_Energy_farm_and_the_Cenotaphs,_the_ancient_and_the_modern,_Jaisalmer_Rajasthan_India.jpg

Wikimedia. (n.d.).https://commons.wikimedia.org/wiki/File:Kachch.png

Wimhurst, J. J., Greene, J. S., & Koch, J. (2023). Predicting commercial wind farm site suitability in the conterminous United States using a logistic regression model. *Applied Energy*, *352*, 121880. DOI: 10.1016/j.apenergy.2023.121880

Wimhurst, J. J., Nsude, C. C., & Greene, J. S. (2023). Standardizing the factors used in wind farm site suitability models: A review. *Heliyon*, *9*(5), e15903. DOI: 10.1016/j.heliyon.2023.e15903 PMID: 37168883

Wood Mackenzie, https://www.woodmac.com/press-releases/2024-press-releases/global-wind-oem-marketshare/#:~:text=China%20accounted%20for%2065%%20of%20global%20wind,global%20rankings%2C%20a%20first%20for%20the%20sector.&text=The%20country%27s%20record%20installations%2C%20backed%20by%20more,not%20shielded%20local%20OEMs%20from%20reducing%20profitability.
World Wind Energy Association, https://wwindea.org/AnnualReport2023

Yadav, B. K., Yadav, R., Jahangiri, M., Priya, S. S., Bajracharya, T. R., & Sudhakar, K. (2024). Decarbonizing airport using solar and wind farm: A case of Biratnagar, Nepal. *e-Prime - Advances in Electrical Engineering. Electronics and Energy*, *8*, 100583. DOI: 10.1016/j.prime.2024.100583

Yildiz, S. S. (2024). Spatial multi-criteria decision making approach for wind farm site selection: A case study in Balıkesir, Turkey. *Renewable & Sustainable Energy Reviews*, *192*, 114158. DOI: 10.1016/j.rser.2023.114158

Yousefi, H., Moradi, S., Zahedi, R., & Ranjbar, Z. (2024). Developed analytic hierarchy process and multi criteria decision support system for wind farm site selection using GIS: A regional-scale application with environmental responsibility. *Energy Conversion and Management: X*, Volume 22, 100594. DOI: 10.1016/j.ecmx.2024.100594

Ziemba, P. (2022). Uncertain Multi-Criteria analysis of offshore wind farms projects investments – Case study of the Polish Economic Zone of the Baltic Sea. *Applied Energy*, *309*, 118232. DOI: 10.1016/j.apenergy.2021.118232

Chapter 8
Reliability, Market Dynamics, and Energy Trading With AI in Wind Energy

Francis Mwasilu
http://orcid.org/0000-0001-7540-4198
University of Dar es Salaam, Tanzania

Jackson Justo
http://orcid.org/0000-0003-0638-0713
ITMO University, Russia

Ramesh C. Bansal
University of Sharjah, UAE

ABSTRACT

This chapter explores artificial intelligence's transformative impact on wind energy, addressing intermittency challenges to enhance reliability, market integration, and trading efficiency. As wind power expands to meet decarbonization goals, supplying over one-third of global electricity by 2050, its variability strains grids and markets, incurring balancing costs and penalties. The analysis dissects reliability issues, highlighting AI's role in predictive maintenance via machine learning on SCADA data for failure anticipation, deep learning for real-time fault detection using convolutional neural networks, and adaptive controls to minimize downtime and extend asset life. These advancements shift operations from reactive to proactive, leveraging vast datasets to predict and prevent disruptions. It then examines electricity market dynamics, where forecast errors lead to imbalances

DOI: 10.4018/979-8-2600-0106-6.ch008

1. INTRODUCTION

1.1 Background and Motivation

The global shift towards a sustainable energy future is accelerating, with wind energy at the forefront of this transition. Driven by ambitious decarbonization goals and precipitously declining technology costs, wind power is projected to supply over one-third of global electricity needs by 2050 (Bouckaert et al., 2021). However, the intermittent and unpredictable nature of wind presents a major hurdle to achieving this potential. The wind doesn't always blow when energy is needed, and its output can fluctuate dramatically, creating instability for power grids designed for predictable, dispatchable generation. This variability not only complicates real-time grid management, requiring increased balancing reserves, but also introduces significant financial risks for wind farm operators who face penalties for deviating from scheduled production in competitive electricity markets (Odeh & Watts, 2019).

These operational and market challenges create a critical barrier to the further integration and investment in wind energy. To mitigate these issues, a paradigm shift from reactive to intelligent, predictive management is essential. This is where artificial intelligence (AI) emerges as a transformative enabler. The core motivation for this chapter is to systematically explore how cutting-edge AI technologies are being deployed to overcome these dual challenges. By leveraging vast datasets from turbines, weather models, and market signals, AI offers a pathway to enhance the reliability and economic competitiveness of wind power (Hassan, 2025).

This chapter will therefore investigate the specific AI methodologies that are reshaping the sector. It will first detail how machine learning algorithms facilitate predictive maintenance and fault detection, directly improving turbine availability and operational stability (Udo & Muhammad, 2021). Subsequently, it will analyze the role of advanced forecasting models in increasing market responsiveness and enabling strategic participation in energy trading. Finally, it will discuss how AI-driven tools for portfolio optimization and risk management are creating a more robust financial case for wind energy investments, solidifying its role in a clean and reliable energy system.

1.2 The Role of Wind Energy in the Global Energy Transition

Wind energy has cemented its status as a cornerstone of the global energy transition, experiencing exponential growth driven by precipitously falling levelized costs of energy (LCOE) and ambitious international climate targets outlined in agreements like the Paris Accord. The technology has evolved from a niche alternative to a mainstream power source, with global capacity surpassing 1 terawatt in 2023 and projected to grow substantially in the coming decades. This remarkable expansion is a direct response to the urgent need for decarbonization, as wind power offers a clean, renewable, and domestically sourced alternative to fossil fuels, directly displacing carbon-intensive generation and enhancing national energy security (IRENA, 2020).

The sheer scale of modern wind energy deployment underscores its critical importance. This is exemplified by the development of gigawatt-scale offshore wind farms in the North Sea and the rapid proliferation of both large-scale onshore wind plants and distributed turbines for community-level generation. This diversified deployment strategy allows nations to leverage their unique geographical advantages, maximizing the utilization of the resource. The continued innovation in turbine technology, particularly the increase in rotor diameter and hub height, has been a key driver in capturing more energy and further reducing costs, making wind power economically competitive with conventional sources in many regions (Wiser et al., 2020).

However, the very characteristic that defines wind power, its dependence on variable meteorological conditions, also presents the primary barrier to its full integration and value realization. This variability and partial unpredictability create systemic challenges, including grid stability concerns, congestion management issues, and price cannibalization in electricity markets during periods of high wind and low demand (H. Lund & Mathiesen, 2009). To fully realize wind energy's potential and ensure it can provide firm, reliable capacity rather than just intermittent energy, the industry must transcend these inherent challenges. This necessitates the development and deployment of sophisticated technologies and strategies, a frontier where artificial intelligence is proving to be a pivotal tool for optimizing both the operations and market participation of wind assets (IRENA, 2020). Figure 1 presents an overview of an offshore wind farm.

Figure 1. Offshore wind farm

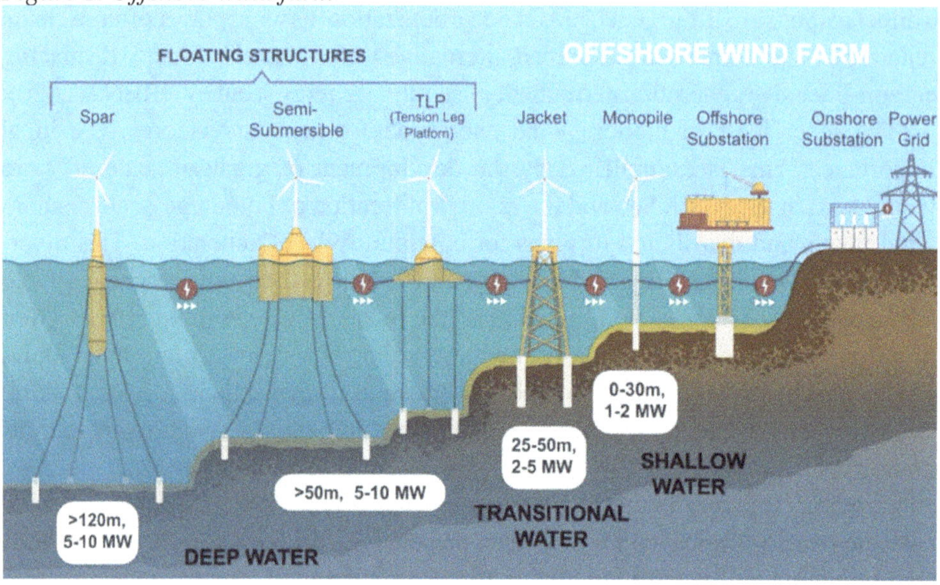

1.3 Challenges: Variability, Market Complexity, and Operational Uncertainty

The primary challenges in wind energy stem from its inherent characteristics, with variability posing the most fundamental obstacle. Variability refers to the unpredictable, often rapid changes in wind speed and direction, which directly cause large fluctuations in power generation output. This intermittency makes it exceptionally difficult for grid operators to balance electricity supply with real-time demand, leading to frequency deviations, voltage instability, and an increased reliance on costly fossil-fueled spinning reserves to maintain grid reliability (P. D. Lund et al., 2015). The economic consequence of this is price cannibalization, where high wind penetration during periods of low demand drives wholesale electricity prices to zero or even negative values, severely eroding the revenue potential of wind assets (Liebensteiner et al., 2025).

Compounding the physical challenge of variability is the market complexity wind operators face. In liberalized electricity markets, wind energy must compete with dispatchable and predictable power sources. Operators are required to submit generation forecasts for day-ahead and intraday markets, and are financially penalized for any deviations between their scheduled and actual delivery (Odeh & Watts, 2019). This creates a complex strategic bidding problem fraught with uncertainty.

Alongside this, operational uncertainty presents a significant financial risk, as unscheduled maintenance and catastrophic component failures, such as gearbox or blade damage, can lead to prolonged downtime and exorbitant repair costs, directly impacting the levelized cost of energy (LCOE) (Tchakoua et al., 2013). These combined challenges create a high-stakes environment where forecasting inaccuracies and equipment failures translate directly into financial losses (Bouckaert et al., 2021).

1.4 Opportunities Presented by Artificial Intelligence

Artificial Intelligence offers a powerful and synergistic set of tools to systematically address the triad of challenges facing wind energy. By leveraging machine learning (ML) and deep learning (DL) algorithms to analyze vast, high-dimensional datasets, including historical and real-time SCADA data, numerical weather predictions (NWP), satellite imagery, and market price signals, AI can generate highly accurate, probabilistic forecasts and actionable insights that were previously unattainable with traditional physical or statistical models (Hong & Rioflorido, 2019). This capability forms the foundation for transforming wind energy from an unpredictable resource into a more manageable and valuable one.

Specifically, AI-driven predictive maintenance models can analyze subtle patterns in turbine sensor data to anticipate component failures weeks or months in advance, enabling condition-based maintenance and drastically reducing unplanned downtime and operational expenditures (OPEX) (Z. Wang & Bu, 2024). In the market domain, AI-powered forecasting models, such as convolutional neural networks (CNNs) and long short-term memory (LSTM) networks, provide the high-fidelity power predictions necessary for strategic bidding, minimizing imbalance costs and maximizing market revenue (Z. Wang & Bu, 2024). Furthermore, reinforcement learning algorithms are now being deployed to autonomously develop and execute complex energy trading strategies, dynamically optimizing bids across multiple market horizons to capture arbitrage opportunities and effectively manage portfolio risk (Vázquez-Canteli & Nagy, 2019). The convergence of these AI applications is creating a new paradigm of intelligent, data-driven wind farm operations.

1.5 Chapter Objectives and Structure

This chapter aims to provide a comprehensive and structured overview of the transformative impact of artificial intelligence on the wind energy sector, bridging the gap between technical potential and practical market application. We will synthesize the latest research and industry case studies to demonstrate how AI is not merely an incremental improvement but a foundational shift in managing wind assets. The objective is to equip researchers, engineers, and market participants with a clear understanding of the methodologies, benefits, and implementation pathways of these advanced AI tools, thereby fostering greater confidence in and adoption of these technologies (Qamsane et al., 2019).

To achieve this, the chapter is structured to follow a logical progression from operational challenges to market integration. We begin by detailing the specific reliability and operational hurdles, followed by an in-depth exploration of AI applications for predictive maintenance, fault detection, and performance optimization. The subsequent section analyzes the evolving dynamics of electricity markets and provides a thorough examination of how AI is enabling more strategic and profitable participation through enhanced forecasting and algorithmic trading. Finally, the chapter will consolidate these concepts by presenting illustrative real-world case studies of successful AI implementation and conclude with a forward-looking discussion on the emerging challenges, such as data quality and model interpretability, and the future opportunities for AI in the global wind energy ecosystem (Aghdam et al., 2019).

2. WIND ENERGY RELIABILITY AND OPERATIONAL CHALLENGES

2.1 Understanding Wind Variability and Grid Integration Issues

The core challenge of wind energy is its fundamental dependence on stochastic weather conditions. Wind speed is not constant, and phenomena such as wind gusts, shear, and lulls can cause rapid and significant fluctuations in power output over timescales ranging from seconds to hours. This inherent intermittency creates profound problems for grid operators, whose primary task is to maintain a precise and stable balance between electricity supply and demand to ensure grid frequency and voltage remain within strict operational limits. The unpredictable nature of wind makes it difficult to schedule power generation reliably, forcing system operators to rely on more flexible, and often more expensive and carbon-intensive, backup power sources, such as natural gas peaker plants or rapidly dispatchable hydroelectricity (Nghitevelekwa & Bansal, 2018).

The integration of large-scale wind power thus necessitates a fundamental redesign of grid operation protocols. Traditional power systems were built around dispatchable generation, where output can be controlled to follow load. Wind energy inverts this paradigm, requiring the grid's load and other generation sources to follow its variable output to a significant degree. This can lead to significant technical issues, including voltage deviations, thermal overloading of transmission lines during periods of high wind, and reduced system inertia, which is critical for maintaining stability during faults (Makolo et al., 2021). The economic impact is equally significant, manifesting as the need for costly grid reinforcement, increased spending on balancing services, and the phenomenon of wind curtailment when generation exceeds the grid's capacity to absorb it.

Addressing these integration issues requires a multi-faceted approach. This includes investing in grid infrastructure, such as high-voltage interconnectors to pool resources over wider geographical areas, and deploying complementary technologies like energy storage systems (ESS) to time-shift wind energy (IRENA, 2020). Furthermore, the development of advanced forecasting tools and market mechanisms that value flexibility are crucial to managing the variability. Without these mitigating strategies, the cost and complexity of integrating wind energy at high penetration levels could hinder its ability to contribute effectively to decarbonization goals.

2.2 Equipment Reliability and Maintenance Concerns

Wind turbines are complex electro-mechanical systems operating under continuous dynamic loads and in harsh, often remote environments, making them highly susceptible to accelerated wear and tear. Components such as gearboxes, blades, and generators are subject to extreme stresses from turbulent winds, gravity, and temperature cycles, leading to failures of bearings, cracks in blades, and insulation breakdowns in generators (Tchakoua et al., 2013). Such equipment failures, particularly of major components located in the nacelle, can be extraordinarily costly due to the expenses associated with crane rentals, complex logistics, and extended downtime required for repair or replacement.

Traditional maintenance strategies, primarily based on periodic scheduled inspections (time-based) or reactive repairs after a breakdown (corrective), are increasingly seen as inefficient and economically unsustainable for modern wind farms. Scheduled maintenance can lead to unnecessary component replacements when life remains, while reactive strategies risk catastrophic failures that result in massive revenue loss and repair costs (Besnard & Bertling, 2010). This "too early or too late" dilemma highlights the critical need for a more intelligent, data-driven approach to improve asset reliability and optimize operational expenditures (OPEX), which constitute a significant portion of the levelized cost of energy (LCOE) for wind power.

The shift towards a proactive maintenance paradigm is therefore a key focus for the industry. This involves moving from a calendar-based schedule to a condition-based approach, where maintenance decisions are informed by the actual health of the turbine. The foundation of this approach is the extensive data collected by Supervisory Control and Data Acquisition (SCADA) systems and additional condition monitoring systems (CMS), which track hundreds of parameters like vibration, temperature, and oil quality. Analyzing this data to predict failures before they occur is the cornerstone of predictive maintenance, a capability powerfully enabled by artificial intelligence, which can detect subtle anomaly patterns indicative of incipient faults (Dao, 2022).

2.3 Impact of Uncertainty on System Stability and Forecasting

The uncertainty in wind power generation creates a ripple effect that compromises the stability and economic efficiency of the entire energy system. From a grid operator's perspective, this uncertainty complicates the crucial task of balancing generation and load in real-time. Large, unexpected ramps in wind power, either up or down, can cause frequency deviations, and if not quickly compensated for by other resources, can lead to load-shedding or even cascading blackouts (Ummels et al., 2008). This requires system operators to hold substantial spinning reserves, which are typically provided by fossil-fuel plants, thereby offsetting some of the carbon savings from wind energy itself.

Accurate forecasting is the primary tool to mitigate this uncertainty. Forecasts are required across different time horizons: day-ahead for market scheduling, intraday for adjusting schedules, and very short-term (minutes to hours) for real-time grid balancing. However, traditional forecasting models, including physical models based on Numerical Weather Prediction (NWP) and simpler statistical methods like ARIMA, often struggle to capture the complex, non-linear, and spatially correlated relationships between meteorological variables and turbine power output (Giebel et al., 2011). This leads to significant prediction errors, which directly translate into financial penalties for wind farm operators in the form of imbalance costs and lost revenue opportunities.

The consequences of poor forecasting extend beyond immediate grid stability and individual project economics. At a systemic level, persistent forecast errors increase the overall cost of integrating renewable energy, which can influence energy policy and market design (Hirth, 2013). Furthermore, the inability to accurately predict output can deter investment, as it increases the perceived risk of wind projects for financiers. Therefore, improving forecasting accuracy is not merely a technical exercise but a critical enabler for the continued growth and bankability of wind energy, paving the way for the development of more advanced AI-driven forecasting methodologies that can learn directly from data and continuously improve their performance.

2.4 Traditional vs. AI-Driven Reliability Approaches

Historically, wind farm reliability has been managed using a combination of rigid, time-based scheduled maintenance and a reactive "run-to-failure" approach for non-critical components. This philosophy is rooted in established industrial practices but is inherently inefficient for complex assets like wind turbines (Wiser et al., 2020). Scheduled maintenance often results in "over-maintenance," where healthy components are serviced or replaced prematurely, wasting resources and potentially introducing infant mortality failures. Conversely, the run-to-failure model inevitably leads to unexpected breakdowns, causing extensive downtime and catastrophic repair costs, especially for major components like gearboxes and blades.

In stark contrast, AI-driven reliability represents a paradigm shift towards a proactive and predictive operational philosophy. This approach leverages the vast streams of sensor data generated by turbines, including vibration spectra, temperature trends, oil debris counts, and acoustic emissions, to create a digital twin of the asset's health. Machine learning models are trained on this historical data to recognize subtle, multivariate patterns that are precursors to specific failure modes (Zhao et al., 2016). This allows for a transition from scheduled to condition-based and ultimately predictive maintenance, where work orders are generated based on the actual remaining useful life (RUL) of a component, thereby maximizing productive uptime and optimizing spare parts logistics.

The benefits of this AI-driven approach are transformative. By accurately predicting failures weeks or even months in advance, operators can plan interventions during low-wind seasons, thus minimizing lost production revenue. This directly reduces operational expenditures (OPEX) by avoiding unnecessary maintenance and preventing expensive catastrophic failures. Furthermore, the diagnostic capabilities of AI enhance root cause analysis, providing valuable feedback to turbine manufacturers for improving future designs. As the industry matures, this data-driven intelligence is becoming a key competitive advantage, enabling asset owners to maximize their return on investment and improve the bankability of their wind energy portfolios (Udo & Muhammad, 2021).

3. AI APPLICATIONS FOR RELIABILITY IN WIND ENERGY

3.1 Predictive Maintenance Using Machine Learning

Predictive maintenance (PdM) is a game-changer for wind farm operations, moving the industry beyond traditional time-based and reactive strategies. At its core, PdM uses machine learning (ML) models trained on historical Supervisory Control and Data Acquisition (SCADA) and Condition Monitoring System (CMS) data to identify subtle, often multivariate patterns that precede a component failure. These models learn the normal operational "fingerprint" of a healthy turbine and can detect deviations from this baseline that are imperceptible to human operators or simple threshold alarms (Sun et al., 2016). By continuously monitoring real-time sensor data, these ML systems can issue early warnings, allowing maintenance teams to intervene proactively before a minor anomaly escalates into a major, costly repair.

The technical implementation of PdM involves several ML approaches. Supervised learning algorithms, such as Support Vector Machines (SVMs) and Random Forests, can be trained on labeled historical data (e.g., "normal" vs. "impending bearing failure") to classify the current state of a component (Zin et al., 2013). More advanced deep learning architectures, like Long Short-Term Memory (LSTM) networks, are particularly effective as they can model the temporal dependencies in time-series sensor data, capturing how a fault develops over time. These models can predict the Remaining Useful Life (RUL) of critical components, providing a quantitative basis for scheduling maintenance.

The operational and economic impact of successfully deploying predictive maintenance is dramatic. It directly increases turbine availability and annual energy production (AEP) by minimizing unplanned downtime. Furthermore, it extends the operational lifespan of assets by preventing catastrophic damage and allows for more efficient use of maintenance crews and resources, significantly reducing operational expenditures (OPEX) (Tchakoua et al., 2013). By transforming maintenance from a cost center into a strategic function, PdM powered by ML strengthens the business case for wind energy, enhancing its reliability and making it a more dependable and bankable source of power (Udo & Muhammad, 2021).

3.2 Fault Detection and Diagnosis With AI

Fault detection and diagnosis (FDD) is a critical capability where AI algorithms, particularly deep learning and neural networks, demonstrate superior performance over traditional methods. While conventional FDD often relies on physics-based models or simple statistical process control limits, AI models excel at identifying and classifying complex, non-linear anomalies in high-dimensional datasets. Techniques such as convolutional neural networks (CNNs) can process vibration spectrograms or thermal images to detect faults like blade cracks or rotor imbalances with a level of accuracy and speed unattainable by manual analysis (Janssens et al., 2016). This enables a highly sensitive and automated surveillance system for entire wind farms.

The diagnostic power of AI extends beyond mere detection to precise fault isolation and classification. Deep learning models can be trained to not only flag an anomaly but also to diagnose its root cause, for instance, distinguishing between a misaligned gearbox, a faulty bearing, and an electrical imbalance in the generator (Zhirong et al., 2016). Recurrent neural networks (RNNs), adept at handling sequential data, can analyze the evolution of sensor readings to pinpoint the originating component of a fault. This precise diagnosis is invaluable for maintenance planning, as it tells technicians exactly what to expect and which parts and tools will be required, thereby reducing the mean time to repair (MTTR).

The implementation of AI-driven FDD creates a more resilient and safer operational environment. By providing early and accurate warnings, these systems allow operators to take corrective action, such as de-rating a turbine or shutting it down, to prevent a minor fault from triggering a cascading failure that could damage multiple subsystems (Tamilselvan & Wang, 2013). This proactive containment of faults minimizes the risk of catastrophic events like fires or structural collapses. Ultimately, the integration of robust AI-based FDD systems is a critical step towards achieving fully autonomous wind farm operations, where the health of each asset is continuously managed by intelligent algorithms, ensuring maximum productivity and system integrity.

The core of this process is a deep neural network that acts as a sophisticated pattern recognition system. It takes in a multitude of data points from a wind turbine's condition monitoring system (CMS) and processes them to determine the health of the turbine's components. A typical architecture for such a network is illustrated in Figure 2.

Figure 2. Neural network architecture for wind turbine fault detection

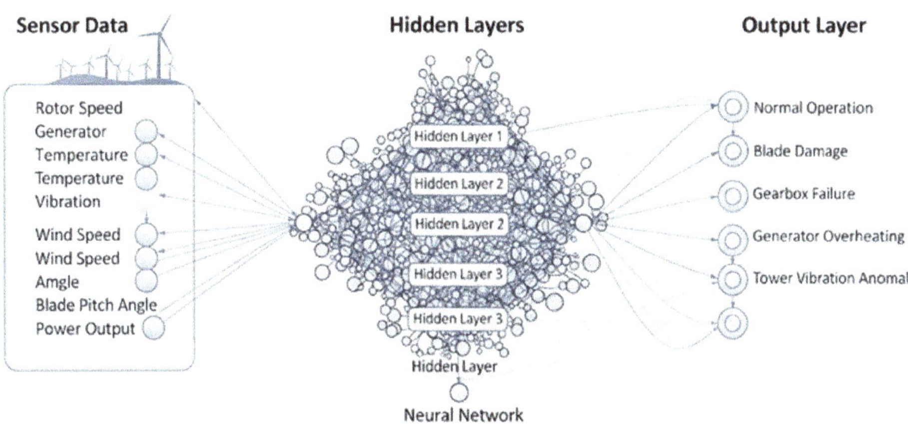

3.3 AI in Wind Turbine Condition Monitoring Systems

Modern wind turbines are equipped with sophisticated Condition Monitoring Systems (CMS) that continuously collect vast amounts of high-frequency data from an array of sensors, including accelerometers on the gearbox and generator, acoustic emission sensors, and temperature and oil particle counters. While this data stream is invaluable, its volume and complexity make manual analysis impractical for detecting subtle, incipient faults. Artificial Intelligence acts as the intelligent brain behind these systems, processing this multivariate data to transform raw signals into actionable insights. By applying machine learning models, operators can move beyond simple threshold alarms to a nuanced understanding of asset health, enabling a truly proactive and continuous monitoring paradigm (Hou & Wang, 2023).

AI models are particularly adept at analyzing complex patterns in specific data types that are indicative of component degradation. For instance, deep learning architectures like Convolutional Neural Networks (CNNs) can process vibration spectrograms to identify frequency patterns associated with specific bearing or gear faults with a high degree of accuracy, often detecting anomalies long before they are apparent in overall vibration levels (Janssens et al., 2016). Similarly, Recurrent Neural Networks (RNNs) and Long Short-Term Memory (LSTM) networks can model temporal sequences in SCADA data parameters, such as the relationship between power output, generator temperature, and ambient conditions, to flag performance deviations that suggest issues like cooling system failures or blade surface erosion (Dou et al., 2022). This multi-modal analysis provides a comprehensive digital health signature for each turbine.

The operational shift enabled by AI-driven CMS is transformative. By providing a continuous, automated, and highly sensitive assessment of component health, it allows maintenance to be scheduled based on the actual condition and predicted remaining useful life (RUL) of assets, rather than on a fixed calendar (Tamilselvan & Wang, 2013). This condition-based approach directly maximizes turbine availability and energy production by preventing unexpected failures and minimizing unnecessary maintenance downtime. Furthermore, the diagnostic precision of AI helps in root cause analysis, providing critical feedback to OEMs for improving future turbine designs and solidifying AI's role as an indispensable tool for optimizing the lifecycle management and profitability of wind assets (Ukoba et al., 2024).

3.4 Case Studies: Real-World Implementations

AI's potential in wind energy reliability transitions from theory to practice via global deployments, yielding measurable gains in efficiency, cost savings, and risk reduction.

A leading European operator allied with a data analytics firm to pioneer machine learning for gearbox failure prediction. Trained on SCADA and vibration datasets from hundreds of turbines, the model discerned early degradation signals. Deployment slashed annual maintenance costs by 20% and boosted turbine availability 15%, enabling proactive replacements during off-peak winds instead of crisis responses, delivering swift ROI through predictive analytics (Stetco et al., 2019).

In North America, a utility harnessed deep learning for blade integrity via drone imagery and convolutional neural networks (CNNs). The system detected subtle flaws like micro-cracks and erosion invisible to inspectors. A pivotal intervention flagged root cracking, averting a blade detachment that risked multimillion-dollar damages and lost output, underscoring AI's prowess in preempting rare, high-impact failures (Kong et al., 2014).

Scaling up, an Asian asset owner rolled out a cloud AI platform aggregating portfolio-wide data. Ensemble models benchmarked turbines against peers, flagging laggards and root causes. This optimized resource allocation, codified best practices, and secured better warranties/insurance via proven metrics. Collectively, these cases affirm AI as a mature force, fortifying resilience and economics in wind projects (Tautz-Weinert & Watson, 2017).

4. MARKET DYNAMICS IN MODERN ELECTRICITY MARKETS

4.1 Structure of Energy Markets

Modern electricity markets form intricate, multi-tiered frameworks to deliver reliable, cost-effective power, balancing supply and demand through temporal layers. At the core is the day-ahead market (DAM), the hub for wholesale trading, where generators and retailers bid hourly for the next day's output. A centralized auction employs security-constrained unit commitment (SCUC) and economic dispatch (SCED) algorithms to clear trades, setting locational marginal prices (LMPs) and dispatch schedules for conventional assets, enabling advance planning for dispatchable resources (Badesa et al., 2020). Yet, forecasting imperfections necessitate the real-time market (RTM), which operates sub-hourly to rectify deviations between planned and actual conditions, often yielding volatile prices amid unforeseen shifts.

Complementing energy-only venues, ancillary services markets safeguard grid integrity and quality. These procure vital supports wind traditionally underdelivers: frequency regulation deploys rapid-response units to stabilize 50/60 Hz; spinning/non-spinning reserves hold capacity for outages or errors (IRENA, 2020). Voltage control via reactive power maintains stability. Regionally varied, these markets offer revenue and mandates. As inverter-based renewables like wind and solar proliferate, reforms recalibrate designs to reward flexibility and novel stability, posing integration hurdles yet opening pathways for wind's enhanced role (IRENA, 2020).

4.2 Participation of Wind Energy in Competitive Markets

Wind producers' market success pivots on precise forecasting, transforming inherent uncertainties into strategic assets. In day-ahead markets, operators submit hourly generation bids based on wind predictions, committing financially for the next day. Deviations exact steep tolls: under-delivery from weak winds forces real-time purchases at premiums, racking up imbalance costs (Gonzalez-Salazar et al., 2018). Over-generation dumps surplus into volatile real-time markets, often at depressed or negative prices, eroding gains.

This setup poses intricate optimization challenges for portfolio managers, prioritizing value over volume. They navigate by aligning forecasts with price spreads between day-ahead and real-time arenas, crafting bids to capture differentials (Hirth, 2013). Risk-averse tactics like derated bidding—submitting conservative outputs, curb under-delivery exposures.

Beyond energy trading, horizons expand: storage-augmented farms tap ancillary services, offering fast-ramp frequency regulation to diversify revenues and affirm grid value. Thus, AI-enhanced foresight not only mitigates perils but unlocks wind's multifaceted market prowess, solidifying its economic foothold in renewables (Lipu et al., 2023).

4.3 Regulatory and Policy Considerations

Wind energy's market participation and viability are deeply influenced by regulatory and policy frameworks. Early adoption was spurred by mandates like Renewable Portfolio Standards (RPS), requiring renewable quotas, and incentives such as Production Tax Credits (PTCs) or Feed-in Tariffs (FITs), which lowered barriers to deployment (IRENA, 2020). Carbon pricing, via emissions trading or taxes, further levels the field by embedding fossil fuel externalities, enhancing wind's competitiveness.

As wind scales from niche to dominant, legacy regulations, crafted for dispatchable sources, impede progress. Hurdles include protracted grid connections, tariffs ignoring true costs, and markets undervaluing renewables' flexibility and ramping speed (Brown & Reichenberg, 2020). To adapt, authorities pursue reforms: intraday/continuous trading for agile bids, capacity payments recognizing wind's reliability contributions, and redefined ancillary services accommodating inverters. These evolutions cultivate supportive ecosystems, easing integration and accelerating the shift to low-carbon grids.

4.4 Price Volatility and Revenue Risk for Wind Operators

Wind operators grapple with acute revenue risks from resource variability clashing with price volatility. Unlike dispatchable generators that time output for peaks, wind's weather-driven production lacks price correlation, yielding mismatches. This fosters "revenue cannibalization": high wind eras flood markets with cheap renewables, crashing wholesale prices and eroding all wind margins during optimal generation (Hirth, 2013). Abrupt drops, from slack demand, renewable surges, or low fossil costs, can nullify profits despite full-capacity turbines.

Sustaining project viability demands robust risk hedging. Financially, operators secure Power Purchase Agreements (PPAs) with reliable buyers, fixing prices to offload merchant exposure. Derivatives like futures and options further buffer against swings (Gora et al., 2025).

Operationally, hybridizing wind with battery storage gains traction. Batteries time-shift output: capturing surplus in low-price wind gluts for discharge at premiums, arbitraging differentials. This steadies revenues, mitigates intermittency, and yields ancillary grid services. Together, these tactics transform volatility into opportunity, bolstering wind's bankability amid energy transitions.

5. AI-DRIVEN FORECASTING AND OPTIMIZATION

5.1 Wind Power Forecasting Techniques: Statistical vs. AI Models

Conventional wind forecasting leans on Numerical Weather Prediction (NWP) models, which computationally simulate atmospheric physics, and statistical tools like Auto-Regressive Integrated Moving Average (ARIMA), extrapolating linearly from past data. Though foundational, these falters: NWP's coarse grids overlook site-specific topography and turbulence, while ARIMA grapples with non-linear, multivariate weather-turbine interactions, yielding errors in extremes or ramps (Giebel et al., 2011). Such imprecision caps wind assets' economic viability, inflating imbalances and costs.

Conversely, AI and machine learning herald a revolution, deriving non-linear insights from expansive datasets. From Random Forests to deep architectures, models ingest NWP, SCADA, and mast readings to unearth subtle variable entanglements, like wind shear, density, and turbulence synergies, beyond physical or linear confines. This empowers granular, probabilistic forecasts spanning minutes to days, slashing errors and bolstering market bids (Hong & Rioflorido, 2019; Wu

& (Hong & Rioflorido, 2019), 2007). Ultimately, AI elevates wind from erratic to predictable, unlocking operational and financial frontiers.

5.2 Deep Learning, Reinforcement Learning, and Hybrid Models

Deep learning excels in wind power forecasting by adeptly modeling temporal dynamics. Long Short-Term Memory (LSTM) networks and Gated Recurrent Units (GRUs) detect long-range dependencies in time-series data, adeptly tracing wind's persistence, diurnal cycles, and weather inertia. Fed with historical power and meteorological sequences, these models yield not just point estimates but probabilistic outputs, quantifying uncertainty to inform risk-sensitive trading decisions (Sheng et al., 2023).

AI's scope broadens to market optimization via Reinforcement Learning (RL). RL agents evolve bidding tactics through iterative market simulations, earning profits as rewards or imbalance fees as penalties, honing policies for sustained revenue maximization (Vázquez-Canteli & Nagy, 2019).

For peak efficacy, hybrid architectures fuse Numerical Weather Prediction's (NWP) physics with deep learning's pattern acuity, channeling refined NWP inputs into LSTMs. This blend harnesses complementary strengths, delivering unmatched forecast precision and resilient trading outcomes over standalone approaches (Hewamalage et al., 2021).

5.3 Market Price Forecasting With AI

Complementing generation predictions, electricity price forecasting is vital for wind farm profitability amid market volatility. AI thrives here, unraveling non-linear price drivers through multifaceted inputs: anticipated loads, rival generation schedules, fuel costs (notably natural gas), renewable forecasts inducing cannibalization, and grid bottlenecks. By discerning intricate factor interplay, AI surpasses legacy time-series methods, foreseeing spikes, plunges, and day-ahead/real-time spreads with unmatched finesse (Wozabal et al., 2016).

This foresight arms operators with strategic edge. Precise price outlooks guide bid derating or aggregation to evade crash-time sales. Paired with power forecasts, it unlocks nuanced tactics: intraday engagements leveraging updates, or joint energy/ancillary bids (Lago et al., 2021). For storage-integrated sites, it anchors arbitrage, timing battery charges in low-price wind surpluses and discharges at peaks, converting volatility into lucrative differentials. Thus, AI reframes market turbulence as a revenue catalyst, fortifying wind's economic resilience.

5.4 Forecasting Accuracy and Its Impact on Trading and Dispatch

The accuracy of wind and price forecasts is the single most critical factor determining a wind farm's profitability in liberalized markets. A more accurate power forecast allows an operator to submit a day-ahead bid that closely matches actual delivery, thereby minimizing the volume of energy that must be settled in the real-time market at a potential loss. Each percentage point reduction in forecast error directly translates to a reduction in imbalance costs and an increase in net revenue, as it prevents the need to buy high (in case of a shortfall) or sell low (in case of a surplus) (Lago et al., 2021). This precision is equally vital for grid operators, who rely on accurate aggregate wind forecasts to schedule the optimal mix of conventional generation and reserves, enhancing overall grid stability, reducing reliance on peaking plants, and lowering system-wide operating costs.

The choice of forecasting model is therefore a strategic business decision. Different AI models offer varying trade-offs between accuracy, computational complexity, data requirements, and interpretability. The following table compares prominent forecasting approaches, highlighting their applicability across different time horizons and operational contexts. This enables wind farm operators and grid planners to select the most appropriate tool for their specific needs, whether it be for long-term planning, day-ahead market bidding, or real-time grid balancing.

Table 1. Comparison of Wind Power Forecasting Models

Model Type	Time Horizon	Strengths	Limitations	Accuracy (RMSE) Example
NWP	Day-ahead	Physics-based, large-scale	Coarse resolution	15-20%
ARIMA	Short-term	Simple, fast computation	Linear assumptions	12-18%
Random Forest	Intraday	Handles non-linearity, feature importance	No temporal modeling	8-12%
LSTM	All	Temporal dependencies, sequences	Data-intensive, training time	5-10%
Hybrid (NWP+DL)	Day-ahead	Synergistic, robust	Complex integration	4-8%

6. AI IN ENERGY TRADING STRATEGIES

6.1 Algorithmic Bidding and Energy Portfolio Optimization

Algorithmic bidding marks a paradigm shift in energy trading, supplanting manual heuristics with AI-orchestrated, data-centric precision. These systems harness advanced optimization and machine learning to autonomously craft and dispatch bids across day-ahead, intraday, and balancing markets. Ingesting real-time feeds, probabilistic wind forecasts, price projections, congestion visuals, and rival analyses, they forge profit-optimizing curves, outpacing human limits to seize arbitrage and adapt to signals, slashing imbalances and amplifying revenues for wind operators (Vázquez-Canteli & Nagy, 2019).

AI's reach amplifies in portfolio realms, as operators blend wind with solar and batteries. Drawing from modern portfolio theory and stochastic programming, algorithms dissect asset synergies and correlations, apportioning energy/capacity to equilibrate yields against perils. Exemplars include timing battery charges amid surplus wind/low prices and discharges at peaks, transmuting intermittent output into premium slots. This mitigates cannibalization, smooths profiles, and stacks gains from arbitrage plus ancillaries (Mantar Gundogdu et al., 2019).

Deployment fuses predictive DL for foresight with reinforcement learning (RL) agents for execution. RL evolves policies via market simulations or live trials, honing on profit rewards and penalty feedbacks (Mocanu et al., 2018). Yielding counterintuitive tactics attuned to dynamics, rules, and physics, these architectures eclipse legacy methods, fortifying wind's fiscal resilience in volatile arenas.

6.2 Risk Assessment and Decision Support Tools

Electricity markets' volatility imperils wind operators with multifaceted risks: volume from forecast inaccuracies, price from fluctuations, and balancing from grid penalties. AI mitigates these through advanced risk assessment tools that craft probabilistic scenarios. Employing Monte Carlo simulations and generative adversarial networks (GANs), these systems simulate myriad future states from historical and predictive data, yielding outcome distributions over deterministic forecasts (Dou et al., 2022). This empowers traders to gauge Value at Risk (VaR) or Conditional Value at Risk (CVaR), fostering granular hedging and exposure management.

Such analytics anchor decision support systems (DSS) for traders. AI-driven DSS integrate risk models, forecasts, and live feeds to deliver tailored advice, like securing a Power Purchase Agreement (PPA) for downturn hedging or holding merchant exposure for spikes (Wozabal et al., 2016). Operationally, it optimizes turbine maintenance timing to curb revenue dips or prompts battery pre-charging amid congestion forecasts.

AI's true prowess lies in augmentation, not substitution. By offloading intricate computations and synthesis, it liberates humans for strategic focus and anomalies. Recommendations arrive with transparent rationales, key assumptions, and confidence metrics (Aghdam et al., 2019), blending AI's impartiality with human acumen, intuition, and foresight. This symbiotic "human-in-the-loop" paradigm yields agile, resilient trading amid market intricacies, fortifying wind's economic viability in turbulent renewables landscapes.

6.3 Dynamic Pricing and Demand Response Integration

AI propels dynamic pricing models, vital for mirroring real-time electricity costs, supplanting static or time-of-use tariffs. Prices oscillate every 5-15 minutes per grid realities, with AI forecasting local supply (variable renewables like wind) and demand to compute marginal loads, dispatching precise economic cues to participants (Chen et al., 2022). For wind generators, this sharpens price signals, rewarding scarcity while curbing surpluses, fostering fair compensation.

Seamlessly, this meshes with AI-orchestrated demand response (DR), amplifying grid agility. In low-wind, high-price lulls, AI alerts commercial, industrial, or smart-home users to curtail usage, forming "virtual power plants" rivaling gas peakers for balance (Stecca et al., 2020). Wind operators gain buffers against curtailment in congested or oversupplied grids, birthing markets for their output.

This wind-DR nexus signals a interactive power evolution. AI platforms co-optimize supply and demand in holistic services: aggregators bid fused wind-flex-load portfolios into capacity/ancillary arenas, yielding firmer, reliable offerings (Siano, 2014). Outcomes? Grid fortification, fossil backup diminishment, novel revenues for producers/consumers, and hastened decarbonization.

6.4 AI in Peer-to-Peer (P2P) and Blockchain-Based Energy Markets

The rise of peer-to-peer (P2P) energy trading, powered by blockchain, is reshaping the energy sector by decentralizing distribution and empowering small-scale wind producers. Prosumers, those with micro-turbines, can now directly sell surplus power to nearby consumers, sidestepping utilities. AI serves as the platform's brain, automating intricate matching of supply and demand based on location, eco-preferences, and pricing, all while honoring local grid limits (Hamdan et al., 2024).

Blockchain underpins these exchanges with immutable security, transparency, and trustlessness, eliminating intermediaries. Yet, AI infuses vitality: embedding logic into smart contracts for conditional auto-execution, forecasting participant generation/consumption, suggesting dynamic prices, and orchestrating community storage to prioritize local use over grid exports (Mengelkamp et al., 2018). This duo fosters resilient microgrids, directly monetizing distributed wind output.

The model's ripple effects are transformative for wind energy. It democratizes markets, inviting farmers and communities to monetize assets previously uneconomic. Local balancing eases transmission burdens, bolstering grid robustness. Evolving platforms will leverage AI for advanced feats: aggregating "energy communities" for wholesale bids or delivering granular services to operators. This AI-blockchain fusion heralds a consumer-led, equitable paradigm, centering distributed wind in a transparent, efficient, and decentralized energy future (Lim et al., 2025).

7. CASE STUDIES AND REAL-WORLD APPLICATIONS

7.1 AI-Enhanced Wind Farm Management

AI is revolutionizing wind farm operations worldwide, boosting efficiency, reliability, and sustainability through targeted innovations. In Europe, Vestas collaborates with Microsoft on AI-driven wake steering, training controllers to yaw turbines and deflect turbulent wakes, recapturing energy for 1-3% output gains at scale without hardware upgrades. EDP's ASTER® platform integrates machine learning and IoT for data analytics, streamlining large-scale operations and signaling a data-centric paradigm shift (Madeswaran et al., 2024).

Asia leads in smart turbine tech. China's Envision employs its EnOS™ platform, embedding hundreds of sensors for real-time adjustments like blade pitch, yielding up to 15% production hikes, complemented by AI digital twins for simulations and predictive maintenance. In India, Suzlon's ML-based PdM analyzes sensor and historical data to forecast gearbox/bearing failures accurately, averting outages, extending intervals, and slashing costs.

In the US, AES leverages manufacturer data and AI cloud tools for >90% accurate failure predictions, curbing repairs and trips to optimize OPEX. Environmentally, a Skkynet partnership automates turbine shutdowns during bat migrations using wind/temperature data, balancing conservation with minimal yield loss and averting penalties (Holdmann & Wies, 2025).

Australia excels in forecasting and integration. Hornsdale Wind Farm processes multi-stream data for 48-hour predictions, refining positioning and low-wind maintenance for yield surges and cost cuts (Operator, 2024). ZendIT streamlines construction via AI risk modeling, forecasting delays, weather-optimized cranes, and supply fixes, in remote sites. The SA VPP employs AI to forecast demand from household/solar data, stabilizing grids and curbing fossil reliance.

These cases illustrate AI's versatility, from yield optimization to eco-compliance, driving wind's scalability in the energy transition.

7.2 AI-Driven Energy Trading Platforms: Examples and Outcomes

Major energy trading entities are harnessing AI platforms to automate bidding and portfolio management, yielding profit gains through refined strategies and minimized forecast error penalties. Reinforcement learning exemplifies this, surpassing human traders in simulations by discerning nuanced price signals and balancing risks in real time (Vázquez-Canteli & Nagy, 2019). A prime case is Statkraft, Europe's premier renewable generator, partnering with Kahoona. Their system ingests vast weather, market, and asset datasets for autonomous decisions, boosting trading margins 5-10% via imbalance cost reductions and arbitrage capture, outpacing conventional tactics.

These platforms transcend profitability, revolutionizing risk mitigation. Deep learning enables probabilistic forecasting of prices and generation, facilitating precise exposure hedging (Dou et al., 2022). Notably, Enbala (now Ormat) deploys predictive analytics for wind-storage hybrids, co-optimizing across energy, capacity, and regulation markets. This elevates annual revenues over 20% while curbing volatility, enhancing investor appeal.

This surge ushers in algorithmic energy markets. MIT analysis reveals AI traders amplify liquidity and efficiency through persistent, data-adaptive bidding, refining price discovery. US firm Gridmatic's AI-as-a-service democratizes elite strategies for independent operators, delivering 99%+ schedule accuracy and eradicating deviation fines, fortifying AI's commercial viability (Lipu et al., 2023).

7.3 Industry Collaborations and Pilot Projects

Intersectoral partnerships between technology giants and energy firms are pivotal in propelling AI innovations for wind energy, bridging computational prowess with operational realities. A landmark collaboration is Google and DeepMind's application of machine learning to a 700-megawatt wind dataset in the central US. Employing neural networks, the initiative delivered 36-hour-ahead power output predictions, optimizing day-ahead market commitments and yielding a 20% uplift in energy value, highlighting AI's economic leverage in forecasting. Paralleling this, Siemens Gamesa teamed with IBM, harnessing Watson AI to scrutinize operational data for preempting failures in blades and bearings. This predictive maintenance curbed downtime and costs, exemplifying AI's role in asset longevity (Lipu et al., 2023; Paidi et al., 2020).

Pilot ventures are pioneering AI at scale for grid resilience and autonomy. Denmark's "Energy Cluster Denmark" spearheads a expansive project deploying AI for integrated grid oversight, fusing forecasts from thousands of dispersed turbines to sustain stability under surging renewables (H. Lund & Mathiesen, 2009). In the EU, the "AutoGEMS" initiative probes reinforcement learning for self-governing distributed energy resources, including wind farms, enabling instantaneous balancing and morphing turbine fleets into agile grid responders.

These endeavors forge pathways to smarter infrastructure. The UK's Orwell project, uniting National Grid ESO with Carbon Trust, crafts an AI digital twin of Britain's electricity grid. By modeling myriad wind-variable scenarios, it refines planning, cuts expenses, and boosts efficiency (Tamilselvan & Wang, 2013). Globally, the IEA's Digitalization for Wind Energy Task fosters knowledge exchange on pilots and practices, expediting AI adoption to hasten the energy transition. Collectively, these synergies underscore AI's transformative potential, democratizing advanced tools for sustainable, adaptive wind ecosystems.

8. CHALLENGES AND FUTURE DIRECTIONS

8.1 Data Quality and Infrastructure Limitations

AI's effectiveness in wind energy hinges on the garbage in, garbage out axiom, elevating data quality to a cornerstone issue. Wind farms produce terabytes from SCADA systems, sensors, and masts, yet harsh conditions breed flaws: miscalibrated sensors, dropouts, outliers, and gaps. Training on impure data fosters biases, yielding flawed forecasts, overlooked faults, and risky decisions that inflate costs or endanger operations (H. Wang et al., 2019). High data volume and speed further strain resources, demanding robust cloud/edge setups for real-time handling, often a hefty CapEx/Opex burden for owners.

Mitigation demands holistic data governance. Automated pipelines for validation and cleansing, spotting anomalies on-the-fly, ensure pristine feeds for (Tchakoua et al., 2013). IT/OT fusion is vital, merging silos like logs, curves, and feeds into unified lakes for seamless access. Edge computing shines here: on-site preprocessing at turbines cuts latency for controls, trims bandwidth for cloud uploads, and bolsters resilience (Ukoba et al., 2024).

By fortifying these pillars, stakeholders craft a durable ecosystem, unleashing AI's promise without pitfalls. This proactive stance not only curbs inefficiencies but propels wind's scalability in renewables, harmonizing data deluge with actionable intelligence.

8.2 Ethical and Regulatory Challenges in AI Deployment

As AI gains autonomy in energy trading and grid control, ethical and regulatory hurdles intensify. A core issue is the black box opacity of deep learning models, obscuring decision logic and eroding accountability. When AI-driven bids spark market disruptions or controls cause equipment failures, pinpointing responsibility proves elusive, complicating liability (Song et al., 2023). Worse, AI agents might exploit market inefficiencies or collude algorithmically, evading outdated oversight and risking systemic manipulation.

Regulators are responding with innovative governance. Standards for Explainable AI (XAI) are emerging to demystify models in critical infrastructure, enabling human audits and comprehension. Cybersecurity demands escalate, as AI vulnerabilities invite attacks disrupting supply or skewing markets; fortified protocols and resilience tests are imperative. Forward-thinking policies must codify AI safety, data privacy, and fairness, aligning deployments with societal benefits.

In wind energy, these imperatives ensure AI's efficiency doesn't compromise trust or equity. By mandating transparency and safeguards, frameworks can harness AI's potential while mitigating perils, fostering a secure, ethical transition to intelligent grids. Proactive collaboration between industry, regulators, and ethicists will bridge gaps, turning challenges into robust safeguards for a sustainable energy future.

8.3 Future Trends: Explainable AI, Digital Twins, and Autonomous Energy Systems

AI's evolution in wind energy heralds an era of enhanced transparency, simulation, and autonomy, propelled by pivotal advancements. At the forefront is Explainable AI (XAI), a burgeoning domain dedicated to demystifying opaque models for human scrutiny. Tools like LIME (Local Interpretable Model-agnostic Explanations) and SHAP (Shapley Additive exPlanations) dissect decision pathways, spotlighting pivotal inputs, such as anomalous vibration frequencies or abrupt wind shears, that sway failure predictions. This interpretability fosters trust, empowering engineers to scrutinize and refine AI outputs with assurance.

Parallelly, high-fidelity Digital Twins, real-time virtual mirrors of wind farms, revolutionize risk management. These dynamic simulacra enable preemptive testing of control tactics, severe weather scenarios, or upkeep protocols in a virtual sandbox, averting real-world pitfalls.

The synergy of XAI, Digital Twins, and sophisticated AI controllers heralds fully autonomous ecosystems. Envision integrated platforms perpetually fine-tuning fleet operations: dynamically tweaking yaw and blade pitch, preemptively calendaring maintenance, and self-executing trades attuned to market fluxes and grid exigencies (Lipu et al., 2023). This self-sustaining paradigm slashes costs, curtails human toil, and amplifies dependability, transmuting wind assets into resilient, value-maximizing entities. Far from supplanting humans, these innovations exalt operators to visionary stewards, charting goals and tackling anomalies, igniting a renaissance of decentralized, ultra-efficient energy paradigms.

8.4 Integrating AI With Other Renewable Energy Sources

AI techniques honed for wind energy seamlessly extend to hybrid renewables, optimizing integrations of wind, solar, and storage. Solar's variability, tied to diurnal cycles, clouds, and weather, gains from AI forecasting via satellite imagery and sky cameras, accurately predicting irradiance and output. Merging these with wind forecasts yields stable hybrid profiles, capitalizing on complementary patterns: solar peaks midday, wind often at night or seasonally offset (Paidi et al., 2020). This complementarity curbs overall intermittency, slashing reliance on expensive grid balancing.

Beyond co-location, AI anchors sophisticated multi-asset portfolios incorporating battery energy storage systems (BESS). Intelligent energy management systems (EMS) enable real-time co-optimization, dynamically allocating surplus wind/solar to charge batteries, discharge during peaks, or deliver frequency regulation. Such decisions enhance economic efficiency, stacking revenues from energy arbitrage, capacity, and services.

This AI orchestration transmutes volatile resources into firm, dispatchable assets, amplifying grid value. As decarbonization surges, AI emerges as the linchpin for harmonizing diverse renewables, fortifying stability, and realizing a 100% renewable grid, transforming challenges into scalable, resilient power paradigms.

9. CONCLUSION

This chapter has systematically illuminated the profound transformative role of artificial intelligence (AI) in the wind energy sector, weaving through its operational, market, and economic dimensions to forge a resilient pathway for renewables. Far from a supplementary tool, AI stands as a foundational force, redefining wind power's integration into global energy systems. Central findings underscore AI's capacity to bolster reliability through predictive maintenance and fault detection, leveraging machine learning on SCADA data to foresee failures and extend asset lifespans, thereby slashing downtime and operational costs (H. Wang et al., 2019). Equally pivotal is the surge in forecasting precision from deep learning hybrids, which demystify wind's variability, curbing imbalance penalties and empowering strategic bidding in day-ahead and real-time markets (Sun et al., 2016). At the trading frontier, reinforcement learning-driven algorithms and portfolio optimizations navigate volatility, hedging risks via storage synergies and unlocking arbitrage, thus elevating revenue streams and financial viability (Vázquez-Canteli & Nagy, 2019).

Collectively, these innovations propel wind from a passive, erratic contributor to an active, intelligent grid ally. By distilling terabytes of sensor, weather, and market data into proactive insights and automated decisions, AI mitigates intermittency's legacy burdens, fostering applications like wake steering for yield boosts and wildlife-sensitive curtailments that yield dual economic and ecological gains. This fusion of technical fortitude and market acumen cements wind as an indispensable, dispatchable mainstay in decarbonized grids, accelerating the shift to sustainable power.

The ripple effects demand strategic recalibration across stakeholders. Wind operators must pivot from output-centric models to AI-infused ecosystems, prioritizing digital upskilling in data analytics and algorithmic trading to sustain edges in a competitive arena (Udo & Muhammad, 2021). Grid stewards gain indispensable levers for stability amid inverter-dominated networks, harnessing AI's foresight to trim balancing expenses, refine dispatch, and blueprint resilient infrastructure. Policymakers bear the onus of visionary reforms: evolving market architectures to remunerate AI-enabled flexibility and ancillary contributions, while mandating cybersecurity, data interoperability, and algorithmic transparency in vital infrastructure. Public-private synergies will propel R&D, temper ethical quandaries like bias and opacity, and democratize AI's dividends, ensuring equitable advancement in a digitized energy paradigm.

Ultimately, AI heralds an epochal advance for wind energy, transmuting nature's caprice into computational mastery. As paradigms mature, embracing explainable AI for trust-building audits, digital twins for virtual stress-testing, and autonomous fleets for seamless orchestration, the horizon brims with amplified efficiencies. Envision self-regulating wind arrays, interwoven with solar, batteries, and demand orchestration, birthing a robust, 100% renewable tapestry. This odyssey to sustainability is inexorably bound to digital ingenuity; AI, as linchpin, unlocks wind's latent prowess, not merely clean electrons, but a steadfast, prosperous bedrock for humanity's verdant tomorrow. In this intelligent renaissance, wind energy transcends aspiration, manifesting as a tangible, tech-empowered reality.

REFERENCES

Aghdam, H. H., Gonzalez-Garcia, A., van de Weijer, J., & López, A. M. (2019). Active learning for deep detection neural networks. *Proceedings of the IEEE/CVF International Conference on Computer Vision*, 3672–3680, DOI: 10.1109/ICCV.2019.00377

Badesa, L., Teng, F., & Strbac, G. (2020). Pricing inertia and frequency response with diverse dynamics in a mixed-integer second-order cone programming formulation. *Applied Energy*, *260*, 114334. DOI: 10.1016/j.apenergy.2019.114334

Besnard, F., & Bertling, L. (2010). An approach for condition-based maintenance optimization applied to wind turbine blades. *IEEE Transactions on Sustainable Energy*, *1*(2), 77–83. DOI: 10.1109/TSTE.2010.2049452

Bouckaert, S., Pales, A. F., McGlade, C., Remme, U., Wanner, B., Varro, L., D'Ambrosio, D., & Spencer, T. (2021). *Net zero by 2050: A roadmap for the global energy sector*, .DOI: 10.1787/c8328405-en

Brown, T., & Reichenberg, L. (2020). Decreasing market value of variable renewables is a result of policy, not variability. *ArXiv Preprint ArXiv:2002.05209*, 13, https://doi.org//arXiv.2002.05209.DOI: 10.48550

Chen, X., Huang, L., Liu, J., Song, D., & Yang, S. (2022). Peak shaving benefit assessment considering the joint operation of nuclear and battery energy storage power stations: Hainan case study. *Energy*, *239*, 121897. DOI: 10.1016/j.energy.2021.121897

Dao, P. B. (2022). Condition monitoring and fault diagnosis of wind turbines based on structural break detection in SCADA data. *Renewable Energy*, *185*, 641–654. DOI: 10.1016/j.renene.2021.12.051

Dou, Z., Zhang, C., Wang, W., Wang, D., Zhang, Q., Cai, Y., & Fan, R. (2022). Review on key technologies and typical applications of multi-station integrated energy systems. *Global Energy Interconnection*, *5*(3), 309–327. DOI: 10.1016/j.gloei.2022.06.009

Giebel, G., Brownsword, R., Kariniotakis, G., Denhard, M., & Draxl, C. (2011). The state-of-the-art in short-term prediction of wind power: A literature overview, DOI: 10.11581/DTU:00000017

Gonzalez-Salazar, M. A., Kirsten, T., & Prchlik, L. (2018). Review of the operational flexibility and emissions of gas-and coal-fired power plants in a future with growing renewables. *Renewable & Sustainable Energy Reviews*, *82*, 1497–1513. DOI: 10.1016/j.rser.2017.05.278

Gora, M., Makolo, P., Mwasilu, F., & Justo, J. J. (2025). Dynamic LQR control for improved DC link voltage stability in grid-integrated solar PV system. *Tanzania Journal of Science*, *51*(3), 735–753. DOI: 10.65085/2507-7961.1060

Hamdan, A., Ibekwe, K. I., Ilojianya, V. I., Sonko, S., & Etukudoh, E. A. (2024). AI in renewable energy: A review of predictive maintenance and energy optimization. *International Journal of Science and Research Archive*, *11*(1), 718–729. DOI: 10.30574/ijsra.2024.11.1.0112

Hassan, M. (2025). *AI-Based Conditional Monitoring & Predictive Maintenance for Offshore Wind Farms*, .DOI: 10.30574/ijsra.2025.14.1.0028

Hewamalage, H., Bergmeir, C., & Bandara, K. (2021). Recurrent neural networks for time series forecasting: Current status and future directions. *International Journal of Forecasting*, *37*(1), 388–427. DOI: 10.1016/j.ijforecast.2020.06.008

Hirth, L. (2013). The market value of variable renewables: The effect of solar wind power variability on their relative price. *Energy Economics*, *38*, 218–236. DOI: 10.1016/j.eneco.2013.02.004

Holdmann, G., & Wies, R. (2025). Harnessing the Arctic Wind: How Alaska became a leader in wind energy microgrids. *IEEE Electrification Magazine*, *13*(2), 6–15. DOI: 10.1109/MELE.2025.3558678

Hong, Y.-Y., & Rioflorido, C. L. P. P. (2019). A hybrid deep learning-based neural network for 24-h ahead wind power forecasting. *Applied Energy*, *250*, 530–539. DOI: 10.1016/j.apenergy.2019.05.044

Hou, Y., & Wang, Q. (2023). Big data and artificial intelligence application in energy field: A bibliometric analysis. *Environmental Science and Pollution Research International*, *30*(6), 13960–13973. DOI: 10.1007/s11356-022-24880-y

IRENA. R. E. S. (2020). International renewable energy agency. *Abu Dhabi, 2020*, https://www.irena.org/publications/2020/Jul/Renewable-energy-statistics-2020

Janssens, O., Slavkovikj, V., Vervisch, B., Stockman, K., Loccufier, M., Verstockt, S., Van de Walle, R., & Van Hoecke, S. (2016). Convolutional neural network based fault detection for rotating machinery. *Journal of Sound and Vibration*, *377*, 331–345. DOI: 10.1016/j.jsv.2016.05.027

Kong, X., Zhang, Z., Yin, X., & Wen, M. (2014). Study of fault current characteristics of the DFIG considering dynamic response of the RSC. *IEEE Transactions on Energy Conversion*, *29*(2), 278–287. DOI: 10.1109/TEC.2014.2309694

Lago, J., Marcjasz, G., De Schutter, B., & Weron, R. (2021). Forecasting day-ahead electricity prices: A review of state-of-the-art algorithms, best practices and an open-access benchmark. *Applied Energy*, *293*, 116983. DOI: 10.1016/j.apenergy.2021.116983

Liebensteiner, M., Ocker, F., & Abuzayed, A. (2025). High electricity price despite expansion in renewables: How market trends shape Germany's power market in the coming years. *Energy Policy*, *198*, 114406. DOI: 10.1016/j.enpol.2024.114448

Lim, S., Seo, K., Park, J.-W., & Lee, K. Y. (2025). New frequency stability assessment based on contribution rates of wind power plants. *International Journal of Electrical Power & Energy Systems*, *164*, 110388. DOI: 10.1016/j.ijepes.2024.110388

Lipu, M. S. H., Miah, M. S., Jamal, T., Rahman, T., Ansari, S., Rahman, M. S., Ashique, R. H., Shihavuddin, A. S. M., & Shakib, M. N. (2023). Artificial intelligence approaches for advanced battery management system in electric vehicle applications: A statistical analysis towards future research opportunities. *Vehicles (Basel)*, *6*(1), 22–70. DOI: 10.3390/vehicles6010002

Lund, H., & Mathiesen, B. V. (2009). Energy system analysis of 100% renewable energy systems—The case of Denmark in years 2030 and 2050. *Energy*, *34*(5), 524–531. DOI: 10.1016/j.energy.2008.04.003

Lund, P. D., Lindgren, J., Mikkola, J., & Salpakari, J. (2015). Review of energy system flexibility measures to enable high levels of variable renewable electricity. *Renewable & Sustainable Energy Reviews*, *45*, 785–807. DOI: 10.1016/j.rser.2015.01.057

Madeswaran, A., Bisht, D., Yuvaraj, S., Reedy, M. U., Al-Attabi, K., & Dhablia, A. (2024). AI-controlled wind turbine systems: Integrating IoT and machine learning for smart grids. *E3S Web of Conferences, 540*, 03008, .DOI: 10.1051/e3sconf/202454003008

Makolo, P., Zamora, R., & Lie, T.-T. (2021). The role of inertia for grid flexibility under high penetration of variable renewables-A review of challenges and solutions. *Renewable & Sustainable Energy Reviews*, *147*, 111223. DOI: 10.1016/j.rser.2021.111223

Mantar Gundogdu, B., Gladwin, D. T., Nejad, S., & Stone, D. A. (2019). Scheduling of grid-tied battery energy storage system participating in frequency response services and energy arbitrage. *IET Generation, Transmission & Distribution*, *13*(14), 2930–2941. DOI: 10.1049/iet-gtd.2018.6690

Mengelkamp, E., Gärttner, J., Rock, K., Kessler, S., Orsini, L., & Weinhardt, C. (2018). Designing microgrid energy markets: A case study: The Brooklyn Microgrid. *Applied Energy*, *210*, 870–880. DOI: 10.1016/j.apenergy.2017.06.054

Mocanu, E., Mocanu, D. C., Nguyen, P. H., Liotta, A., Webber, M. E., Gibescu, M., & Slootweg, J. G. (2018). On-line building energy optimization using deep reinforcement learning. *IEEE Transactions on Smart Grid*, *10*(4), 3698–3708. DOI: 10.1109/TSG.2018.2834219

Nghitevelekwa, K., & Bansal, R. C. (2018). A review of generation dispatch with large-scale photovoltaic systems. *Renewable & Sustainable Energy Reviews*, *81*, 615–624. DOI: 10.1016/j.rser.2017.08.035

Odeh, R. P., & Watts, D. (2019). Impacts of wind and solar spatial diversification on its market value: A case study of the Chilean electricity market. *Renewable & Sustainable Energy Reviews*, *111*, 442–461. DOI: 10.1016/j.rser.2019.01.015

Operator, A. E. M. (2024). *Australian Energy Market Operator. 2023.* January.

Paidi, E. S. N. R., Marzooghi, H., Yu, J., & Terzija, V. (2020). Development and Validation of Artificial Neural Network-Based Tools for Forecasting of Power System Inertia With Wind Farms Penetration. *IEEE Systems Journal*, *14*(4), 4978–4989. DOI: 10.1109/JSYST.2020.3017640

Qamsane, Y., Chen, C.-Y., Balta, E. C., Kao, B.-C., Mohan, S., Moyne, J., Tilbury, D., & Barton, K. (2019). A unified digital twin framework for real-time monitoring and evaluation of smart manufacturing systems. *2019 IEEE 15th International Conference on Automation Science and Engineering (CASE)*, 1394–1401, DOI: 10.1109/COASE.2019.8842880

Sheng, Y., Wang, H., Yan, J., Liu, Y., & Han, S. (2023). Short-term wind power prediction method based on deep clustering-improved Temporal Convolutional Network. *Energy Reports*, *9*, 2118–2129. DOI: 10.1016/j.egyr.2023.01.015

Siano, P. (2014). Demand response and smart grids—A survey. *Renewable & Sustainable Energy Reviews*, *30*, 461–478. DOI: 10.1016/j.rser.2013.10.022

Song, J., Zhou, X., Zhou, Z., Wang, Y., Wang, Y., & Wang, X. (2023). Review of Low Inertia in Power Systems Caused by High Proportion of Renewable Energy Grid Integration. *Energies*, *16*(16), 6042. DOI: 10.3390/en16166042

Stecca, M., Elizondo, L. R., Soeiro, T. B., Bauer, P., & Palensky, P. (2020). A comprehensive review of the integration of battery energy storage systems into distribution networks. *IEEE Open Journal of the Industrial Electronics Society*, *1*, 46–65. DOI: 10.1109/OJIES.2020.2981832

Stetco, A., Dinmohammadi, F., Zhao, X., Robu, V., Flynn, D., Barnes, M., Keane, J., & Nenadic, G. (2019). Machine learning methods for wind turbine condition monitoring: A review. *Renewable Energy*, *133*, 620–635. DOI: 10.1016/j.renene.2018.10.047

Sun, P., Li, J., Wang, C., & Lei, X. (2016). A generalized model for wind turbine anomaly identification based on SCADA data. *Applied Energy*, *168*, 550–567. DOI: 10.1016/j.apenergy.2016.01.133

Tamilselvan, P., & Wang, P. (2013). Failure diagnosis using deep belief learning based health state classification. *Reliability Engineering & System Safety*, *115*, 124–135. DOI: 10.1016/j.ress.2013.02.022

Tautz-Weinert, J., & Watson, S. J. (2017). Using SCADA data for wind turbine condition monitoring–a review. *IET Renewable Power Generation*, *11*(4), 382–394. DOI: 10.1049/iet-rpg.2016.0248

Tchakoua, P., Wamkeue, R., Tameghe, T. A., & Ekemb, G. (2013). A review of concepts and methods for wind turbines condition monitoring. *2013 World Congress on Computer and Information Technology (WCCIT)*, 1–9, DOI: 10.1109/WCCIT.2013.6618706

Udo, W., & Muhammad, Y. (2021). Data-driven predictive maintenance of wind turbine based on SCADA data. *IEEE Access : Practical Innovations, Open Solutions*, *9*, 162370–162388. DOI: 10.1109/ACCESS.2021.3132684

Ukoba, K., Olatunji, K. O., Adeoye, E., Jen, T.-C., & Madyira, D. M. (2024). Optimizing renewable energy systems through artificial intelligence: Review and future prospects. *Energy & Environment*, *35*(7), 3833–3879. DOI: 10.1177/0958305X241256293

Ummels, B. C., Pelgrum, E., & Kling, W. L. (2008). Integration of large-scale wind power and use of energy storage in the Netherlands' electricity supply. *IET Renewable Power Generation*, *2*(1), 34–46. DOI: 10.1049/iet-rpg:20070056

Vázquez-Canteli, J. R., & Nagy, Z. (2019). Reinforcement learning for demand response: A review of algorithms and modeling techniques. *Applied Energy*, *235*, 1072–1089. DOI: 10.1016/j.apenergy.2018.11.002

Wang, H., Lei, Z., Zhang, X., Zhou, B., & Peng, J. (2019). A review of deep learning for renewable energy forecasting. *Energy Conversion and Management*, *198*, 111799. DOI: 10.1016/j.enconman.2019.111799

Wang, Z., & Bu, S. (2024). Probabilistic frequency stability analysis considering dynamics of wind power generation with different control strategies. *IEEE Transactions on Power Systems, 39*(5), 6412–6425. DOI: 10.1109/TPWRS.2024.3362056

Wiser, R. H., Bolinger, M., Hoen, B., Millstein, D., Rand, J., Barbose, G. L., Darghouth, N. R., Gorman, W., Jeong, S., & Mills, A. D. (2020). *Wind energy technology data update: 2020 edition*, https://emp.lbl.gov/publications/wind-energy-technology-data-update

Wozabal, D., Graf, C., & Hirschmann, D. (2016). The effect of intermittent renewables on the electricity price variance. *OR-Spektrum, 38*(3), 687–709. DOI: 10.1007/s00291-015-0395-x

Zhao, J., Lyu, X., Fu, Y., Hu, X., & Li, F. (2016). Coordinated microgrid frequency regulation based on DFIG variable coefficient using virtual inertia and primary frequency control. *IEEE Transactions on Energy Conversion, 31*(3), 833–845. DOI: 10.1109/TEC.2016.2537539

Zhirong, Z., Hao, Y., Feng, W., Fang, Z., & Zhenxiong, W. (2016). A novel control strategy of photovoltaic-battery system for restraining the photovoltaic power fluctuations and suppressing the low frequency oscillations of power system. *2016 IEEE 8th International Power Electronics & Motion Control Conference (IPEMC-ECCE Asia)*, 2978, .DOI: 10.1109/IPEMC.2016.7512770

Zin, A. A. B. M. HA, M. P., Khairuddin, A. B., Jahanshaloo, L., & Shariati, O. (2013). An overview on doubly fed induction generators′ controls and contributions to wind based electricity generation. *Renewable and Sustainable Energy Reviews, 27*, 692–708, DOI: 10.1016/j.rser.2013.07.010

Chapter 9
HVDC Transmission for Wind Energy Systems in Microgrids and Virtual Power Plants

Francis Mwasilu
http://orcid.org/0000-0001-7540-4198
University of Dar es Salaam, Tanzania

ABSTRACT

A significant logistical challenge inherent to wind energy is that the most abundant and consistent wind resources are frequently situated in remote, inaccessible areas, such as deep offshore locations or isolated plains, far from the urban centers that constitute primary electricity load centers. This geographical disparity necessitates a highly efficient, reliable, and economically viable method for evacuating large volumes of power over extended distances. This task is often proven inadequate for conventional alternating current (AC) transmission systems due to substantial power losses and stability issues.

1. INTRODUCTION

1.1 Overview of Wind Energy Systems

Wind energy has solidified its position as a cornerstone of the global transition to a clean energy portfolio, gathering immense attention for its potential to provide a sustainable and virtually carbon-free power source. The fundamental principle involves converting the kinetic energy of moving air into electrical energy through

DOI: 10.4018/979-8-2600-0106-6.ch009

wind turbines, which are deployed in wind farms. These installations, both onshore and offshore, now contribute significantly to the energy mix of numerous national and regional electricity grids (IRENA, 2023; Wu et al., 2024) Continuous technological advancements have been pivotal in this expansion, leading to remarkable increases in turbine efficiency, scalability, and grid integration capacity. Modern turbines, with larger rotor diameters and taller towers, can capture energy more effectively, while sophisticated power electronics have improved their ability to support grid stability. This evolution has transformed wind energy conversion systems from a niche alternative into a mainstream, cost-competitive energy source essential for meeting global climate targets (Jiang, 2021; Li et al., 2017; Souza et al., 2022). Figure 1 shows the current wind evolution trend with rotor size and power extraction capability. The most striking trend is that power extraction capability grows significantly faster than rotor diameter. While the rotor diameter has roughly doubled from 2005 (~90m) to 2020 (220m), the power output has increased by a factor of over five (from 2.5 MW to 13 MW). This visually demonstrates the aerodynamic principle where power capture is proportional to the swept area of the rotor, meaning a doubling of the diameter leads to a quadrupling of the potential power capture. An inflection point is noticeable around 2010-2015. The move from land-based to offshore wind farms removed logistical constraints on transportation and installation, allowing for a rapid escalation in turbine size. Turbines with rotor diameters exceeding 150m became practical only in offshore environments. The trend indicates that by 2030, state-of-the-art offshore turbines are projected to have rotor diameters surpassing 300 meters and power ratings exceeding 25 MW. This massive scaling is a primary driver in reducing the Levelized Cost of Energy (LCOE) for offshore wind, making it increasingly competitive with traditional energy sources.

Figure 1. Current wind evolution trend with rotor size and power extraction capability.

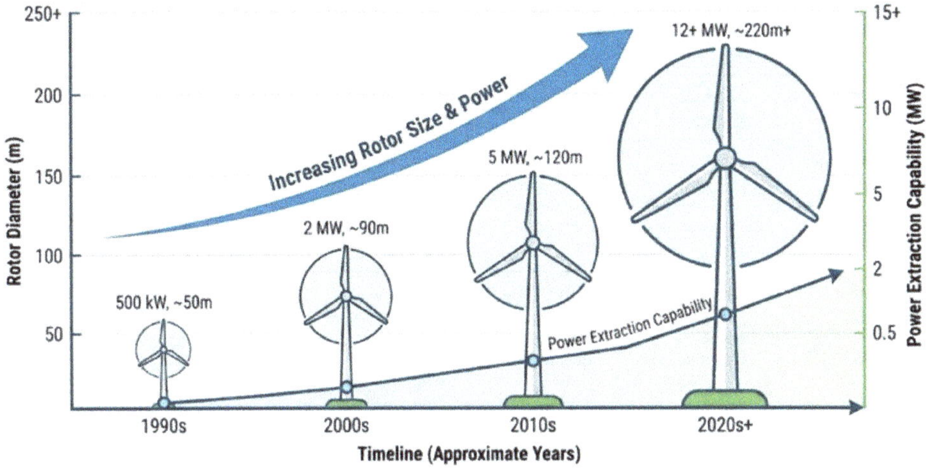

Therefore, the focus has increasingly shifted towards offshore wind resources, which present a vastly untapped potential. Offshore wind farms, situated in deep-sea environments like the Hornsea wind farm, off the east coast of England and Hywind Tampen floating wind farm in the Norwegian North Sea, benefit from stronger, more consistent, and less turbulent wind profiles compared to their onshore counterparts. This results in significantly higher capacity factors, meaning they can generate electricity closer to their maximum potential output more consistently. As summarized in Table 1, current North Sea projects demonstrate a clear correlation between distance from shore and technological complexity. It highlights the rapid scaling of turbine dimensions, contrasting the 8.0 MW turbines used at Hornsea 2 with the massive 13 MW GE Haliade-X units at Dogger Bank, while also distinguishing between standard fixed-bottom monopiles in shallow waters and floating spar solutions required for deep-water sites like Hywind Tampen (GWEC, 2023).

Table 1. Comparative analysis of major North Sea wind farms

Wind Farm	Location	Distance from shore / Water Depth	Foundation Type	Turbine Technology	Rotor / Power
Hornsea 2	United Kingdom	89 km / 30–40 m	Fixed (Monopile)	Siemens Gamesa SG 8.0-167 DD	167 m / 8.0 MW
Dogger Bank A, B & C	United Kingdom	130 km+ / 20–35 m	Fixed (Monopile)	GE Haliade-X	220 m / 13–14 MW
Hollandse Kust Zuid	Netherlands	~20 km / 18–28 m	Fixed (Monopile)	Siemens Gamesa SG 11.0-200 DD	200 m / 11.0 MW
Hywind Tampen	Norway	140 km / 260–300 m	Floating (Spar)	Siemens Gamesa SG 8.0-167 DD	167 m / 8.6 MW*
Gode Wind 3	Germany	~40 km / 30 m	Fixed (Monopile)	Siemens Gamesa SG 11.0-200 DD	200 m / 11.0 MW

However, this immense promise is counterbalanced by a set of formidable logistical and technical challenges. The geographical isolation of these prime generation sites often dozens to hundreds of kilometers from major load centers and the existing grid infrastructure, complicates the reliable and economic delivery of power. Transporting massive turbine components, performing maintenance in harsh marine environments, and, most critically, establishing a viable connection to the onshore grid present substantial hurdles (GWEC, 2023). The conventional use of High-Voltage Alternating Current (HVAC) transmission for these long distances becomes increasingly inefficient due to high capacitive losses, necessitating a more robust technological solution for successful power evacuation. This push for efficient long-distance power transmission from remote wind resources has catalyzed the integration of High Voltage Direct Current (HVDC) technology into wind energy systems. HVDC transmission is uniquely suited to address the limitations of HVAC over extended distances, offering lower electrical losses, enhanced controllability of power flow, and the ability to connect asynchronous grids (Korompili et al., 2016; Mohammadi et al., 2024; Yu et al., 2024). Its application is particularly critical in the context of modern grid architectures like microgrids and Virtual Power Plants (VPPs), which aim to aggregate and optimize diverse distributed energy resources. By providing a stable and controllable backbone, HVDC links enable the reliable incorporation of large-scale, variable wind power into these decentralized systems, thereby mitigating intermittency issues and improving overall grid resilience (Mohammadi et al., 2024).

1.2 Challenges of Integrating Wind Power into Decentralized Grids

Integrating large-scale wind energy into decentralized or weak electrical grids presents a paradigm shift from traditional, centralized power system operation. These grids, often characterized by lower short-circuit strength and less rotational inertia, are highly susceptible to the unique characteristics of wind power, leading to a host of technical and operational challenges that must be addressed to ensure reliability and stability (Dreidy et al., 2017; Gabriel et al., 2025; Heylen et al., 2021). Modern HVDC transmission system with Voltage Sourced Converter (VSC) can complement most of the challenges exemplified in the following subsections, intrinsically realized in wind energy conversion systems integrated into grid network today (Korompili et al., 2016; Li et al., 2017; Pfeifenberger et al., 2023).

1.2.1 Intermittency and Variability

The fundamental challenge of wind energy integration stems from its inherent intermittency and variability. Unlike conventional power plants such as thermal power plants or hydro power, the electrical output of a wind farm is not dispatchable but is entirely dependent on meteorological conditions, which can change rapidly over short time scales. This stochastic nature introduces significant fluctuations in both active and reactive power injection into the grid (IRENA, 2019). In a decentralized grid with limited generation resources, these power swings can lead to severe voltage sags and swells, directly impacting power quality for connected consumers and potentially causing protective device malfunctions. More critically, on a system-wide level, a sudden drop in wind generation creates an instantaneous generation-load deficit. In traditional systems, this deficit is initially compensated for by the rotational inertia stored in the massive spinning rotors of synchronous generators, which slows the rate of frequency decline and gives time for other resources to respond (Gabriel et al., 2025). However, wind turbines, particularly those using power electronics-based Type 3 with doubly-fed induction generators (DFIG) and Type 4 with permanent magnet synchronous generators (PMSG) technologies, are decoupled from the grid frequency and provide little to no inherent inertial response (Ratnam et al., 2020; Ullah et al., 2025). This lack of natural damping effect means that frequency deviations in a wind-heavy grid can be faster and more severe, increasing the risk of under-frequency load shedding or even cascading blackouts if not properly managed with fast-responding reserves and advanced forecasting (Dreidy et al., 2017; Heylen et al., 2021). Figure 2 shows different types of wind energy conversion systems available in practical wind farms today.

Figure 2. Different types of wind energy conversion systems available in practical wind farms.

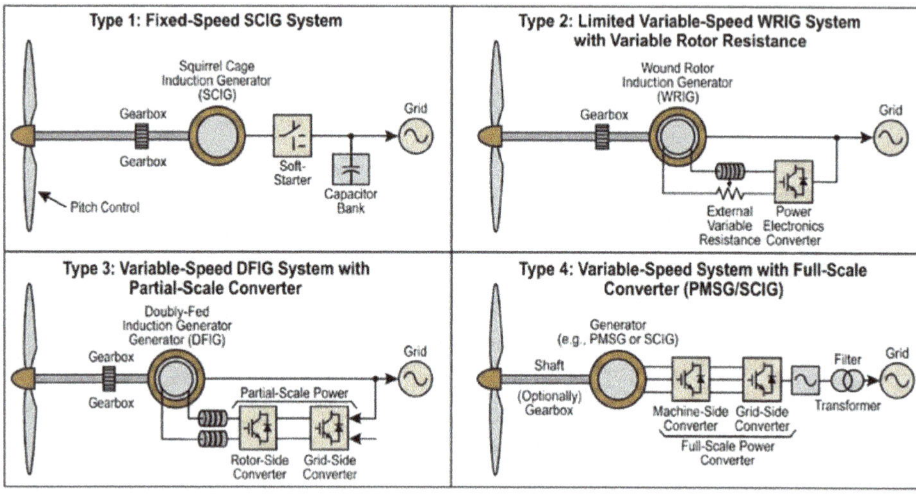

1.2.2 Long Transmission Distance of Offshore Wind Farms to Load Centers

The most promising sites for wind energy, particularly for maximizing capacity factor, are often located in remote areas far from the urban and industrial centers that constitute the major loads. This is particularly the case for offshore wind farms, which are increasingly being developed in deep-sea locations tens to hundreds of kilometers from the coastline (Enevoldsen & Xydis, 2019). This geographical disparity creates a daunting power transmission challenge. Using conventional HVAC transmission technology for such long distances becomes highly problematic due to the phenomenon of capacitive charging current. In AC submarine cables, the conductor acts as a capacitor against the ground, and over long distances, the current required just to charge this capacitance becomes significant, consuming the cable's thermal current-carrying capacity (Van Hertem et al., 2016). This results in substantial reactive power losses, requiring the installation of large and costly shunt reactors at regular intervals to compensate, and ultimately limits the feasible transmission distance and power capacity. For distances beyond approximately 80-100 kilometers for submarine cables, HVAC systems become technically and economically not viable. This distance barrier necessitates an alternative transmission solution to successfully evacuate the vast amounts of power generated by these remote wind resources and deliver it efficiently to the onshore grid, a challenge that has become a central focus of offshore wind development plans worldwide (Dakic et al., 2020; Marzbnai et al., 2025).

1.2.3 Grid Congestion in AC Transmission Corridors

As the penetration of wind and other renewable energy sources increases, existing AC transmission corridors often become bottlenecks, leading to grid congestion. This occurs when the power scheduled to flow through a particular transmission line or transformer exceeds its thermal or stability limits. Congestion is a particular issue in decentralized grids where the network was originally designed for a one-way flow of power from centralized conventional thermal power plants to load centers. The integration of large-scale wind farms in peripheral locations forces a reconfiguration of these power flows, often overloading lines that were not designed for such purposes. For instance, a powerful wind farm in a coastal region may generate more power than the local load can consume, forcing the surplus to be exported over HVDC interconnectors to neighboring regions. If these interconnectors lack sufficient capacity, they become congested. This not only poses a physical risk of overheating and damaging assets but also has significant economic consequences. System operators are forced to curtail wind power generation, even when it is available and cost-effective, and dispatch more expensive local generation to manage the congestion, leading to higher electricity prices and inefficiencies that undermine the economic and environmental benefits of renewable energy (Pfeifenberger et al., 2023).

1.2.4 Grid Network Stability Issues

The large-scale replacement of conventional synchronous generators with inverter-based resources (IBRs) like wind and solar power leads to inertia reduction in a particular grid network, which fundamentally undermines the stability of AC power systems. In a traditional grid, the kinetic energy stored in the spinning masses of synchronous generators provides a crucial buffer against disturbances. When a generator trips or a large load is suddenly connected, this inertia automatically opposes the change in system frequency, slowing its rate of change (RoCoF) and providing a critical window, typically several seconds, for primary frequency response or automatic generation control (AGC) to activate (Gabriel et al., 2025; Pfeifenberger et al., 2023; Ullah et al., 2025). Wind turbines, however, are connected via power electronics that decouple their rotor speed from the grid frequency. Consequently, they do not naturally contribute to this inertial response. As the share of wind power keep on growing, the total system inertia decreases, making the grid stiffer and more fragile (Heylen et al., 2021). In such a low-inertia system, any imbalance between generation and load results in a much faster and more severe frequency deviation, increasing the risk of reaching protection thresholds and triggering widespread disconnections. This necessitates the development of synthetic or virtual inertia solutions from wind turbines and other IBRs, where control algorithms detect frequency changes and temporarily inject additional power into the grid, mimicking the behavior of physical inertia to preserve grid stability (Sang et al., 2022; Wu et al., 2024).

1.2.5 Limited Reactive Power Control From Wind Farms

Reactive power is essential for maintaining voltage levels within statutory limits throughout the power system. It is required to energize magnetic fields in transformers and motors and is directly related to voltage control. Traditional induction generator-based wind turbines (e.g., Type 1 and some Type 2, refer to Figure 2 above) are known for being net consumers of reactive power, exacerbating voltage drops on weak grids, especially during start-up. While modern full-scale converter wind turbines (e.g., Type 4) have the inherent capability to control their reactive power output, their operational mandates have often prioritized other functions. Typically, these turbines are operated at a unity power factor to maximize active power export, meaning they neither consume nor produce reactive power. This represents a missed opportunity for grid support. In decentralized grids, where voltage stability is a primary concern, the ability of distributed generators to provide dynamic reactive power support is invaluable. It can help maintain voltage profiles during contingencies, reduce transmission losses, and increase the grid's hosting capacity for more renewables. The evolving grid codes are now mandating such capabilities, requiring wind farms to behave like grid-forming assets that can regulate voltage and frequency, not just follow it (IRENA, 2022; Wu et al., 2024). Effectively harnessing this potential requires sophisticated plant-level controllers that can coordinate the reactive power output of dozens of turbines to provide a unified and robust response to grid disturbances.

1.3 Microgrids and Virtual Power Plants (VPPs) in Wind Integration

Microgrids represent a paradigm shift from centralized power generation to localized, decentralized power networks capable of operating either in tandem with the main grid or autonomously as an island during main grid outages. Their core function is to integrate a diverse portfolio of distributed energy resources (DERs), including wind turbines, solar photovoltaics, and battery energy storage systems (BESS) within a defined electrical boundary to serve local loads (Farrokhabadi et al., 2019). This architecture significantly enhances energy resilience and reliability, making it particularly valuable for critical infrastructure, remote communities, and areas with unstable grid connections. By managing generation and consumption locally, microgrids can mitigate the impact of disturbances on the main grid and optimize the use of local renewables. However, integrating a variable resource like wind power as a primary generation source within a microgrid introduces complex stability challenges. The microgrid energy management system (EMS) must continuously balance the stochastic output of the wind turbines with the state of charge of the batteries and the local load demand, a task that requires sophisticated forecasting and real-time control to prevent voltage or frequency violations, especially in islanded mode where no external grid supports stability (Bidram & Davoudi, 2012; Hirsch et al., 2018).

Virtual Power Plants (VPPs) represent a more abstract but equally transformative approach to managing the energy transition. Rather than being a physically connected network like a microgrid, a VPP is an advanced aggregation of numerous, geographically dispersed DERs from residential solar-plus-storage systems and commercial backup generators to utility-scale wind farms coordinated through a cloud-based digital platform (Esfahani et al., 2025). This platform uses sophisticated software and communications technology to aggregate the capacity and flexibility of these diverse assets, enabling them to function as a single, unified, and remotely dispatchable power plant. By pooling resources, a VPP can effectively smooth out the aggregate output of intermittent renewables; for example, a drop in wind generation at one site might be compensated by solar generation or discharged batteries elsewhere in the portfolio. This aggregation dramatically enhances the dispatchability, predictability, and market value of wind energy, allowing VPP operators to participate in wholesale energy markets, offer ancillary services like frequency regulation, and defer costly traditional grid upgrades by providing non-wire alternatives. This turns a collection of variable generators into a reliable and market-competitive utility asset (Abdelkader et al., 2024; Esfahani et al., 2025; Kaiss et al., 2025). Figure 3 illustrates the VPP conceptual framework, which describes the transition from decentralized distributed generation to a unified, dispatchable power house. It is observed that diverse and dispersed generation sets are aggregated into a more coordinated aand ochestrated power plant. As a result renwable energy resources such as wind farmss become reliable, markert compitive utility asset.

Figure 3. Vitual power plant conceptual framework from decentralized distributed generation system to disptachable powerhouse.

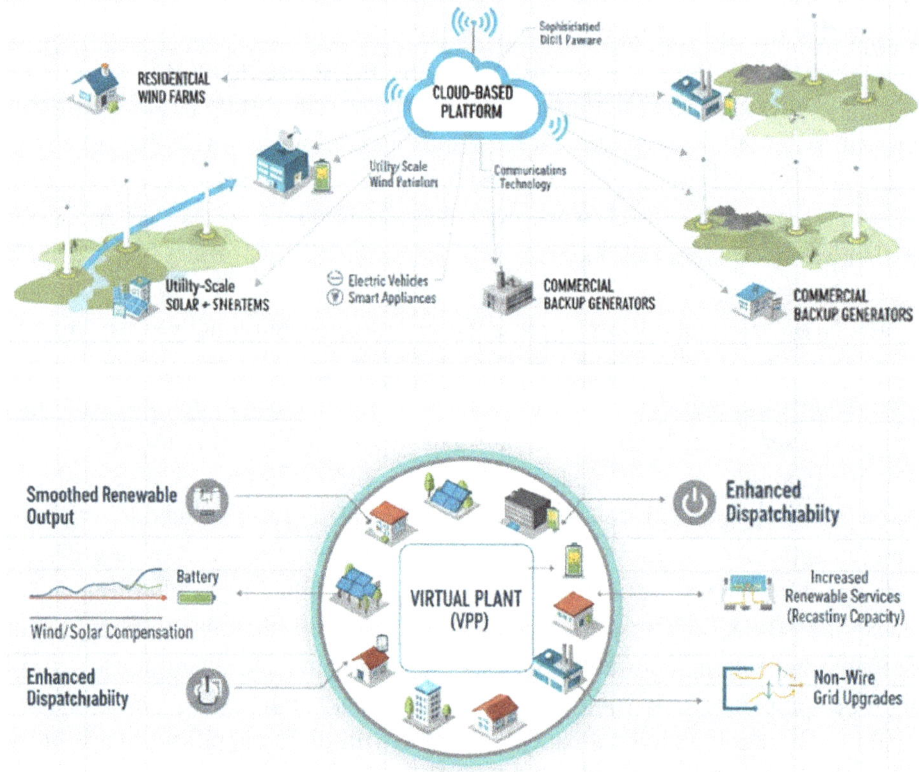

While both microgrids and VPPs are instrumental in enhancing grid flexibility and integrating renewables, their effectiveness is critically challenged by the fundamental characteristics of large-scale wind energy, inherent variability and its remote location. Conventional AC networks, which form the backbone of most existing distribution and transmission systems, struggle with the long-distance transmission of bulk power from these remote wind sites to the load centers where microgrids and VPP assets are often concentrated. The technical limitations of HVAC, including significant reactive power losses and voltage instability over long ac cables, can create a bottleneck, preventing the efficient and reliable transfer of wind energy to load centers. This is where HVDC technology creates a synergetic relationship. HVDC systems offer a superior solution for bulk power transmission over long distances, with lower losses and no distance-related charging current issues. By providing a stable, efficient, and controllable highway for wind power, HVDC links can deliver large blocks of energy from remote wind farms directly to the points of interconnection for microgrids or into the portfolio of a VPP. This enables the reliable injection of wind energy at a scale that would be impossible with AC networks, thereby strengthening the business case and operational feasibility of both decentralized microgrids and widely aggregated VPPs in a future dominated by renewable energy. As shown in Figure 4, HVDC provides a stable, efficient link for bulk wind power integration, enabling reliable integration for microgrids and VPPs.

Figure 4. HVDC-enabled wind farm integration for decentralized microgrid and aggregated VPP

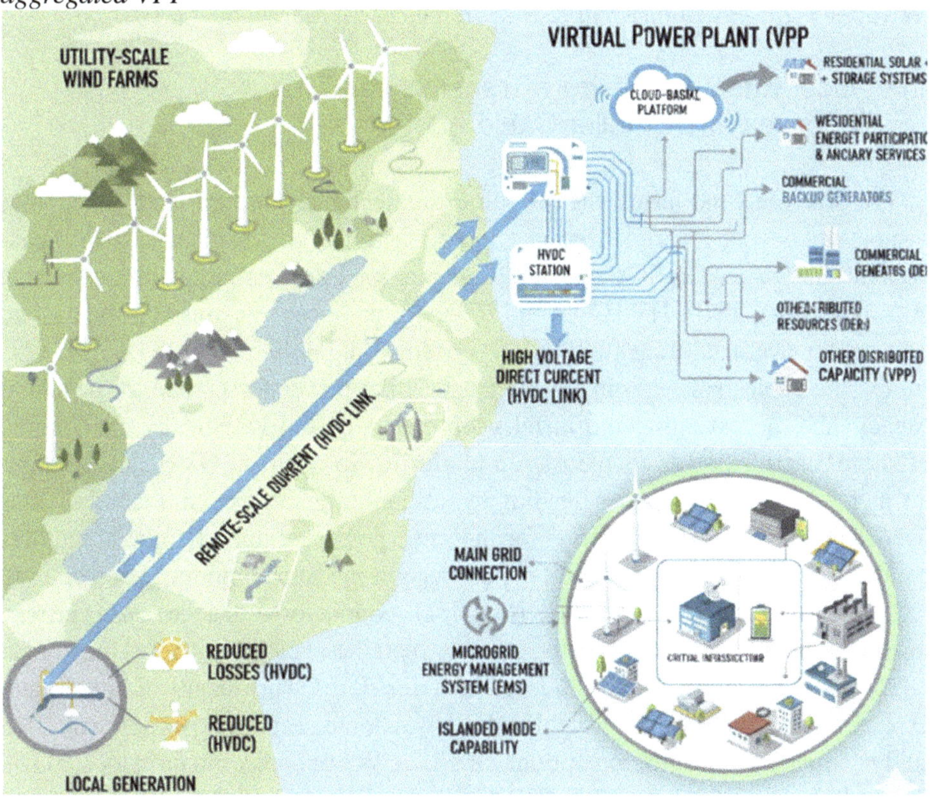

1.4 Introduction to HVDC Technology for Wind Systems

HVDC technology has emerged as a critical enabler for the global integration of renewable energy, addressing fundamental limitations of traditional AC transmission. It facilitates efficient, long-distance, and high-capacity power transfer by converting AC to DC for transmission and back to AC at the point of consumption. Unlike AC systems, HVDC is not limited by capacitive charging currents, experiences lower resistive losses over long distances, and provides operators with full, rapid control over the magnitude and direction of power flow (Van Hertem et al., 2016). A key advantage is its ability to interconnect asynchronous AC grids, such as those of different countries or regions with different frequencies, even voltages without the risk of cascading failures. For wind energy systems, particularly large-scale and remote offshore farms, these characteristics translate into several distinct and powerful advantages. HVDC transmission can be realized by employing either natural commutation converter technology, that is, Line commutated converter (LCC-HVDC) or forced commutation converter technology, that is, voltage sourced converter (VSC-HVDC) (Khan & Bhowmick, 2022; Marzbnai et al., 2025).

1.4.1 Cost-Effective Offshore Transmission With Lower Losses and Smaller Cable Footprints

For offshore wind farms, the economic and technical case for HVDC becomes compelling beyond a certain distance and power rating, often referred to as break-even distance. Specifically, the HVDC transmission line become economical viable in comparison with the HVAC beyond 50 km for undersea transmission and beyond 600 km for overhead transmission line system (Van Hertem et al., 2016). HVAC submarine cables suffer from significant reactive power losses due to their high capacitance, necessitating the installation of costly and logistically challenging offshore reactive power compensation platforms. In contrast, HVDC cables have no such reactive power issues, leading to substantially lower overall energy losses over long distances. Furthermore, for the same power rating, a DC cable requires a much smaller physical footprint; it can transmit the same amount of power as an AC cable using only two conductors (or a single conductor with a sea return) instead of three, and without the need for a heavy metallic sheath (Khan & Bhowmick, 2022; Marzbnai et al., 2025; Van Hertem et al., 2016). This results in lighter, less expensive, and easier-to-install cables, directly reducing capital expenditure. A 2023 study comparing transmission options for a 1.2 GW offshore wind farm located 150 km from shore found that a VSC-HVDC solution offered a 15-20% lower levelized cost of transmission compared to an HVAC system with compensation, primarily due to lower losses and the avoidance of multiple offshore platforms (Lyu et al., 2024). This makes HVDC the only technically and economically viable solution for gigawatt-scale wind farms in deep sea locations.

1.4.2 Enhanced Grid Stability With HVDC Links

Modern HVDC systems, particularly those based on VSC technology, are equipped with advanced power electronics that allow them to act as powerful stability assets for the AC grids to which they are connected (Li et al., 2017). A VSC-HVDC terminal can independently control active power flow and reactive power at each connection point. This means it can not only import or export real power but also dynamically inject or absorb reactive power to regulate AC voltage, much like a static synchronous compensator (STATCOM). This capability is crucial for weak onshore grid connection points often associated with large wind farms. Moreover, since the power electronics can respond in milliseconds, HVDC links can provide rapid frequency support. If a frequency dip is detected in the onshore grid, the HVDC control system can instantly increase the power injection (using the stored energy in the DC link and the wind farm available power) to mitigate the frequency collapse. This fast frequency response is a critical service in grids with reduced inertia capability. Research and practical site results have demonstrated that the intelligent control of HVDC links can significantly improve the transient stability and damping of low-frequency oscillations in power systems, making the entire grid network more resilient to disturbances (Pfeifenberger et al., 2023).

1.4.3 Scalable Infrastructure Suited for Large-Scale Wind Farms and Multi-Terminal DC Systems

The modular nature of VSC-HVDC technology makes it inherently scalable, perfectly aligning with the trend towards larger offshore wind farms and interconnected offshore grids. While early projects were point-to-point links connecting a single wind farm to shore, the development of Multi-Terminal HVDC (MTDC) systems enabler to bulk power transfer to interconnected regional grid network or microgrids (Chaudhuri et al., 2014). An MTDC network functions like an offshore energy highway, where multiple wind farms and multiple onshore connection points can be interconnected via a common DC grid. This architecture offers tremendous advantages as enhances security of supply (if one onshore converter fails, power can be rerouted through another), allows for the pooling of wind resources from a large geographical area to smooth overall power output, and provides access to electricity markets in different countries or regions (Van Hertem et al., 2016). The scalability of HVDC is evident in projects like the North Sea Wind Power Hub concept, which envisions a meshed DC grid connecting tens of gigawatts of wind capacity across several nations (Lüth & Keles, 2024). Furthermore, the standardized, modular design of VSC stations allows for capacity to be increased incrementally as more wind farms are built, providing a future-proof transmission infrastructure. Figure 5 summarizes the capabilities of HVDC technology for wind power, enabling long-distance, high-capacity integration of renewable energy sources.

Figure 5. HVDC technology capability for wind power integration

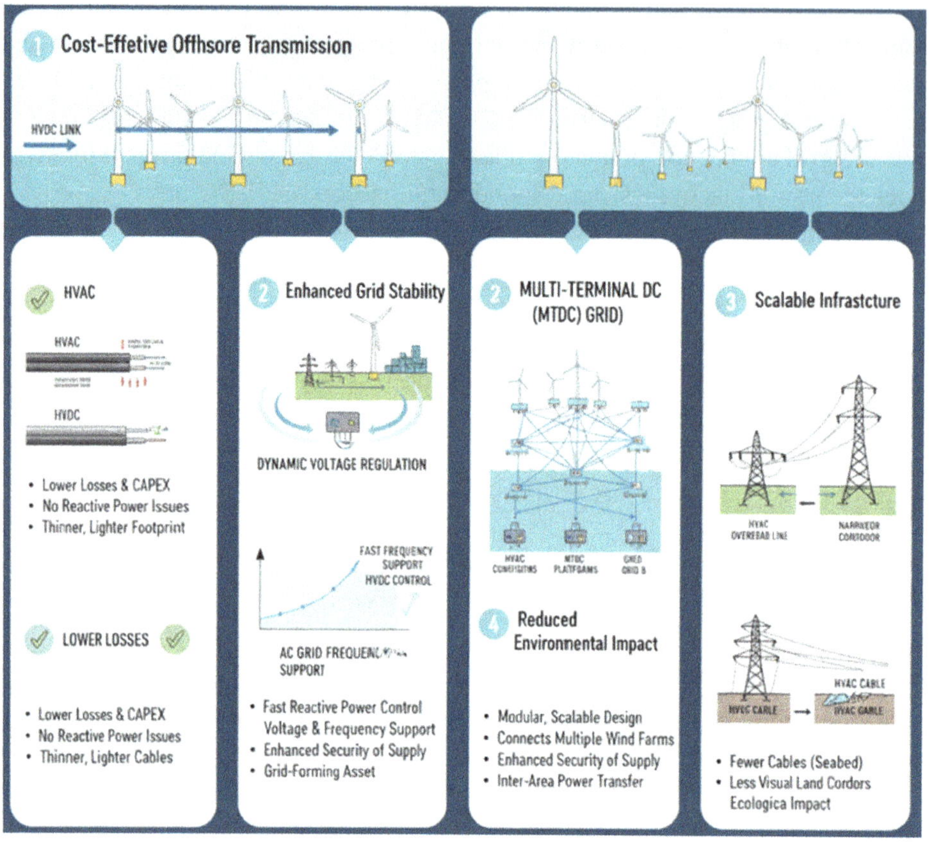

1.4.4 Reduced Environmental Impact Compared to HVAC

HVDC transmission offers a clear advantage in minimizing the environmental footprint of energy infrastructure, both at sea and on land. As mentioned, an HVDC cable system requires fewer conductors than an equivalent AC system to transmit the same amount of power. This translates directly into a reduced physical footprint on the seabed, which is a significant consideration in environmentally sensitive marine areas. On land, the advantage is even more pronounced. HVDC overhead lines can transmit more power per right-of-way than AC lines. For the same voltage level and power capacity, HVDC lines require a narrower corridor and have a lower visual impact. A comparative analysis often shows that a bipolar HVDC line can transmit power equivalent to a double-circuit AC line while using only about half the corridor width (Kalair et al., 2016; Pfeifenberger et al., 2023). This reduction in land use is a critical factor in gaining regulatory approval and public acceptance for new transmission projects needed to bring renewable energy from remote resource centers to load areas. By enabling more efficient use of space and resources, HVDC technology not only solves a technical problem but also contributes to the broader sustainability goals of the energy transition.

2. HVDC SYSTEM ARCHITECTURES FOR WIND INTEGRATION

The effective integration of wind power into modern energy systems is heavily dependent on selecting the appropriate HVDC architecture. These architectures are not merely about connecting point A to point B; they represent sophisticated solutions tailored to the scale, location, and grid requirements of wind generation assets. This section investigates the core technologies and topological configurations that define how HVDC enables the large-scale harnessing of wind energy.

2.1 HVDC Transmission Technologies

The choice of converter technology is foundational to the capabilities and limitations of any HVDC system. Two primary technologies dominate the landscape, each with distinct operational principles and suitability for wind integration.

2.1.1 Line-Commutated Converter (LCC-HVDC)

Based on thyristor technology, LCC-HVDC is a mature and robust solution renowned for its high efficiency and capacity, making it well-suited for bulk power transfer over very long distances. However, its application in wind energy is constrained by several inherent characteristics. LCC converters require a synchronous voltage source for commutation, the process of transferring current from one valve to another. This means they depend on a strong and stable AC grid at both the sending and receiving ends. Furthermore, LCC-HVDC consumes significant reactive power over 50% of the rated capacity, necessitating large and costly capacitor banks for compensation, and is prone to commutation failures during AC grid disturbances, which can lead to power interruptions (Arrillaga et al., 2007; Jovcic, 2019). While cost-effective for certain point-to-point applications from strong grids, these limitations have rendered LCC-HVDC less favorable for the majority of modern offshore wind projects.

2.1.2 Voltage-Source Converter (VSC-HVDC)

Utilizing Insulated-Gate Bipolar Transistors (IGBTs), VSC-HVDC represents a more flexible and advanced technology that has become the de facto standard for wind integration, particularly offshore. Its key advantage lies in its ability to independently control active and reactive power at each terminal. A VSC-HVDC can generate its own stable AC voltage waveform (grid forming capability), meaning it does not rely on the connected AC grid for commutation. This makes it perfectly suited for connecting to weak grids or even entirely passive networks, such as an offshore windfarm where the converters must establish the grid frequency and voltage (i.e., grid-forming capability), providing a vital black-start capability (Pfeifenberger et al., 2023). VSC-HVDC offers features like rapid power reversal without requiring voltage polarity change and provides inherent active filtering, resulting in superior power quality (Flourentzou et al., 2009). The modular Multilevel Converter (MMC) topology, a leading VSC technology recently, further reduces harmonic distortion and losses, solidifying the role of VSC-HVDC as the enabling technology for complex wind integration architectures like multi-terminal systems (Sharifabadi et al., 2016).

2.2 Point-to-Point HVDC Links

The point-to-point configuration is the simplest and most widely deployed HVDC architecture for wind energy. In this topology, a single offshore converter station is connected to a wind farm (or a cluster of turbines) and linked via a DC cable to a single onshore converter station, which interfaces with the main AC grid. This one-to-one approach is highly effective and economically viable for medium to large-scale offshore wind projects located beyond the economic distance of HVAC transmission. Its primary advantages are simplicity in control, protection, and project execution. The dedicated link provides efficient, low-loss transmission and, when using VSC technology, can offer critical grid support services to the onshore connection point, such as voltage stability and frequency regulation. Numerous projects, such as the BorWin series in the German North Sea, exemplify this model, where individual converter platforms transmit gigawatts of power from specific wind farm clusters directly to the German coast (Pfeifenberger et al., 2023). While straightforward, this architecture lacks redundancy; a fault on the single DC cable or a converter outage results in a complete loss of the transmission path (Jovcic, 2019).

2.3 Multi-Terminal HVDC Systems

To enhance security, flexibility, and economic efficiency, Multi-Terminal HVDC (MT-HVDC) systems are emerging as the next evolutionary step. These systems interconnect three or more converter stations through a common DC network, which can be arranged in a radial star topology or a more resilient meshed configuration. For wind integration, this means multiple offshore wind farms, potentially owned by different developers, can be connected to a shared HVDC transmission infrastructure, which then delivers power to multiple onshore connection points, possibly in different countries. This architecture offers significant benefits such as providing redundancy, as power can be re-routed if one line or converter fails; it allows for the pooling of wind generation from a wide geographical area, smoothing out aggregate power output; and it enables access to different electricity markets, optimizing revenue. The development of efficient and reliable HVDC circuit breakers is a key enabler for meshed MTDC grids, as they allow for the selective isolation of faulty sections without collapsing the entire network (Chaudhuri et al., 2014). The Zhangbei project in China is a pioneering example of a four-terminal VSC-based system that integrates large-scale wind and solar energy, demonstrating the operational feasibility of this advanced architecture (An et al., 2025; Liu et al., 2022; Pang & Wei, 2018).

2.4 HVDC Grids for Microgrid and VPP Integration

The concept of HVDC extends beyond massive offshore transmission to the distribution level, where HVDC microgrids and regional DC grids present a paradigm shift for local energy management. In areas with high penetration of distributed renewables, such as industrial parks, remote islands, or university campuses, an internal HVDC grid can be established. This architecture connects local wind and solar generation, battery storage, and DC loads (like electric vehicle charging stations and data centers) directly via a DC bus, eliminating repeated AC/DC conversion losses. Such a grid can operate in a highly flexible and resilient manner, either interconnected with the main AC grid or autonomously in islanded mode. For Virtual Power Plants (VPPs), a regional HVDC grid provides the physical backbone to aggregate diverse and geographically dispersed resources at a much larger scale and with greater controllability than a conventional AC network would allow. The HVDC grid ensures that the power from a distant wind farm, a local solar park, and a distributed fleet of batteries can be pooled and dispatched as a single, firm resource to provide grid services or participate in energy markets, thereby maximizing the value and reliability of the aggregated portfolio.

3. CONTROL STRATEGIES IN HVDC-ENABLED MICROGRIDS AND VPPS

The sophisticated integration of HVDC links into microgrids and Virtual Power Plants (VPPs) necessitates a multi-layered control hierarchy. VSC-HVDC are These strategies ensure that the superior hardware capabilities of HVDC are fully leveraged to deliver stable, efficient, and economically optimal operation. The control architecture spans from the rapid internal management of individual converters to the high-level, market-driven coordination of an entire aggregated portfolio.

3.1 HVDC Converter-Level Control

At the heart of any VSC-HVDC system are the converter-level controls, which govern the fundamental operation of each terminal. These fast-acting algorithms, typically implemented in a cascaded loop structure, are the first line of defense for system stability and are crucial for integrating variable wind power.

3.1.1 Active Power Control

This function is responsible for precisely regulating the amount of wind power injected into the grid. The control system receives a power reference, which can be set by a higher-level controller or an operator. The inner current control loops then manipulate the IGBT switches to ensure the converter output matches this reference. In a wind integration context, this is vital for following dispatch instructions, implementing power curtailment if needed, and ensuring that the power generated by the wind farm is accurately transmitted through the DC link. Advanced schemes can also use active power control for frequency support, where the power reference is temporarily increased upon detecting a drop in the onshore grid frequency (Li et al., 2017; Lin & Wu, 2021). Figure 6 shows coordinated control approach for VSC-HVDC connecting an offshore windfarm to onshore power grid.

Figure 6. VSC-HVDC interconnecting offshore windfarm with control details (Lin & Wu, 2021).

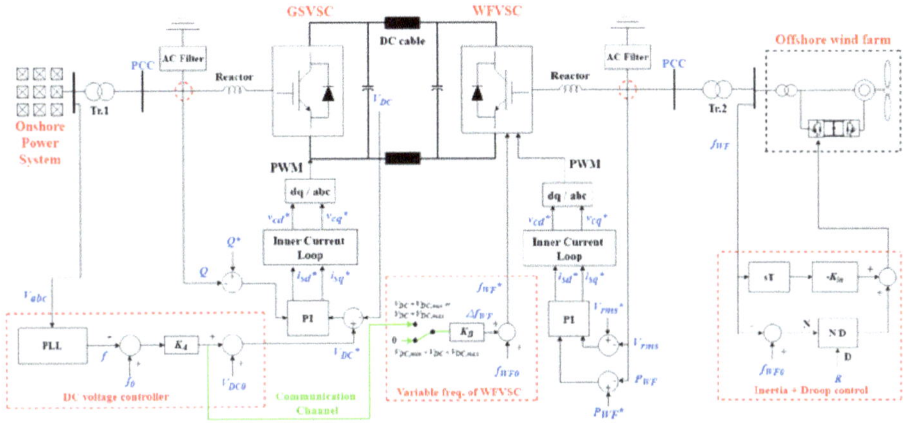

3.1.2 Reactive Power Support

A key advantage of VSC-HVDC is its independent control of reactive power. Regardless of the active power flow, the converter can be commanded to inject or absorb reactive power to manage the voltage profile at its point of connection (PoC). This is especially critical for weak grids or remote connection points where voltage stability is a concern. The converter can operate in various modes, such as fixed reactive power injection (Q-control) or, more commonly, AC voltage control (V-control), where it automatically adjusts its reactive power output to maintain the AC voltage at the PoC within a tight band, effectively acting as a large STATCOM (Jovcic, 2019).

3.1.3 Fault Ride-Through (FRT)

Grid codes mandate that generation assets, including HVDC links, must remain connected and support the grid during and after short-duration faults. Fault ride-through (FRT) control strategies ensure the HVDC system can handle voltage dips on the AC side without tripping. The FRT may include low voltage FRT and high voltage FRT requirements. During a fault, the controllers may temporarily alter the power flow priorities, utilizing the converter's current capacity to inject reactive current to support grid voltage recovery, even if it means briefly reducing active power transmission (IRENA, 2022; Wu et al., 2024). This capability is essential for preventing wide spread generation loss during disturbances and maintaining overall grid resilience (Justo et al., 2018).

3.2 System-Level Coordination in Microgrids

In a microgrid setting, the HVDC link must be seamlessly coordinated with all other distributed energy resources (DERs) and loads. This is achieved through a hierarchical control structure that distributes functions across different time scales and system levels (Zhang et al., 2026).

3.2.1 Primary Control

This is the fastest, localized control layer, typically operating in milliseconds. Its main goal is to maintain instantaneous power balance and stable voltage/frequency following a disturbance or fluctuation. In an HVDC-based microgrid, primary control often uses droop characteristics, where the DC voltage is used as an indicator of power balance. For example, if generation exceeds load, the DC voltage will rise, and the droop control will signal sources to reduce output or storage to charge faster. This decentralized approach ensures immediate stabilization without the need for communication, ensuring the microgrid's survival during islanded operation (Lin & Wu, 2021; Zhang et al., 2026).

3.2.2 Secondary Control

Operating on a timescale of seconds to minutes, secondary control acts to restore the system voltage and frequency to their nominal values after a primary control action has stabilized it. It compensates for the steady-state deviations inherent in droop control. The secondary controller, often a centralized Energy Management System (EMS), calculates the required power set-point adjustments for the HVDC link and other dispatchable resources to eliminate accumulated errors and resynchronize the microgrid with the main grid before reconnection (Lin & Wu, 2021).

3.2.3 Tertiary Control

This is the highest control layer, functioning on a timescale of minutes to hours. Its focus is on economic optimization and interaction with the main grid. Tertiary control considers factors such as generation costs, load forecasts, wholesale electricity prices, and grid tariff structures. It calculates the most economically efficient power schedule, determining optimal power import/export through the HVDC link, the charging/discharging cycles of batteries, and the dispatch of other local generators to minimize operational cost or maximize revenue (Lin & Wu, 2021).

3.3 Aggregator Control in Virtual Power Plants (VPP)

For a VPP, control is less about physical stability and more about the intelligent, data-driven aggregation and dispatch of a distributed portfolio. The HVDC link is a critical asset within this portfolio, managed by the VPP central intelligence platform (Bao et al., 2022). The Australian Energy Market Operator (AEMO) has demonstrated various VPP trials to understand the capability and operational limitations of VPP solutions (AEMO, 2020). In this case, numerous real world VPP projects such as FENIX project in Europe, Next Kraftwerke in Gemarny, Statkraf in Norway, South Australia's Virtual Power Plant (SA VPP) and the Hornsdale Power Reserve in Australia validate its applicability and flexibility in serving the next generation power grid (Abdelkader et al., 2024; Marzbnai et al., 2025).

3.3.1 Forecast-Based Dispatching

The VPP platform relies heavily on accurate forecasts for wind generation, load demand, and market prices. Using sophisticated weather models and machine learning algorithms, it predicts the available power from the HVDC-connected wind farm and other resources hours or days in advance. These forecasts are then used to create a day-ahead and intra-day dispatch plan that aligns available generation with committed load and market positions, ensuring the VPP can reliably meet its obligations (Marzbnai et al., 2025).

3.3.2 Market Participation

The core business of a VPP is to act as a single, market-facing entity. The aggregator control platform bundles the capacity of the HVDC-linked wind farm with other DERs (solar, storage, flexible demand) to form a bid that can be offered into wholesale energy, capacity, and ancillary service markets. The controllability and reliability of HVDC link are key to this, as they allow the VPP to offer firm, dispatchable power, enhancing its credibility and market value compared to a single, variable wind farm (Abdelkader et al., 2024; Marzbnai et al., 2025).

3.3.3 Grid Services

Beyond energy, VPPs can deliver critical ancillary services to grid operators. By coordinating the HVDC link fast power control with other assets, a VPP can provide services like frequency regulation (automatically adjusting power output to match grid frequency), voltage support, and congestion management. For instance, upon request from a grid operator, the VPP controller can send a signal to the HVDC link to rapidly increase or decrease its power flow, using the wind farm's available power and the portfolio's storage to deliver a precise and timely response (AEMO, 2020; Bao et al., 2022).

4. CASE STUDIES AND REAL-WORLD APPLICATIONS

Real-world projects provide critical validation for the theoretical and technical advantages of HVDC-wind integration. The following case studies, spanning different scales and geographies, exemplify the practical application and proven benefits of this technology in enhancing grid capacity, stability, and sustainability.

4.1 BorWin and DolWin HVDC Projects (Germany)

The cluster of HVDC projects in the German North Sea, including the BorWin and DolWin series, represents one of the world's most ambitious and extensive applications of VSC-HVDC technology for offshore wind integration. Driven by Germany's *Energiewende* (energy transition), these projects are responsible with transmitting multiple gigawatts of wind power from distant offshore sites to the mainland. The BorWin and DolWin platforms are essentially massive offshore converter stations that collect AC power from dozens of surrounding wind turbines, convert it to DC, and transmit it via submarine cables over distances of 100-200 kilometers to onshore converter stations.

The choice of VSC-HVDC was critical for several reasons. First, the long transmission distances made HVAC technically and economically unviable due to excessive reactive power losses. Second, the weak AC grid at the onshore connection points in northern Germany required the dynamic reactive power support and black-start capability that only VSC technology can provide. These converter stations act as giant grid-forming assets, stabilizing the voltage at the point of common coupling and providing essential short-circuit power. For instance, the BorWin3 project, with a capacity of 900 MW, uses VSC technology with voltage level ±320 kV to not only transmit power but also to regulate grid voltage without the need for additional compensation equipment, a feat impossible with older LCC-HVDC. These projects have been instrumental in proving the reliability of VSC-HVDC at a multi-gigawatt scale, though they have also highlighted challenges such as the complexity of offshore maintenance and the need for robust protection systems. The success of this cluster has established a blueprint for large-scale offshore wind integration worldwide, demonstrating that HVDC is not just an option but a necessity for achieving global renewable energy targets.

4.2 Zhangbei VSC Multi-Terminal HVDC Grid (China)

The Zhangbei VSC-HVDC project in Hebei Province is a landmark achievement, recognized as the world's first true large-scale, multi-terminal, VSC-based HVDC grid. Unlike the point-to-point links in the German North Sea, Zhangbei connects four converter stations in a flexible ring configuration, integrating 4500 MW of renewable energy, predominantly from vast wind farms and large-scale solar PV plants in the resource-rich but remote Zhangjiakou area.

This project is a monumental step beyond simple transmission; it is a functioning laboratory for the supergrid concepts essential for a 100% renewable future. Its multi-terminal architecture allows for power to be routed flexibly between different generation sites and load centers. For example, surplus wind power from one area can be directed to feed load in another or to pump-hydro storage facilities, all through the DC grid, minimizing congestion on the existing AC network. The project showcases HVDC's unparalleled ability to manage the variability and geographic diversity of renewables. When the wind is not blowing at one wind farm, power can be imported from another station connected to a solar farm or a different wind zone, smoothing the aggregate power output delivered to the Beijing-Tianjin-Hebei load center. The Zhangbei grid employs advanced DC power flow controllers and protection systems to maintain stability and manage fault currents, addressing key technical hurdles for future MTDC grids. Its successful operation provides invaluable data and operational experience, proving that complex, meshed HVDC grids are a practical and powerful solution for balancing large-scale renewable energy integration with grid resilience and reliability.

4.3 Microgrid HVDC Pilot in Alaska (USA)

While large-scale projects capture headlines, the application of HVDC principles at the microgrid level is equally transformative, as demonstrated by a pioneering pilot project in a remote Alaskan community. This community, like many isolated settlements, historically relied almost entirely on expensive and polluting diesel generators. The pilot project integrated a local wind farm with a battery energy storage system (BESS) using a low-voltage DC (LVDC) microgrid architecture, significantly reducing diesel consumption.

The core of this system is a DC bus that directly connects the wind turbines (via AC/DC converters), the battery storage (via a bi-directional DC/DC converter), and the critical loads (many of which, like LED lighting and electronics, are inherently

DC). This architecture provides several key advantages over a traditional AC microgrid. First, it eliminates multiple power conversion stages, reducing energy losses by an estimated 10-15% and lowering system costs. Second, the DC microgrid offers superior stability and power quality when integrating highly variable wind generation. The battery storage system, directly connected to the DC bus, can absorb or inject power almost instantaneously to balance fluctuations in wind output, ensuring a stable voltage without the frequency stability challenges of an AC island grid. The diesel generators can now be switched off for extended periods or operated at their most efficient set-points, with the DC microgrid managing the real-time balance. This project serves as a scalable model for the tens of thousands of remote communities worldwide, demonstrating that HVDC/LVDC technology is a key enabler for achieving high-penetration renewable energy, enhancing energy security, and reducing the environmental footprint of off-grid and weak-grid power systems.

5. FUTURE TRENDS

The integration of HVDC with wind energy systems is a rapidly evolving field. To fully realize its potential in building a resilient, sustainable, and efficient global power network, several key trends must be accelerated and supported. The future lies not in choosing between AC and DC, but in intelligently combining their strengths and creating the frameworks for their seamless operation.

5.1 Hybrid AC/DC Grids

The electrical grid of the future will not be a wholesale replacement of AC with DC, but rather an optimized, layered hybrid architecture that leverages the unique advantages of each technology. Hybrid AC/DC grids represent a paradigm where long-distance HVDC transmission corridors act as the continent-spanning energy superhighways, efficiently delivering bulk power from remote renewable resource centers such as offshore wind hubs, to major load centers. Meanwhile, the existing AC infrastructure will continue to serve as the robust local roads for regional and distribution-level power delivery, leveraging its well-understood protection schemes and infrastructure. The synergy occurs at the interface, where VSC-HVDC terminals provide enhanced stability and controllability to the AC grid. Furthermore, at the distribution level, hybrid microgrids will use local DC nanogrids to efficiently connect DC-based renewables, storage, and loads, while maintaining an AC bus for traditional infrastructure. This hybrid approach allows for the optimal integration of diverse energy resources, minimizes conversion losses, and creates a more flexible and resilient network capable of managing complex power flows and isolating disturbances.

5.2 Standardization of Multi-Terminal HVDC

The vision of a meshed, multi-terminal HVDC grid, analogous to the current AC transmission system, is currently hindered by a lack of universal standards. For MT-HVDC systems to be developed by multiple vendors and operated interoperable across national borders, international collaboration is paramount. Standardization efforts must focus on several critical areas such as harmonized voltage levels, standardized dynamic performance and control interfaces (e.g., for power flow controllers), and, most critically, universal protection schemes and HVDC circuit breaker specifications. Without standardized protection, a fault in one part of a multi-vendor DC grid could be impossible to isolate without collapsing the entire network. Recently, organizations like CIGRE and IEC are leading this effort, but the process must be accelerated through government-funded pilot projects and international treaties that mandate interoperability. This will reduce costs, increase competition, and ensure that future HVDC grids are secure, reliable, and open-access, much like the internet.

5.3 Advanced Control and AI Integration

The complexity of managing HVDC-enabled systems with high wind penetration exceeds the capability of traditional, deterministic control algorithms. The future lies in integrating Artificial Intelligence (AI) and Machine Learning (ML) to create cognitive, self-optimizing grid management systems. AI can revolutionize forecasting by analyzing vast datasets of weather, satellite imagery, and historical generation to produce highly accurate short-term predictions of wind power output, drastically reducing balancing costs. In real-time operation, ML-based dynamic control can optimize the power set-points of HVDC links, storage systems, and other DERs within a VPP or microgrid to achieve multiple objectives simultaneously such as minimizing losses, avoiding congestion, and providing ancillary services. Furthermore, AI-powered digital twins can simulate the entire system in real-time, enabling predictive maintenance and advanced fault detection. For instance, an AI model can analyze subtle patterns in VSC converter sensor data to predict an IGBT failure weeks before it occurs, preventing costly downtime or outage. This transition from reactive to predictive and prescriptive control is essential for the reliable and economic operation of future power networks.

5.4 Policy and Market Integration

Technological solutions cannot succeed without supportive policy and market structures. Policymakers must recognize HVDC infrastructure as a critical enabler of decarbonization and create regulatory frameworks that de-risk investment, streamline permitting processes, and potentially offer incentives for projects that enhance grid resilience and cross-border energy exchange. Concurrently, electricity market designs must be modernized to properly value the services that HVDC systems and VPPs provide. Current energy-only markets often fail to compensate assets for their capacity or fast-responding grid support. Markets need to be designed to explicitly reward speed, accuracy, and location that is creating robust revenue streams for frequency regulation, voltage support, and congestion relief. This would allow HVDC links and the VPPs that enable to monetize their rapid controllability and reactive power capabilities, improving their business case and encouraging private investment. Regulatory bodies should also mandate technology-neutral performance standards (e.g., for frequency response) to ensure a level playing field where HVDC and aggregated renewables can compete fairly with traditional generators.

6. CONCLUSION

The integration of High Voltage Direct Current technology into wind energy systems marks a transformative and indispensable step toward building a resilient, efficient, and truly low-carbon global energy future. This synergy addresses one of the most pressing challenges of the clean energy transition, which is the geographical mismatch between prime renewable resources and major load centers. By fundamentally overcoming the distance and stability limitations inherent in traditional AC transmission, HVDC unlocks the potential for large-scale wind power harvesting from the world's most remote and wind-rich offshore and onshore sites. It ensures that this clean electricity can be delivered reliably and with minimal losses to the load centers where it is desperately needed, thereby turning vast, untapped wind potential into a firm, dispatchable pillar of the grid.

Within the evolving architectures of microgrids and virtual power plants, HVDC moves beyond simple transmission to become a core enabling technology for intelligent energy management. It provides the precise controllability, inherent scalability, and operational flexibility required to seamlessly coordinate a diverse portfolio of distributed energy resources, from fluctuating wind turbines and solar arrays to fast-responding battery storage and flexible demand. Through strategic deployment and the application of sophisticated, hierarchical control strategies, HVDC systems do more than just move power. HVDC actively enhances overall power system performance. They dampen instability, mitigate congestion, provide critical grid-forming services, and improve the economic viability of renewable investments by ensuring their output can be reliably and profitably brought to market.

While the transition to an HVDC-enabled grid presents significant challenges including the need for standardized multi-terminal systems, advanced protection schemes, and substantial capital investment; the path forward is clear. These hurdles can be overcome through sustained technological innovation in power electronics and artificial intelligence, the establishment of supportive policy and market frameworks that value the unique services HVDC provides, and collaborative international efforts to harmonize standards. By embracing this path, HVDC technology is poised to form the robust, intelligent, and adaptive backbone of tomorrow's sustainable power infrastructure, seamlessly connecting a world of distributed clean energy and powering a secure and prosperous future for generations to come.

REFERENCES

Abdelkader, S., Amissah, J., & Abdel-Rahim, O. (2024). Virtual power plants: An in-depth analysis of their advancements and importance as crucial players in modern power systems. *Energy, Sustainability and Society*, *14*(1), 52. DOI: 10.1186/s13705-024-00483-y

AEMO. (2020). *AEMO Virtual Power Plant Demonstration*. Australian Energy Market Operator.

An, T., Liu, D., Li, F., & Zhu, L. (2025). Experience from the world's first HVDC grid. In *High Voltage DC Transmission Systems: HVDC* (pp. 2285–2326). Springer. DOI: 10.1007/978-3-031-27353-7_53

Arrillaga, J., Liu, Y. H., & Watson, N. R. (2007). *Flexible power transmission: the HVDC options*. John Wiley & Sons., DOI: 10.1002/9780470511862

Bao, P., Zhang, W., & Zhang, Y. (2022). Secondary frequency control considering optimized power support from virtual power plant containing aluminum smelter loads through VSC-HVDC link. *Journal of Modern Power Systems and Clean Energy*, *11*(1), 355–367. DOI: 10.35833/MPCE.2021.000072

Bidram, A., & Davoudi, A. (2012). Hierarchical structure of microgrids control system. *IEEE Transactions on Smart Grid*, *3*(4), 1963–1976. DOI: 10.1109/TSG.2012.2197425

Chaudhuri, N., Chaudhuri, B., Majumder, R., & Yazdani, A. (2014). *Multi-terminal direct-current grids: Modeling, analysis, and control*. John Wiley & Sons., DOI: 10.1002/9781118960486

Dakic, J., Cheah-Mane, M., Gomis-Bellmunt, O., & Prieto-Araujo, E. (2020). HVAC transmission system for offshore wind power plants including mid-cable reactive power compensation: Optimal design and comparison to VSC-HVDC transmission. *IEEE Transactions on Power Delivery*, *36*(5), 2814–2824. DOI: 10.1109/TPWRD.2020.3027356

Dreidy, M., Mokhlis, H., & Mekhilef, S. (2017). Inertia response and frequency control techniques for renewable energy sources: A review. *Renewable & Sustainable Energy Reviews*, *69*, 144–155. DOI: 10.1016/j.rser.2016.11.170

Enevoldsen, P., & Xydis, G. (2019). Examining the trends of 35 years growth of key wind turbine components. *Energy for Sustainable Development*, *50*, 18–26. DOI: 10.1016/j.esd.2019.02.003

Esfahani, M., Alizadeh, A., Cao, B., Kamwa, I., & Xu, M. (2025). Bridging theory and practice: A comprehensive review of virtual power plant technologies and their real-world applications. *Renewable & Sustainable Energy Reviews, 222*, 115929. DOI: 10.1016/j.rser.2025.115929

Farrokhabadi, M., Canizares, C. A., Simpson-Porco, J. W., Nasr, E., Fan, L., Mendoza-Araya, P. A., Tonkoski, R., Tamrakar, U., Hatziargyriou, N., Lagos, D., Wies, R. W., Paolone, M., Liserre, M., Meegahapola, L., Kabalan, M., Hajimiragha, A. H., Peralta, D., Elizondo, M. A., Schneider, K. P., & Reilly, J. (2019). Microgrid stability definitions, analysis, and examples. *IEEE Transactions on Power Systems, 35*(1), 13–29. DOI: 10.1109/TPWRS.2019.2925703

Flourentzou, N., Agelidis, V. G., & Demetriades, G. D. (2009). VSC-based HVDC power transmission systems: An overview. *IEEE Transactions on Power Electronics, 24*(3), 592–602. DOI: 10.1109/TPEL.2008.2008441

Gabriel, S., Mwasilu, F., & Mwinyiwiwa, B. M. M. (2025). The Impacts of Renewable Energy Sources: A Review on Grid Inertia and Frequency Regulation. *Tanzania Journal of Engineering & Technology, 44*(2), 171–180. Advance online publication. DOI: 10.52339/tjet.v44i1.1275

GWEC. (2023). *Global Wind Report 2023.*

Heylen, E., Teng, F., & Strbac, G. (2021). Challenges and opportunities of inertia estimation and forecasting in low-inertia power systems. *Renewable & Sustainable Energy Reviews, 147*, 111176. DOI: 10.1016/j.rser.2021.111176

Hirsch, A., Parag, Y., & Guerrero, J. (2018). Microgrids: A review of technologies, key drivers, and outstanding issues. *Renewable & Sustainable Energy Reviews, 90*, 402–411. DOI: 10.1016/j.rser.2018.03.040

IRENA. (2019). *Solutions to integrate high shares of variable renewable energy.*

IRENA. (2022). *Grid codes for renewable powered systems.*

IRENA. (2023). *World Energy Transitions Outlook 2023: 1.5°C Pathway,* Abu Dhabi.

Jiang, Z. (2021). Installation of offshore wind turbines: A technical review. *Renewable & Sustainable Energy Reviews, 139*, 110576. DOI: 10.1016/j.rser.2020.110576

Jovcic, D. (2019). *High voltage direct current transmission: converters, systems and DC grids.* John Wiley & Sons., DOI: 10.1002/9781119566632

Justo, J. J., Mwasilu, F., & Jung, J. W. (2018). Effective protection for doubly fed induction generator-based wind turbines under three-phase fault conditions. *Electrical Engineering, 100*(2), 543–556. DOI: 10.1007/s00202-017-0528-0

Kaiss, M., Wan, Y., Gebbran, D., Vila, C. U., & Dragičević, T. (2025). Review on virtual power plants/virtual aggregators: Concepts, applications, prospects and operation strategies. *Renewable & Sustainable Energy Reviews, 211*, 115242. DOI: 10.1016/j.rser.2024.115242

Kalair, A., Abas, N., & Khan, N. (2016). Comparative study of HVAC and HVDC transmission systems. *Renewable & Sustainable Energy Reviews, 59*, 1653–1675. DOI: 10.1016/j.rser.2015.12.288

Khan, S., & Bhowmick, S. (2022). *Power-flow Modelling of HVDC transmission systems*. CRC Press., DOI: 10.1201/9781003252078

Korompili, A., Wu, Q., & Zhao, H. (2016). Review of VSC HVDC connection for offshore wind power integration. *Renewable & Sustainable Energy Reviews, 59*, 1405–1414. DOI: 10.1016/j.rser.2016.01.064

Li, Y., Xu, Z., Østergaard, J., & Hill, D. J. (2017). Coordinated control strategies for offshore wind farm integration via VSC-HVDC for system frequency support. *IEEE Transactions on Energy Conversion, 32*(3), 843–856. DOI: 10.1109/TEC.2017.2663664

Lin, C.-H., & Wu, Y.-K. (2021). Overview of frequency-control technologies for a VSC-HVDC-integrated wind farm. *IEEE Access : Practical Innovations, Open Solutions, 9*, 112893–112921. DOI: 10.1109/ACCESS.2021.3102829

Liu, C., Yao, W., Xiang, C., Cheng, X., & Wang, Q. (2022). Design of±535 kV hybrid high voltage direct current (HVDC) circuit breaker. *IET Generation, Transmission & Distribution, 16*(19), 4017–4026. DOI: 10.1049/gtd2.12568

Lüth, A., & Keles, D. (2024). Risks, strategies, and benefits of offshore energy hubs: A literature-based survey. *Renewable & Sustainable Energy Reviews, 203*, 114761. DOI: 10.1016/j.rser.2024.114761

Lyu, Y., Luo, Y., Jia, Z., Fu, S., Tang, Y., Wang, Q., & Zeng, Y. (2024). Benefit evaluation of HVAC and HVDC for offshore wind power transmission system under the multidimensional index. *Electric Power Systems Research, 237*, 111018. DOI: 10.1016/j.epsr.2024.111018

Marzbnai, F., Osman, A., & Hassan, M. S. (2025). Advances in virtual power plant operations: A review of optimization models. *IEEE Access : Practical Innovations, Open Solutions, 13*, 131525–131548. Advance online publication. DOI: 10.1109/ACCESS.2025.3591697

Mohammadi, F., Danapour, N., & Rouzbehi, K. (2024). Integrated Wind Turbines and HVDC Power Transmission Lines. *IEEE Transactions on Sustainable Energy, 15*(3), 2134–2137. DOI: 10.1109/TSTE.2024.3402010

Pang, H., & Wei, X. (2018). Research on key technology and equipment for Zhangbei 500kV DC grid. *2018 International Power Electronics Conference (IPEC-Niigata 2018-ECCE Asia)*, 2343–2351. DOI: 10.23919/IPEC.2018.8507575

Pfeifenberger, J. P., Bai, L., Levitt, A., Plet, C. A., Sonnathi, C. M., Alliance, C. G., United, G., & Energy, P. (2023). *The operational and market benefits of HVDC to system operators*. The Brattle Group and DNV.

Ratnam, K. S., Palanisamy, K., & Yang, G. (2020). Future low-inertia power systems: Requirements, issues, and solutions-A review. *Renewable & Sustainable Energy Reviews, 124*, 109773. DOI: 10.1016/j.rser.2020.109773

Sang, W., Guo, W., Dai, S., Tian, C., Yu, S., & Teng, Y. (2022). Virtual synchronous generator, a comprehensive overview. *Energies, 15*(17), 6148. DOI: 10.3390/en15176148

Sharifabadi, K., Harnefors, L., Nee, H.-P., Norrga, S., & Teodorescu, R. (2016). *Design, control, and application of modular multilevel converters for HVDC transmission systems*. John Wiley & Sons., DOI: 10.1002/9781118851555

Souza, M. E. T.Junior, & Freitas, L. C. G. (2022). Power electronics for modern sustainable power systems: Distributed generation, microgrids and smart grids—A review. *Sustainability (Basel), 14*(6), 3597. DOI: 10.3390/su14063597

Ullah, M., Guan, Y., Yu, Y., Chaudhary, S. K., Vasquez, J. C., & Guerrero, J. M. (2025). Dynamic Performance and Power Quality of Large-Scale Wind Power Plants: A Review on Challenges, Evolving Grid Code, and Proposed Solutions. *IEEE Open Journal of Power Electronics, 6*, 1148–1173. Advance online publication. DOI: 10.1109/OJPEL.2025.3582012

Van Hertem, D., Gomis-Bellmunt, O., & Liang, J. (2016). *HVDC grids: for offshore and supergrid of the future*. John Wiley & Sons. DOI: 10.1002/9781119115243

Wu, D., Seo, G.-S., Xu, L., Su, C., Kocewiak, L., Sun, Y., & Qin, Z. (2024). Grid integration of offshore wind power: Standards, control, power quality and transmission. *IEEE Open Journal of Power Electronics, 5*, 583–604. DOI: 10.1109/OJPEL.2024.3390417

Yu, Z., Song, Q., Yuan, Z., Zou, C., Qiao, X., & Liu, W. (2024). Transmission topology and control for ultra-large offshore wind bases integrating multiple offshore low-frequency AC links and onshore HVDC link. *IEEE Transactions on Power Delivery*, *39*(5), 2797–2809. DOI: 10.1109/TPWRD.2024.3432897

Zhang, L., Sun, L., Xu, W., & Gong, R. (2026). Voltage droop control optimization in microgrid integration with VSCHVDC systems and wind power plants for enhanced dynamic stability. *Electric Power Systems Research*, *252*, 112363. DOI: 10.1016/j.epsr.2025.112363

Compilation of References

A. W. Energy. (2024). *Edge driven digital twins in distributed energy systems.*

Abdelkader, S., Amissah, J., & Abdel-Rahim, O. (2024). Virtual power plants: An in-depth analysis of their advancements and importance as crucial players in modern power systems. *Energy, Sustainability and Society*, *14*(1), 52. DOI: 10.1186/s13705-024-00483-y

Abdullahi, I., Perinpanayagam, S., & Hamidu, I. (2022, September). A fog computing based approach towards improving asset management and performance of wind turbine plants using digital twins. In *2022 27th International Conference on Automation and Computing (ICAC)* (pp. 1–5). IEEE. DOI: 10.1109/ICAC55051.2022.9911170

Abubakr, H., Vasquez, J. C., Mahmoud, K., Darwish, M. M. F., & Guerrero, J. M. (2022). Comprehensive review on renewable energy sources in Egypt—Current status, grid codes and future vision. *IEEE Access : Practical Innovations, Open Solutions*, *10*, 4081–4101. DOI: 10.1109/ACCESS.2022.3140385

Ackermann, T. (2012). Wind Power in Power Systems. In T. Ackermann (Ed.), *Wind Power in Power Systems* (2nd ed.). wiley. DOI: 10.1002/9781119941842

Adefarati, T., & Bansal, R. C. (2019). Energizing Renewable Energy Systems and Distributed Generation. In Tascikaraoglu, A., & Erdinç, O. (Eds.), *Pathways to a Smarter Power System* (pp. 29–65). Academic Press, Elsevier.

AEMO. (2020). *AEMO Virtual Power Plant Demonstration.* Australian Energy Market Operator.

Afridi, Y. S., Ahmad, K., & Hassan, L. (2022). Artificial intelligence based prognostic maintenance of renewable energy systems: A review of techniques, challenges, and future research directions. *International Journal of Energy Research*, *46*(15), 21619–21642. DOI: 10.1002/er.7100

Agency, I. E. (2022). International Energy Agency (IEA) World Energy Outlook 2022. Https://Www.Iea.Org/Reports/World-Energy-Outlook-2022/Executive-Summary

Aghdam, H. H., Gonzalez-Garcia, A., van de Weijer, J., & López, A. M. (2019). Active learning for deep detection neural networks. *Proceedings of the IEEE/ CVF International Conference on Computer Vision*, 3672–3680, DOI: 10.1109/ ICCV.2019.00377

Aitken, M. (2010). Wind power and community benefits: Challenges and opportunities. *Energy Policy*, *38*(10), 6066–6075. DOI: 10.1016/j.enpol.2010.05.062

Al Abdulwahid, A. (2024). Cyber-physical threat mitigation in wind energy systems: A novel secure architecture for industry 4.0 power grids. *Energy Informatics*, *7*(1), 141. DOI: 10.1186/s42162-024-00449-6

Alam, M. S., Chowdhury, T. A., Dhar, A., Al-Ismail, F. S., Choudhury, M. S. H., Shafiullah, M., Hossain, M., Hossain, M., Ullah, A., & Rahman, S. M. (2023). Solar and wind energy integrated system frequency control: A critical review on recent developments. *Energies*, *16*(2), 812. DOI: 10.3390/en16020812

Aledhari, M., Razzak, R., Parizi, R. M., & Saeed, F. (2020). Federated Learning: A Survey on Enabling Technologies, Protocols, and Applications. *IEEE Access : Practical Innovations, Open Solutions*, *8*, 140699–140725. DOI: 10.1109/ACCESS.2020.3013541 PMID: 32999795

Alfaro-Pelico, R. (2025). *A JUST ENERGY TRANSITION FOR COMMUNITIES LARGE-SCALE WIND AND SOLAR PROJECTS IN SUB-SAHARAN AFRICA The report benefited from the review and inputs of IRENA colleagues Babucarr Bittaye, Mirjam Reiner, Ntsebo Sephelane*. www.irena.org

Alghamdi, W. Y., Alshamrani, R. M., Aloufi, R. K., Ba Lhamar, S. O., Altwirqi, R. A., Alotaibi, F. S., & Alazwari, A. S. (2025). Exploring the synergy between digital twin technology and artificial intelligence: A comprehensive survey. *International Journal of Advanced Computer Science and Applications*, *16*(3). Advance online publication. DOI: 10.14569/IJACSA.2025.0160399

Ali, S. W., Sadiq, M., Terriche, Y., Naqvi, S. A. R., Hoang, L. Q. N., Mutarraf, M. U., Hassan, M. A., Yang, G., Su, C.-L., & Guerrero, J. M. (2021). Offshore wind farm-grid integration: A review on infrastructure, challenges, and grid solutions. *IEEE Access : Practical Innovations, Open Solutions*, *9*, 102811–102827. DOI: 10.1109/ACCESS.2021.3098705

Al-Shetwi, A. Q., Atawi, I. E., El-Hameed, M. A., & Abuelrub, A. (2025). Digital twin technology for renewable energy, smart grids, energy storage and vehicle-to-grid integration: Advancements, applications, key players, challenges and future perspectives in modernising sustainable grids. *IET Smart Grid*, *8*(1), e70026. DOI: 10.1049/stg2.70026

Alsmadi, Y. M., Xu, L., Blaabjerg, F., Ortega, A. J. P., Abdelaziz, A. Y., Wang, A., & Albataineh, Z. (2018). Detailed investigation and performance improvement of the dynamic behavior of grid-connected DFIG-based wind turbines under LVRT conditions. *IEEE Transactions on Industry Applications*, *54*(5), 4795–4812. DOI: 10.1109/TIA.2018.2835401

Amuthan, N., M, M. B., Velrajkumar, P., Sivakumar, N., & Jarin, T. (2023). IOT based adjustment mechanism for direct reference model adaptive IMC to support voltage sag in DFIG wind farm. *Measurement. Sensors*, *27*, 100809. DOI: 10.1016/j.measen.2023.100809

An, T., Liu, D., Li, F., & Zhu, L. (2025). Experience from the world's first HVDC grid. In *High Voltage DC Transmission Systems: HVDC* (pp. 2285–2326). Springer. DOI: 10.1007/978-3-031-27353-7_53

Anagnostopoulos, S. J., Bauer, J., Clare, M. C. A., & Piggott, M. D. (2023). Accelerated wind farm yaw and layout optimisation with multi-fidelity deep transfer learning wake models. *Renewable Energy*, *218*, 119293. DOI: 10.1016/j.renene.2023.119293

Ang, T. Z., Salem, M., Kamarol, M., Das, H. S., Nazari, M. A., & Prabaharan, N. (2022). A comprehensive study of renewable energy sources: Classifications, challenges and suggestions. *Energy Strategy Reviews*, *43*, 100939. DOI: 10.1016/j.esr.2022.100939

Anushalini, T., & Sri Revathi, B. (2024). Role of machine learning algorithms for wind power generation prediction in renewable energy management. *Journal of the Institution of Electronics and Telecommunication Engineers*, *70*(4), 4319–4332. DOI: 10.1080/03772063.2023.2205838

Arrillaga, J., Liu, Y. H., & Watson, N. R. (2007). *Flexible power transmission: the HVDC options*. John Wiley & Sons., DOI: 10.1002/9780470511862

Arsal, M., Asad, H., Kamel, T., & Khan, A. (2025). *Cyber-safety assessment of wind turbines: A reachability analysis approach against cyber-attacks*. City, University of London. https://openaccess.city.ac.uk/id/eprint/35350/

Asamoah, S. S., Parbey, J., Yankey, I. K., & Awuah, A. (2023). Techno-economic assessment of a central grid-connected wind farm in Ghana using RETScreen® Expert. *Heliyon*, *9*(1), e12902. DOI: 10.1016/j.heliyon.2023.e12902 PMID: 36685476

Asian Power, https://asian-power.com/videos/asia-pacific-dominates-global-wind-market-71-share

Astolfi, D., Pandit, R., Celesti, L., Lombardi, A., & Terzi, L. (2022). SCADA data analysis for long-term wind turbine performance assessment: A case study. *Sustainable Energy Technologies and Assessments*, *52*, 102357. DOI: 10.1016/j.seta.2022.102357

Atallah, A. M., & El Tantawy, E. S. F. (2015). Direct torque control of machine side multilevel converter for variable speed wind turbines. *Energy*, *90*, 1091–1099. DOI: 10.1016/j.energy.2015.06.065

Atilgan Turkmen, B., & Germirli Babuna, F. (2024). Life Cycle Environmental Impacts of Wind Turbines: A Path to Sustainability with Challenges. *Sustainability 2024, Vol. 16, Page 5365, 16*(13), 5365. DOI: 10.3390/su16135365

Badesa, L., Teng, F., & Strbac, G. (2020). Pricing inertia and frequency response with diverse dynamics in a mixed-integer second-order cone programming formulation. *Applied Energy*, *260*, 114334. DOI: 10.1016/j.apenergy.2019.114334

Bao, X., & Zhang, L. (2022). Review of grid codes and implementation for wind farm frequency regulation. In *E3S Web of Conferences*, 352. https://doi.org/DOI: 10.1051/e3sconf/202235202012

Bao, P., Zhang, W., & Zhang, Y. (2022). Secondary frequency control considering optimized power support from virtual power plant containing aluminum smelter loads through VSC-HVDC link. *Journal of Modern Power Systems and Clean Energy*, *11*(1), 355–367. DOI: 10.35833/MPCE.2021.000072

Barbre, Z., & Li, G. (2025). Enhanced Wind Energy Forecasting Using an Extended Long Short-Term Memory Model. *Algorithms 2025, Vol. 18, Page 206, 18*(4), 206. DOI: 10.3390/a18040206

Basak, D., Nagababu, G., Puppala, H., Patel, J., & Kumar, S. V. V. A. K. (2023). Foreseeing the spatio-temporal offshore wind energy potential of India using a differential weighted ensemble created using CMIP6 datasets. *Regional Studies in Marine Science*, *65*, 103066. DOI: 10.1016/j.rsma.2023.103066

Benbouzid, M., Berghout, T., Sarma, N., Djurović, S., Wu, Y., & Ma, X. (2021). Intelligent condition monitoring of wind power systems: State of the art review. *Energies*, *14*(18), 5967. DOI: 10.3390/en14185967

Beretta, M., Julian, A., Sepulveda, J., Cusidó, J., & Porro, O. (2021). An ensemble learning solution for predictive maintenance of wind turbines main bearing. *Sensors (Basel)*, *21*(4), 1512. https://doi.org/10.3390/s21041512

Berger, R. (2024). *Lazard' S Levelized Cost of Energy Analysis—Version 9.0. 24*, 1–3.

Berger, R. (2025). *Lazards LCOE 2025. June*.

Besnard, F., & Bertling, L. (2010). An approach for condition-based maintenance optimization applied to wind turbine blades. *IEEE Transactions on Sustainable Energy*, *1*(2), 77–83. DOI: 10.1109/TSTE.2010.2049452

Bhattacharjee, P., Jana, R. K., & Bhattacharya, S. (2022). Boosting the annual profit of an offshore wind farm in India with a bio-inspired meta-heuristic scheme. *Environmental Challenges*, *9*, 100642. DOI: 10.1016/j.envc.2022.100642

Bhuj 1 Project India. (n.d.). White-Paper-Poject-Wind-India_2.pdf

Bi, C., & Law, A. W.-K. (2023). Co-locating offshore wind and floating solar farms – Effect of high wind and wave conditions on solar power performance. *Energy*, *266*, 126437. DOI: 10.1016/j.energy.2022.126437

Bidram, A., & Davoudi, A. (2012). Hierarchical structure of microgrids control system. *IEEE Transactions on Smart Grid*, *3*(4), 1963–1976. DOI: 10.1109/TSG.2012.2197425

Biswas, H. S., Kundu, A. K., & Poddar, S. (2024). Releasing the Power of Nature: Exploration of Sustainable Energy for a Flourishing Future. In Yıldırım, S., Yıldırım, D., Demirtaş, I., & Kandpal, V. (Eds.), *Green Transition Impacts on the Economy, Society, and Environment* (pp. 108–124). IGI Global Scientific Publishing., DOI: 10.4018/979-8-3693-3985-5.ch007

Blaabjerg, F., Chen, M., & Huang, L. (2024). Power electronics in wind generation systems. *Nature Reviews Electrical Engineering*, *1*(4), 234–250. DOI: 10.1038/S44287-024-00032-X

Blaabjerg, F., Teodorescu, R., Liserre, M., & Timbus, A. V. (2006). Overview of control and grid synchronization for distributed power generation systems. *IEEE Transactions on Industrial Electronics*, *53*(5), 1398–1409. DOI: 10.1109/TIE.2006.881997

Blodt, M., Chabert, M., Régnier, J., & Faucher, J. (2006). Mechanical load fault detection in induction motors by stator current time-frequency analysis. *IEEE Transactions on Industry Applications*, *42*(6), 1454–1463. DOI: 10.1109/TIA.2006.882631

Bonthu, S., Purvaja, R., Singh, K. S., Ganguly, D., Muruganandam, R., Paul, T., & Ramesh, R. (2024). Offshore wind energy potential along the Indian Coast considering ecological safeguards. *Ocean and Coastal Management, 249*, 107017. DOI: 10.1016/j.ocecoaman.2024.107017

Bošnjaković, M., Martinović, M., & Đokić, K. (2025). Application of artificial intelligence in wind power systems. *Applied Sciences (Basel, Switzerland), 15*(5), 2443. DOI: 10.3390/app15052443

Bouckaert, S., Pales, A. F., McGlade, C., Remme, U., Wanner, B., Varro, L., D'Ambrosio, D., & Spencer, T. (2021). *Net zero by 2050: A roadmap for the global energy sector*, .DOI: 10.1787/c8328405-en

Bouramdane, A. A.. (2023). *Cyberattacks in smart grids: Challenges and solving the multi-criteria decision-making for cybersecurity options. Cybersecurity*. MDPI., https://doi.org/10.3390/jcp3040031

Brøndsted, P., Lilholt, H., & Lystrup, A. (2018). Composite materials for wind power turbine blades. *Renewable Energy: Four Volume Set, 3–4*, 182–216. DOI: 10.4324/9781315793245-94/COMPOSITE-MATERIALS-WIND-POWER-TURBINE-BLADES-POVL-BR

Burton, T., Jenkins, N., Sharpe, D., Bossanyi, E., & Graham, M. (2021). *Wind energy handbook*. https://www.wiley.com/en-us/Wind+Energy+Handbook%2C+3rd+Edition-p-9781119451167

Campoverde-Vilela, L., Feijóo, M. D. C., Vidal, Y., Sampietro, J., & Tutivén, C. (2023). Anomaly-based fault detection in wind turbine main bearings. *Wind Energy Science, 8*, 557–574. https://doi.org/10.5194/wes-8-557-2023

Canan, A. (2023). Offshore wind energy policy paths: A comparative analysis of Denmark and Germany. *Economics and Policy of Energy and the Environment, 1*(1), 35–59. DOI: 10.3280/EFE2023-001003

Carley, S., Baldwin, E., MacLean, L. M., & Brass, J. N. (2017). Global expansion of RE generation: An analysis of policy instruments. *Environmental and Resource Economics, 68*(2), 397–440. DOI: 10.1007/s10640-016-0025-3

Castellazzi, A., Gurpinar, E., Wang, Z., Hussein, A. S., & Fernandez, P. G. (2019). Impact of wide-bandgap technology on renewable energy and smart-grid power conversion applications including storage. *Energies, 12*(23). Advance online publication. DOI: 10.3390/EN12234462

Cavus, M. (2025). Advancing Power Systems with Renewable Energy and Intelligent Technologies: A Comprehensive Review on Grid Transformation and Integration. *Electronics (Basel)*, *14*(6), 1159. DOI: 10.3390/electronics14061159

CEEW. (2022). https://www.ceew.in/publications/assessing-offshore-wind-power-development-for-indias-power-system-in-2030

Çelik, Ö., Yalman, Y., Tan, A., Bayındır, K. Ç., Çetinkaya, Ü., Akdeniz, M., Chaudhary, S. K., Høyer, M., & Guerrero, J. M. (2022). Grid code requirements–A case study on the assessment for integration of offshore wind power plants in Turkey. *Sustainable Energy Technologies and Assessments*, *52*, 102137. DOI: 10.1016/j.seta.2022.102137

Chang, C., Kocar, I., Hu, J., Karaagac, U., Chan, K. W., & Mahseredjian, J. (2022). Coordinated control of DFIG converters to comply with reactive current requirements in emerging grid codes. *Journal of Modern Power Systems and Clean Energy*, *9*(1), 1–14. DOI: 10.35833/MPCE.2021.000191

Chattejee, D., Chakraborty, C., & Dalapati, S. (2023). Pulse-width modulation techniques in two-level voltage source inverters - State of the art and future perspectives. *Power Electronics and Drives*, *8*(1), 335–367. DOI: 10.2478/PEAD-2023-0023

Chaudhuri, N., Chaudhuri, B., Majumder, R., & Yazdani, A. (2014). *Multi-terminal direct-current grids: Modeling, analysis, and control*. John Wiley & Sons., DOI: 10.1002/9781118960486

Chen, B. Q., Liu, K., Yu, T., & Li, R. (2024). Enhancing reliability in floating offshore wind turbines through digital twin technology: A comprehensive review. *Energies*, *17*(8), 1964. DOI: 10.3390/en17081964

Chen, J., Yan, J., Kemmeugne, A., Kassouf, M., & Debbabi, M. (2025). Cybersecurity of distributed energy resource systems in the smart grid: A survey. *Applied Energy*, *383*, 125364. https://doi.org/10.1016/j.apenergy.2025.125364

Chen, X., Huang, L., Liu, J., Song, D., & Yang, S. (2022). Peak shaving benefit assessment considering the joint operation of nuclear and battery energy storage power stations: Hainan case study. *Energy*, *239*, 121897. DOI: 10.1016/j.energy.2021.121897

Chinnappan, C. C. (2025). Integrating data-driven and physics-based approaches for robust wind power prediction: A comprehensive ML-PINN-Simulink framework. *Scientific Reports*, *15*(1), 1–29.

Civera, M., & Surace, C. (2022). Non-Destructive Techniques for the Condition and Structural Health Monitoring of Wind Turbines: A Literature Review of the Last 20 Years. *Sensors 2022, Vol. 22, Page 1627, 22*(4), 1627. DOI: 10.3390/s22041627

Cleveland, F. M. (2008). IEC 61850-7-420 communications standard for distributed energy resources (DER). *IEEE Power Energy, 1–4*, 1–4. Advance online publication. DOI: 10.1109/PES.2008.4596553

Clifton, A., Barber, S., Bray, A., Enevoldsen, P., Fields, J., Sempreviva, A. M., Williams, L., Quick, J., Purdue, M., Totaro, P., & Ding, Y. (2023). Grand challenges in the digitalisation of wind energy. *Wind Energy Science, 8*(6), 947–974. DOI: 10.5194/wes-8-947-2023

Cordova, J. G. L. (2025). The right to energy and the data protection and privacy rights: Towards a coherent enforcement under EU law within the energy transition. *European Journal of Risk Regulation*. Advance online publication. https://doi.org/10.1017/err.2025.30

Council on Energy, Environment and Water. (2022). https://www.ceew.in/publications/indias-expanding-clean-energy-workforce

Culler, M. J. (2021). *Cybersecurity guide for distributed wind* (Technical Report INL/EXT-21-64999). Idaho National Laboratory. https://www.osti.gov/biblio/1826578

Dakic, J., Cheah-Mane, M., Gomis-Bellmunt, O., & Prieto-Araujo, E. (2020). HVAC transmission system for offshore wind power plants including mid-cable reactive power compensation: Optimal design and comparison to VSC-HVDC transmission. *IEEE Transactions on Power Delivery, 36*(5), 2814–2824. DOI: 10.1109/TPWRD.2020.3027356

Danish Energy Agency. (2021). *Cybersecurity annual review 2021*. Danish Ministry of Climate, Energy and Utilities. https://www.scirp.org/reference/referencespapers?referenceid=3884055

Dao, P. B. (2022). Condition monitoring and fault diagnosis of wind turbines based on structural break detection in SCADA data. *Renewable Energy, 185*, 641–654. DOI: 10.1016/j.renene.2021.12.051

Daoudi, M., Mou, A. A. S., Idrissi, A., Ihoume, I., Arbaoui, N., & Benchrifa, M. (2023). New approach to prioritize wind farm sites by data envelopment analysis method: A case study. *Ocean Engineering, 271*, 113820. DOI: 10.1016/j.oceaneng.2023.113820

Daoudi, M., Mou, A. A. S., & Naceur, L. A. (2022). Analysis of the first onshore wind farm installation near the Morocco-United Kingdom green energy export project. *Scientific African, 17*, e01388. DOI: 10.1016/j.sciaf.2022.e01388

Davari, M., & Mohamed, Y. A. R. I. (2017). Robust vector control of a very weak-grid-connected voltage-source converter considering the phase-locked loop dynamics. *IEEE Transactions on Power Electronics*, *32*(2), 977–994. DOI: 10.1109/TPEL.2016.2546341

de Vasconcelos, R. M., Silva, L. L. C., Gonzalez, M. O. A., Santiso, A. M., & de Melo, D. C. (2022). Environmental licensing for offshore wind farms: Guidelines and policy implications for new markets. *Energy Policy*, *171*, 113248. DOI: 10.1016/j.enpol.2022.113248

Debnath, S., Qin, J., Bahrani, B., Saeedifard, M., & Barbosa, P. (2015). Operation, control, and applications of the modular multilevel converter: A review. *IEEE Transactions on Power Electronics*, *30*(1), 37–53. DOI: 10.1109/TPEL.2014.2309937

Desalegn, B., Gebeyehu, D., & Tamirat, B. (2022). Wind energy conversion technologies and engineering approaches to enhancing wind power generation: A review. *Heliyon*, *8*(11), e11263. DOI: 10.1016/J.HELIYON.2022.E11263 PMID: 36345520

Devine-Wright, P., & Wiersma, B. (2020). Understanding community acceptance of a potential offshore wind energy project in different locations: An island-based analysis of 'place-technology fit.'. *Energy Policy*, *137*, 111086. DOI: 10.1016/j.enpol.2019.111086

Díaz-González, F., Hau, M., Sumper, A., & Gomis-Bellmunt, O. (2014). Participation of wind power plants in system frequency control: Review of grid code requirements and control methods. *Renewable & Sustainable Energy Reviews*, *34*, 551–564. DOI: 10.1016/j.rser.2014.03.040

Djalel, D., Abdallah, G., & Hocine, L. (2014). STUDY OF THE LIGHTNING IMPACT ON THE WIND-TURBINE. *Energy Research Journal*, *5*(1), 17–25. DOI: 10.3844/erjsp.2014.17.25

DNV. (2023). *OT cyber security for offshore wind: Joint industry project (JIP)*. DNV. https://www.dnv.com/group/joint-industry-projects/ot-cyber-security-for-offshore-wind/

DOE. U. (2015). *Wind Vision. A New Era for Wind Power in the United States (Executive Summary, Full Report, and Appendices); U.S. Department of Energy (DOE), NREL (National Renewable Energy Laboratory)*. DOI: 10.2172/1215051

Dou, Z., Zhang, C., Wang, W., Wang, D., Zhang, Q., Cai, Y., & Fan, R. (2022). Review on key technologies and typical applications of multi-station integrated energy systems. *Global Energy Interconnection*, *5*(3), 309–327. DOI: 10.1016/j.gloei.2022.06.009

Dreidy, M., Mokhlis, H., & Mekhilef, S. (2017). Inertia response and frequency control techniques for renewable energy sources: A review. *Renewable & Sustainable Energy Reviews*, *69*, 144–155. DOI: 10.1016/j.rser.2016.11.170

Edwards-Jones, A., Watson, S. C. L., Szostek, C. L., & Beaumont, N. J. (2024). Stakeholder insights into embedding marine net gain for offshore wind farm planning and delivery. *Environmental Challenges*, *14*, 100814. DOI: 10.1016/j.envc.2023.100814

Egan, M. (2022). *A retrospective on 2022 cyber incidents in the wind energy sector* (Graduate project). Boise State University. https://scholarworks.boisestate.edu/cyber_gradproj/3/

Ekechukwu, D., & Simpa, P. (2024). The importance of cybersecurity in protecting renewable energy investment: A strategic analysis of threats and solutions. *Engineering Science and Technology Journal*, *5*(6), 1845–1883. DOI: 10.51594/estj.v5i6.1186

El Jaadi, M., Haidi, T., Belfqih, A., Farah, M., & Dailmy, A. (2024). Optimizing wind farm layout for enhanced electricity extraction using a new hybrid PSO-ANN method. *Global Energy Interconnection*, *7*(3), 254–269. DOI: 10.1016/j.gloei.2024.06.006

Elkelawy, M., Atta, Z. A., & Seleem, H. (2024). Technological advances, efficiency optimization, and challenges in wind power plants: A comprehensive review. *Pharos Engineering Science Journal*, *1*(1), 57–65. DOI: 10.21608/pesj.2025.344906.1006

Elkelawy, M., Wael, N., Alm-Eldin Bastawissi, H., & Seleem, H. (n.d.). Cutting-Edge Innovations in Wind Power: Enhancing Efficiency, Sustainability, and Grid Integration. [PESJ]. *Pharos Engineering Science Journal*, *02*, 2025. DOI: 10.21608/pesj.2025.369017.1020

Elkodama, A., Ismaiel, A., Abdellatif, A., Shaaban, S., Yoshida, S., & Rushdi, M. A. (2023). Control Methods for Horizontal Axis Wind Turbines (HAWT): State-of-the-Art Review. *Energies*, *16*(17), 6394. Advance online publication. DOI: 10.3390/en16176394

Enevoldsen, P., & Xydis, G. (2019). Examining the trends of 35 years growth of key wind turbine components. *Energy for Sustainable Development*, *50*, 18–26. DOI: 10.1016/j.esd.2019.02.003

Ennis, S., & Colangelo, G. (2022). *Energy data sharing: The case of EV smart charging*. Centre on Regulation in Europe (CERRE). https://cerre.eu/publications/energy-data-sharing-and-the-case-of-ev-smart-charging/

EQ Magazine. (n.d.) https://www.eqmagpro.com/india-plans-37gw-offshore-wind-energy-power-by-2030-to-issue-bids-eq-mag/#:~:text=Their%20estimate%2C%20made%20at%20a%20Group%20of,and%20Gujarat%20have%20agreed%20to%20purchase%20power

Erge, O., & van Oort, E. (2022). Combining physics-based and data-driven modeling in well construction: Hybrid fluid dynamics modeling. *Journal of Natural Gas Science and Engineering*, *97*, 104348. DOI: 10.1016/j.jngse.2021.104348

Errami, Y., Maaroufi, M., Cherkaoui, M., & Ouassaid, M. (2012). Maximum power point tracking strategy and direct torque control of permanent magnet synchronous generator wind farm. *Proceedings of 2012 International Conference on Complex Systems, ICCS 2012*. https://doi.org/DOI: 10.1109/ICOCS.2012.6458520

Errami, Y., Maaroufi, M., & Ouassaid, M. (2013). Maximum Power Point Tracking of a Wind Power system based on the PMSG using Sliding Mode Direct Torque Control. *Proceedings of 2013 International Renewable and Sustainable Energy Conference, IRSEC 2013*, 218–223. DOI: 10.1109/IRSEC.2013.6529706

Esfahani, M., Alizadeh, A., Cao, B., Kamwa, I., & Xu, M. (2025). Bridging theory and practice: A comprehensive review of virtual power plant technologies and their real-world applications. *Renewable & Sustainable Energy Reviews*, *222*, 115929. DOI: 10.1016/j.rser.2025.115929

ET Energy World. (n.d.) https://energy.economictimes.indiatimes.com/news/renewable/india-identifies-offshore-wind-energy-potential-of-70000-mw-along-gujarat-tn-coasts/73072580

ET Energy World. (n.d.) https://energy.economictimes.indiatimes.com/news/renewable/indias-renewable-power-potential-crosses-2-1-million-mw-wind-energy-leads-with-55-share/119837495

European Union Agency for Cybersecurity (ENISA). (2024). *Threat landscape for the energy sector 2024*. https://www.enisa.europa.eu/publications/enisa-threat-landscape-2024

Fahim, M., Sharma, V., Cao, T. V., Canberk, B., & Duong, T. Q. (2022). Machine learning-based digital twin for predictive modeling in wind turbines. *IEEE Access : Practical Innovations, Open Solutions*, *10*, 14184–14194. DOI: 10.1109/ACCESS.2022.3147602

Farrar, N. O., Ali, M. H., & Dasgupta, D. (2023). Artificial intelligence and machine learning in grid connected wind turbine control systems: A comprehensive review. *Energies*, *16*(3), 1530. DOI: 10.3390/en16031530

Farrokhabadi, M., Canizares, C. A., Simpson-Porco, J. W., Nasr, E., Fan, L., Mendoza-Araya, P. A., Tonkoski, R., Tamrakar, U., Hatziargyriou, N., Lagos, D., Wies, R. W., Paolone, M., Liserre, M., Meegahapola, L., Kabalan, M., Hajimiragha, A. H., Peralta, D., Elizondo, M. A., Schneider, K. P., & Reilly, J. (2019). Microgrid stability definitions, analysis, and examples. *IEEE Transactions on Power Systems*, *35*(1), 13–29. DOI: 10.1109/TPWRS.2019.2925703

Feng, H., Qiao, L., & Lv, Z. (2023). Innovative soft computing-enabled cloud optimization for next-generation IoT in digital twins. *Applied Soft Computing*, *136*, 110082. DOI: 10.1016/j.asoc.2023.110082

Feng, J., & Shen, W. Z. (2017). Design optimization of offshore wind farms with multiple types of wind turbines. *Applied Energy*, *205*, 1283–1297. DOI: 10.1016/j.apenergy.2017.08.107

Fernández-Guillamón, A., Gómez-Lázaro, E., Muljadi, E., & Molina-García, Á. (2019). Power systems with high RE sources: A review of inertia and frequency control strategies over time. *Renewable & Sustainable Energy Reviews*, *115*, 109369. DOI: 10.1016/j.rser.2019.109369

Flourentzou, N., Agelidis, V. G., & Demetriades, G. D. (2009). VSC-based HVDC power transmission systems: An overview. *IEEE Transactions on Power Electronics*, *24*(3), 592–602. DOI: 10.1109/TPEL.2008.2008441

Gabriel, S., Mwasilu, F., & Mwinyiwiwa, B. M. M. (2025). The Impacts of Renewable Energy Sources: A Review on Grid Inertia and Frequency Regulation. *Tanzania Journal of Engineering & Technology*, *44*(2), 171–180. Advance online publication. DOI: 10.52339/tjet.v44i1.1275

Gao, X., Zhou, S., & Luo, T. (2025). Navigating wind energy siting in the U.S.: The role of permitting and siting regulations and public engagement. *Energy Policy*, *205*, 114692. DOI: 10.1016/j.enpol.2025.114692

Garcia Marquez, F. P., & Peinado Gonzalo, A. (2022). A comprehensive review of artificial intelligence and wind energy. *Archives of Computational Methods in Engineering*, *29*(5), 2935–2958. DOI: 10.1007/s11831-021-09678-4

Ghosh, D., Adhikary, A., Nayek, A. K., & Hazra, S. (2025). Renewable energy: Opportunities, applications, and future scope of artificial intelligence and internet of things. *Next Generation Artificial Intelligence Driven Smart and Renewable Energy*, 1–12. https://doi.org/DOI: 10.1201/9781003628279-1/RENEWABLE-ENERGY-DIPANWITA-GHOSH-ARPAN-ADHIKARY-ASIT-KUMAR-NAYEK-SUNANDA-HAZRA

Gidwani, L., Tiwari, H., & Bansal, R. C. (2013). Improving power quality of wind energy conversion system with unconventional power electronic interface. *International Journal of Electrical Power & Energy Systems, 44*(1), 445–453. DOI: 10.1016/j.ijepes.2012.07.037

Giebel, G., Brownsword, R., Kariniotakis, G., Denhard, M., & Draxl, C. (2011). The state-of-the-art in short-term prediction of wind power: A literature overview, DOI: 10.11581/DTU:00000017

Gipe, P. (2016). *Wind energy for the rest of us : a comprehensive guide to wind power and how to use it : introducing electricity rebels and how they are changing the face of wind energy*. 560.

Glasson, J. (2021). Community Benefits and UK Offshore Wind Farms: Evolving Convergence in a Divergent Practice. Https://Doi.Org/10.1142/S1464333221500010, *22*(1–2). DOI: 10.1142/S1464333221500010

Global Energy Review 2025 – Analysis - IEA. (n.d.). Retrieved October 27, 2025, from https://www.iea.org/reports/global-energy-review-2025

Global Wind Report. (2025). https://www.gwec.net/events/2025-global-wind-report-launch

Gonzalez-Salazar, M. A., Kirsten, T., & Prchlik, L. (2018). Review of the operational flexibility and emissions of gas-and coal-fired power plants in a future with growing renewables. *Renewable & Sustainable Energy Reviews, 82*, 1497–1513. DOI: 10.1016/j.rser.2017.05.278

Gora, M., Makolo, P., Mwasilu, F., & Justo, J. J. (2025). Dynamic LQR control for improved DC link voltage stability in grid-integrated solar PV system. *Tanzania Journal of Science, 51*(3), 735–753. DOI: 10.65085/2507-7961.1060

Guerreiro, G. M. G., Martin, F., Dreyer, T., Yang, G., & Andresen, B. (2024). Advancements on grid compliance in wind power: Component & subsystem testing, software-/hardware-in-the-loop, and digital twins. *IEEE Access : Practical Innovations, Open Solutions, 12*, 25949–25966. DOI: 10.1109/ACCESS.2024.3366091

Gu, H., Liu, W., Gao, Q., & Zhang, Y. (2021). A review on wind turbines gearbox fault diagnosis methods. *Journal of Vibroengineering, 23*(1), 26–43. DOI: 10.21595/jve.2020.20178

GWEC. (2023). *Global Wind Report 2023*.

Gyatso, N., Li, Y., Gao, Z., Wang, Q., Li, S., Yin, Q., Chen, J., Jin, P., Liu, Z., Ma, Z., Chen, X., Feng, J., & Dorje, . (2023). Wind power performance assessment at high plateau region: A case study of the wind farm field test on the Qinghai-Tibet plateau. *Applied Energy*, *336*, 120789. DOI: 10.1016/j.apenergy.2023.120789

Habbouche, H., Amirat, Y., & Benbouzid, M. (2025). Leveraging digital twins and AI for enhanced gearbox condition monitoring in wind turbines: A review. *Applied Sciences (Basel, Switzerland)*, *15*(10), 5725. DOI: 10.3390/app15105725

Haghshenas, A., Hasan, A., Osen, O., & Mikalsen, E. T. (2023). Predictive digital twin for offshore wind farms. *Energy Informatics*, *6*(1), 1. DOI: 10.1186/s42162-023-00257-4

Hamdan, A., Ibekwe, K. I., Ilojianya, V. I., Sonko, S., & Etukudoh, E. A. (2024). AI in renewable energy: A review of predictive maintenance and energy optimization. *International Journal of Science and Research Archive*, *11*(1), 718–729. DOI: 10.30574/ijsra.2024.11.1.0112

Hannan, M. A., Al-Shetwi, A. Q., Mollik, M. S., Ker, P. J., Mannan, M., Mansor, M., Al-Masri, H. M. K., & Mahlia, T. I. (2023). Wind energy conversions, controls, and applications: A review for sustainable technologies and directions. *Sustainability (Basel)*, *15*(5), 3986. DOI: 10.3390/su15053986

Hansen, A. D., & Michalke, G. (2007). Fault ride-through capability of DFIG wind turbines. *Renewable Energy*, *32*(9), 1594–1610. DOI: 10.1016/J.RENENE.2006.10.008

Hassan, M. (2025). *AI-Based Conditional Monitoring & Predictive Maintenance for Offshore Wind Farms*, .DOI: 10.30574/ijsra.2025.14.1.0028

Hassan, A., Ahmad, G., Shafiullah, M., Islam, A., & Alam, M. S. (2025). Review of the intelligent frameworks for pitch angle control in wind turbines. *IEEE Access : Practical Innovations, Open Solutions*, *13*, 29864–29885. DOI: 10.1109/ACCESS.2025.3540367

Hau, E. (2006). Wind turbines: Fundamentals, technologies, application, economics. *Wind Turbines: Fundamentals, Technologies, Application, Economics*, 1–783. DOI: 10.1007/3-540-29284-5

Hazra, S., Madhusoodhanan, S., Moghaddam, G. K., Hatua, K., & Bhattacharya, S. (2016). Design considerations and performance evaluation of 1200-V 100-A SiC MOSFET-based two-level voltage source converter. *IEEE Transactions on Industry Applications*, *52*(5), 4257–4268. DOI: 10.1109/TIA.2016.2587098

Hewamalage, H., Bergmeir, C., & Bandara, K. (2021). Recurrent neural networks for time series forecasting: Current status and future directions. *International Journal of Forecasting*, *37*(1), 388–427. DOI: 10.1016/j.ijforecast.2020.06.008

Heylen, E., Teng, F., & Strbac, G. (2021). Challenges and opportunities of inertia estimation and forecasting in low-inertia power systems. *Renewable & Sustainable Energy Reviews*, *147*, 111176. DOI: 10.1016/j.rser.2021.111176

Hiorns, A., Smith, R., & Wright, D. (2010). An integrated approach to offshore electricity transmission. *IET Conference Publications, 2010*(570 CP). DOI: 10.1049/cp.2010.0977

Hirsch, A., Parag, Y., & Guerrero, J. (2018). Microgrids: A review of technologies, key drivers, and outstanding issues. *Renewable & Sustainable Energy Reviews*, *90*, 402–411. DOI: 10.1016/j.rser.2018.03.040

Hirth, L. (2013). The market value of variable renewables: The effect of solar wind power variability on their relative price. *Energy Economics*, *38*, 218–236. DOI: 10.1016/j.eneco.2013.02.004

Holdmann, G., & Wies, R. (2025). Harnessing the Arctic Wind: How Alaska became a leader in wind energy microgrids. *IEEE Electrification Magazine*, *13*(2), 6–15. DOI: 10.1109/MELE.2025.3558678

Hong, S., McMorland, J., Zhang, H., Collu, M., & Halse, K. H. (2024). Floating offshore wind farm installation, challenges and opportunities: A comprehensive survey. *Ocean Engineering*, *304*, 117793. DOI: 10.1016/j.oceaneng.2024.117793

Hong, Y.-Y., & Rioflorido, C. L. P. P. (2019). A hybrid deep learning-based neural network for 24-h ahead wind power forecasting. *Applied Energy*, *250*, 530–539. DOI: 10.1016/j.apenergy.2019.05.044

Horbaty, R., Huber, S., & Ellis, G. (2012). Large-scale wind deployment, social acceptance. *Wiley Interdisciplinary Reviews. Energy and Environment*, *1*(2), 194–205. DOI: 10.1002/wene.9

Hou, Y., & Wang, Q. (2023). Big data and artificial intelligence application in energy field: A bibliometric analysis. *Environmental Science and Pollution Research International*, *30*(6), 13960–13973. DOI: 10.1007/s11356-022-24880-y

Huang, D., Shuu, Y., Ruan, J., & Hu, Y. (2009). Ultra high voltage transmission in China: Developments, current status and future prospects. *Proceedings of the IEEE*, *97*(3), 555–583. DOI: 10.1109/JPROC.2009.2013613

Huang, X., Wang, C., & Zhang, S. (2024). Research and application of a Model selection forecasting system for wind speed and theoretical power generation in wind farms based on classification and wind conversion. *Energy*, *293*, 130606. DOI: 10.1016/j.energy.2024.130606

Hu, J., Shang, L., He, Y., & Zhu, Z. Q. (2011). Direct active and reactive power regulation of grid-connected DC/AC converters using sliding mode control approach. *IEEE Transactions on Power Electronics*, *26*(1), 210–222. DOI: 10.1109/TPEL.2010.2057518

Hussain, S., Sharma, S. K., & Lal, S. (2024). Feasible synergy between hybrid solar PV and wind system for energy supply of a green building in Kota (India): A case study using iHOGA. *Energy Conversion and Management*, *315*, 118783. DOI: 10.1016/j.enconman.2024.118783

Hu, Y., Zhang, Y., Li, Y., Wang, Y., Li, G., & Liu, X. (2024). Wind turbine blade recycling: A review of the recovery and high-value utilization of decommissioned wind turbine blades. *Resources, Conservation and Recycling*, *210*, 107813. DOI: 10.1016/j.resconrec.2024.107813

Iberdola. (2023). https://www.iberdrola.com/sustainability/renewables-energy-wind-power

Idaho National Laboratory. (2024). *Attack surface of wind energy technologies in the United States* (INL Threat Assessment Report). https://inl.gov/content/uploads/2024/02/INL-Wind-Threat-Assessment-v5.0.pdf

IEA Wind TCP. (n.d.). https://iea-wind.org/about-iea-wind-tcp/members/india/

India Climate & Energy Dashboard, NITI Aayog. (n.d.) https://iced.niti.gov.in/energy/fuel-sources/wind/potential

International Electrotechnical Commission (IEC). (2024). *IEC 62443 series – Industrial communication networks – Network and system security*. IEC. https://www.isa.org/standards-and-publications/isa-standards/isa-iec-62443-series-of-standards

International Renewable Energy Agency. (n.d.) https://www.irena.org/Energy-Transition/Technology/Wind-energy

International Society of Automation (ISA). (n.d.). *ISA/IEC 62443 series of standards — Standards overview*. https://www.isa.org/standards-and-publications/isa-standards/isa-iec-62443-series-of-standards

Ioannides, M. G., Stamelos, A. P., Papazis, S. A., Stamataki, E. E., & Stamatakis, M. E. (2024). Internet of things-based control of induction machines: Specifics of electric drives and wind energy conversion systems. *Energies*, *17*(3), 645. DOI: 10.3390/en17030645

IRENA & ILO. (2025). *Renewable energy and jobs: Annual review 2024*. www .irena.org

IRENA. (2019). Future of wind: Deployment, investment, technology, grid integration and socio-economic aspects. In *International Renewable Energy Agency*.

IRENA. (2019). *Future of wind: Deployment, investment, technology, grid integration and socio-economic aspects*. International Renewable Energy Agency.

IRENA. (2019). https://www.irena.org/-/media/Files/IRENA/Agency/Publication/2019/Oct/IRENA_Future_of_wind_2019.pdf#:~:text=This%20entails%20increasing%20the%20global%20cumulative%20installed,to%20installed%20capacity%20in%202018%20(542%20GW)

IRENA. (2019). *Solutions to integrate high shares of variable renewable energy*.

IRENA. (2022). *Grid codes for renewable powered systems*. International RE Agency.

IRENA. (2023). *World Energy Transitions Outlook 2023: 1.5°C Pathway*, Abu Dhabi.

IRENA. R. E. S. (2020). International renewable energy agency. *Abu Dhabi, 2020*, https://www.irena.org/publications/2020/Jul/Renewable-energy-statistics-2020

Islam, M. M., Hossain, E., Padmanaban, S., & Brice, C. W. (2020). A new perspective of wind power grid codes under unbalanced and distorted grid conditions. *IEEE Access : Practical Innovations, Open Solutions*, *8*, 15931–15944. DOI: 10.1109/ACCESS.2020.2966907

Islam, M., Mohammadpour, H. A., Ghaderi, A., & Zare, M. (2015). Time-frequency-based instantaneous power components for transient disturbances according to IEEE standard 1459. *IEEE Transactions on Power Delivery*, *30*(3), 1288–1297. DOI: 10.1109/TPWRD.2014.2361203

Islek, F., & Yuksel, Y. (2022). Evaluation of future wind power potential and their projected changes in the Black Sea and possible stable locations for wind farms. *Ocean Engineering*, *266*(Part 2), 112832. DOI: 10.1016/j.oceaneng.2022.112832

Issa, R., Hamad, M. S., & Abdel-Geliel, M. (2023, February). Digital twin of wind turbine based on Microsoft® Azure IoT platform. In *2023 IEEE Conference on Power Electronics and Renewable Energy (CPERE)* (pp. 1–8). IEEE. DOI: 10.1109/CPERE56564.2023.10119576

Jalili, S., Maheri, A., Ivanovic, A., Neilson, R., Bentin, M., Kotzur, S., May, R., & Sunner, I. (2024). Economic and environmental assessments to support the decision-making process in the offshore wind farm decommissioning projects. *Renewable and Sustainable Energy Reviews*, Volume 190. *Renewable & Sustainable Energy Reviews, 114080*, 114080. Advance online publication. DOI: 10.1016/j.rser.2023.114080

Janet, L. Sawin. (n.d.). *REN21. 2025. Renewables 2025 Global Status Report Collection*. Retrieved October 28, 2025, from https://www.ren21.net/gsr-2025/technologies/wind-power/

Jani, H. K., Singh Kachhwaha, S., Nagababu, G., & Das, A. (2022). A brief review on recycling and reuse of wind turbine blade materials. *Materials Today: Proceedings, 62*(P13), 7124–7130. DOI: 10.1016/j.matpr.2022.02.049

Janssens, O., Slavkovikj, V., Vervisch, B., Stockman, K., Loccufier, M., Verstockt, S., Van de Walle, R., & Van Hoecke, S. (2016). Convolutional neural network based fault detection for rotating machinery. *Journal of Sound and Vibration, 377*, 331–345. DOI: 10.1016/j.jsv.2016.05.027

Jayachitra, S., Prasanth, A., Dhanaraj, R. K., & Pelusi, D. (Eds.). (2025). *Applications of federated learning in technological advancements: Use cases and applications*. CRC Press. DOI: 10.1201/9781003532323

Jenkins, N., Burton, T., Bossanyi, E., Sharpe, D., & Graham, M. (2021). *Wind Energy Handbook* (3rd ed.). Wind Energy Handbook., DOI: 10.1002/9781119451143

Jiang, Z. (2021). Installation of offshore wind turbines: A technical review. *Renewable & Sustainable Energy Reviews, 139*, 110576. DOI: 10.1016/j.rser.2020.110576

Jovcic, D. (2019). *High voltage direct current transmission: converters, systems and DC grids*. John Wiley & Sons., DOI: 10.1002/9781119566632

JSW Energy Limited. (n.d.). LettertoBSEandNSE51MWRenewEnergysigned-261223.pdf

Justo, J. J., Mwasilu, F., & Jung, J. W. (2018). Effective protection for doubly fed induction generator-based wind turbines under three-phase fault conditions. *Electrical Engineering, 100*(2), 543–556. DOI: 10.1007/s00202-017-0528-0

Kabir, M. R., Halder, D., & Ray, S. (2024). Digital twins for IoT-driven energy systems: A survey. *IEEE Access : Practical Innovations, Open Solutions, 12*, 177123–177143. DOI: 10.1109/ACCESS.2024.3506660

Kaiss, M., Wan, Y., Gebbran, D., Vila, C. U., & Dragičević, T. (2025). Review on virtual power plants/virtual aggregators: Concepts, applications, prospects and operation strategies. *Renewable & Sustainable Energy Reviews*, *211*, 115242. DOI: 10.1016/j.rser.2024.115242

Kalair, A., Abas, N., & Khan, N. (2016). Comparative study of HVAC and HVDC transmission systems. *Renewable & Sustainable Energy Reviews*, *59*, 1653–1675. DOI: 10.1016/j.rser.2015.12.288

Kandemir, E., Hasan, A., Kvamsdal, T., & Abdel-Afou Alaliyat, S. (2024). Predictive digital twin for wind energy systems: A literature review. *Energy Informatics*, *7*(1), 68. DOI: 10.1186/s42162-024-00373-9

Kang, J., Schwartz, R., Flickinger, J., & Beriwal, S. (2015). Machine Learning Approaches for Predicting Radiation Therapy Outcomes: A Clinician's Perspective. *International Journal of Radiation Oncology*Biology*Physics*, *93*(5), 1127–1135. DOI: 10.1016/j.ijrobp.2015.07.2286

Karad, S., & Thakur, R. (2021). Efficient monitoring and control of wind energy conversion systems using Internet of Things (IoT): A comprehensive review. *Environment, Development and Sustainability*, *23*(10), 14197–14214. DOI: 10.1007/s10668-021-01267-6

Kardooni, R., Yusoff, S. B., & Kari, F. B. (2016). Renewable energy technology acceptance in Peninsular Malaysia. *Energy Policy*, *88*, 1–10. DOI: 10.1016/j.enpol.2015.10.005

Kathirvel, A., Gobalakrinshan, N., & Naren, A. K. (2025). Cybersecurity in industrial automation using AI. In *AI-enhanced cybersecurity for industrial automation*. IGI Global., https://doi.org/10.4018/979-8-3373-3241-3.ch024

Kavousi-Fard, A., Dabbaghjamanesh, M., Sheikh, M., & Jin, T. (2025). A novel deep learning based digital twin model for mitigating wake effects in wind farms. *Renewable Energy Focus*, *53*, 100686. DOI: 10.1016/j.ref.2025.100686

Khan, B., Josep, M., Guerrero, J. M., Chaudhary, S., Vasquez, J. C., Frederiksen, K. H. B., & Wu, Y. (2022). A review of grid code requirements for the integration of renewable energy sources in Ethiopia. *Energies*, *15*(14), 5197. DOI: 10.3390/en15145197

Khan, K., Khurshid, A., Cifuentes-Faura, J., & Xianjun, D. (2024). Does renewable energy development enhance energy security? *Utilities Policy*, *87*, 101725. DOI: 10.1016/j.jup.2024.101725

Khan, S., & Bhowmick, S. (2022). *Power-flow Modelling of HVDC transmission systems*. CRC Press., DOI: 10.1201/9781003252078

Knack, A., Kam, Y. H. S., & Tam, K. (2024). *Enhancing the cyber resilience of offshore wind* (CETaS Research Report). The Alan Turing Institute. https://cetas.turing.ac.uk/publications/enhancing-cyber-resilience-offshore-wind

Koetsier, T. (2025). 6 Advances in Historical Studies Windmills in The Netherlands. *Advances in Historical Studies, 2025*(1), 16–35. DOI: 10.4236/ahs.2025.141002

Kolli, S., Dammala, P. K., Bhattacharya, S., Fan, C., Wang, T., & Cui, L. (2023). Offshore wind farms as additional coolant power sources to enhance seismic resilience of nuclear power plants– A case study. *Nuclear Engineering and Design, 405*, 112204. DOI: 10.1016/j.nucengdes.2023.112204

Kong, X., Zhang, Z., Yin, X., & Wen, M. (2014). Study of fault current characteristics of the DFIG considering dynamic response of the RSC. *IEEE Transactions on Energy Conversion, 29*(2), 278–287. DOI: 10.1109/TEC.2014.2309694

Korompili, A., Wu, Q., & Zhao, H. (2016). Review of VSC HVDC connection for offshore wind power integration. *Renewable & Sustainable Energy Reviews, 59*, 1405–1414. DOI: 10.1016/j.rser.2016.01.064

Kumar, K., Pandey, N. K., Verre, F., & Shah, R. (2025). Cybersecurity challenges in the digitization and integration of renewable energy systems: A review. *IEEE Transactions on Engineering Management, 72*, 3042–3054. Advance online publication. DOI: 10.1109/TEM.2025.3586557

Kurbatova, T., & Perederii, T. (2020, October). *Global trends in RE development. In 2020 IEEE KhPI Week on Advanced Technology (KhPIWeek), 260-263.*, DOI: 10.1109/KhPIWeek51551.2020.9250098

Lago, J., Marcjasz, G., De Schutter, B., & Weron, R. (2021). Forecasting day-ahead electricity prices: A review of state-of-the-art algorithms, best practices and an open-access benchmark. *Applied Energy, 293*, 116983. DOI: 10.1016/j.apenergy.2021.116983

Land-Based Wind Market Report : 2023 Edition. (2023).

le Maitre, J., Ryan, G., & Power, B. (2024). Do concerns about wind farms blow over with time? Residents' acceptance over phases of project development and proximity. *Renewable and Sustainable Energy Reviews*, Volume 189. *Renewable & Sustainable Energy Reviews, 113839*, 113839. Advance online publication. DOI: 10.1016/j.rser.2023.113839

Lee, H., & Romero, J. (n.d.). *Climate Change 2023 Synthesis Report IPCC, 2023: Sections. In: Climate Change 2023: Synthesis Report. Contribution of Working Groups I, II and III to the Sixth Assessment Report of the Intergovernmental Panel on Climate Change [Core Writing Team.* 35–115. DOI: 10.59327/IPCC/AR6-9789291691647

Lezana, P., Silva, C. A., Rodríguez, J., & Pérez, M. A. (2007). Zero-steady-state-error input-current controller for regenerative multilevel converters based on single-phase cells. *IEEE Transactions on Industrial Electronics, 54*(2), 733–740. DOI: 10.1109/TIE.2007.891994

Liebensteiner, M., Ocker, F., & Abuzayed, A. (2025). High electricity price despite expansion in renewables: How market trends shape Germany's power market in the coming years. *Energy Policy, 198*, 114406. DOI: 10.1016/j.enpol.2024.114448

Li, L., Zhu, D., Zou, X., Hu, J., Kang, Y., & Guerrero, J. M. (2023). Review of frequency regulation requirements for wind power plants in international grid codes. *Renewable & Sustainable Energy Reviews, 187*, 113731. DOI: 10.1016/j.rser.2023.113731

Lim, S., Seo, K., Park, J.-W., & Lee, K. Y. (2025). New frequency stability assessment based on contribution rates of wind power plants. *International Journal of Electrical Power & Energy Systems, 164*, 110388. DOI: 10.1016/j.ijepes.2024.110388

Lin, Y., Eto, J. H., Johnson, B. B., Flicker, J. D., Lasseter, R. H., Villegas Pico, H. N., ... & Ellis, A. (2020). *Research roadmap on grid-forming inverters* (No. NREL/TP-5D00-73476). National Renewable Energy Lab.(NREL), Golden, CO (United States). https://doi.org/DOI: 10.2172/1721727

Lin, C.-H., & Wu, Y.-K. (2021). Overview of frequency-control technologies for a VSC-HVDC-integrated wind farm. *IEEE Access : Practical Innovations, Open Solutions, 9*, 112893–112921. DOI: 10.1109/ACCESS.2021.3102829

Lipu, M. H., Miah, M. S., Hannan, M. A., Hussain, A., Sarker, M. R., Ayob, A., Saad, M. H. M., & Mahmud, M. S. (2021). Artificial intelligence-based hybrid forecasting approaches for wind power generation: Progress, challenges, and prospects. *IEEE Access : Practical Innovations, Open Solutions, 9*, 102460–102489. DOI: 10.1109/ACCESS.2021.3097102

Lipu, M. S. H., Miah, M. S., Jamal, T., Rahman, T., Ansari, S., Rahman, M. S., Ashique, R. H., Shihavuddin, A. S. M., & Shakib, M. N. (2023). Artificial intelligence approaches for advanced battery management system in electric vehicle applications: A statistical analysis towards future research opportunities. *Vehicles (Basel), 6*(1), 22–70. DOI: 10.3390/vehicles6010002

Li, T., Sahu, A. K., Talwalkar, A., & Smith, V. (2020). Federated Learning: Challenges, Methods, and Future Directions. *IEEE Signal Processing Magazine*, *37*(3), 50–60. DOI: 10.1109/MSP.2020.2975749

Liu, C., Yao, W., Xiang, C., Cheng, X., & Wang, Q. (2022). Design of±535 kV hybrid high voltage direct current (HVDC) circuit breaker. *IET Generation, Transmission & Distribution*, *16*(19), 4017–4026. DOI: 10.1049/gtd2.12568

Liu, Y., Zhang, J. M., Min, Y. T., Yu, Y., Lin, C., & Hu, Z. Z. (2023). A digital twin-based framework for simulation and monitoring analysis of floating wind turbine structures. *Ocean Engineering*, *283*, 115009. DOI: 10.1016/j.oceaneng.2023.115009

Liu, Z. (2025). Intelligent monitoring and control of farmland based on edge-cloud collaboration and digital twin for digital energy management: Investment benefit analysis. *Sustainable Energy Research*, *12*(1), 43. DOI: 10.1186/s40807-025-00179-7

Li, Y., Xu, Z., Østergaard, J., & Hill, D. J. (2017). Coordinated control strategies for offshore wind farm integration via VSC-HVDC for system frequency support. *IEEE Transactions on Energy Conversion*, *32*(3), 843–856. DOI: 10.1109/TEC.2017.2663664

Loewe, M. (1966). Science and civilisation in China. Vol. 4: Physics and physical technology. Part II: Mechanical engineering. By J. Needham. pp. lv, 759; 158 plates. Cambridge University Press, 1965. £9 10 s. *Journal of the Royal Asiatic Society of Great Britain & Ireland*, *98*(2), 160–163. DOI: 10.1017/S0035869X0012502X

Lopez, G., Arboleya, P., Nunez, D., Freire, A., & Lopez, D. (2023). Wind resource assessment and influence of atmospheric stability on wind farm design using Computational Fluid Dynamics in the Andes Mountains, Ecuador. *Energy Conversion and Management*, *284*, 116972. DOI: 10.1016/j.enconman.2023.116972

Lundberg, S. M., & Lee, S. I. (2017). *A unified approach to interpreting model predictions* (Vol. 2017). Advances in Neural Information Processing Systems. Decem.

Lund, H., & Mathiesen, B. V. (2009). Energy system analysis of 100% renewable energy systems—The case of Denmark in years 2030 and 2050. *Energy*, *34*(5), 524–531. DOI: 10.1016/j.energy.2008.04.003

Lund, P. D., Lindgren, J., Mikkola, J., & Salpakari, J. (2015). Review of energy system flexibility measures to enable high levels of variable renewable electricity. *Renewable & Sustainable Energy Reviews*, *45*, 785–807. DOI: 10.1016/j.rser.2015.01.057

Luo, N., Vidal, Y., & Acho, L. (2014). *Wind Turbine Control and Monitoring*. DOI: 10.1007/978-3-319-08413-8

Lüth, A., & Keles, D. (2024). Risks, strategies, and benefits of offshore energy hubs: A literature-based survey. *Renewable & Sustainable Energy Reviews, 203*, 114761. DOI: 10.1016/j.rser.2024.114761

Lyu, H., Zhang, Q., & Chen, Y. (2023a). Anomaly detection of wind turbine driveline based on sequence decomposition interactive network (DSI-Net). *Sensors (Basel), 23*, 8964. https://doi.org/10.3390/s23218964

Lyu, J., Rao, Y., Wang, Z., Dai, J., & Cai, X. (2023). Data-driven impedance identification and stability online assessment of wind farm connected with MMC-HVDC. *IEEE Transactions on Industry Applications*, *60*(2), 2567–2576. DOI: 10.1109/TIA.2023.3333760

Lyu, Y., Luo, Y., Jia, Z., Fu, S., Tang, Y., Wang, Q., & Zeng, Y. (2024). Benefit evaluation of HVAC and HVDC for offshore wind power transmission system under the multidimensional index. *Electric Power Systems Research, 237*, 111018. DOI: 10.1016/j.epsr.2024.111018

Madeswaran, A., Bisht, D., Yuvaraj, S., Reedy, M. U., Al-Attabi, K., & Dhablia, A. (2024). AI-controlled wind turbine systems: Integrating IoT and machine learning for smart grids. *E3S Web of Conferences, 540*, 03008, .DOI: 10.1051/e3sconf/202454003008

Magesh, T., Devi, G., & Lakshmanan, T. (2022). Measurement and simulation of power quality issues in grid connected wind farms. *Electric Power Systems Research, 210*, 108142. DOI: 10.1016/j.epsr.2022.108142

Mahata, S., Harsh, P., & Shekher, V. (2024). Comparative study of time-series forecasting models for wind power generation in Gujarat, India. *e-Prime - Advances in Electrical Engineering. Electronics and Energy*, *8*, 100511. DOI: 10.1016/j.prime.2024.100511

Mahmoud, M., Semeraro, C., Abdelkareem, M. A., & Olabi, A. G. (2024). Designing and prototyping the architecture of a digital twin for wind turbine. *International Journal of Thermofluids*, *22*, 100622. DOI: 10.1016/j.ijft.2024.100622

Makolo, P. (2013). *Wind generator co-simulation with fault case analysis* [Master's thesis, Chalmers University of Technology]. Department of Energy and Environment.

Makolo, P., Justo, J. J., Mwasilu, F., & Zamora, R. (2018). Fault ride through technique for DFIG-based wind turbines under grid three-phase faults. In *2018 Australasian Universities Power Engineering Conference (AUPEC)* (pp. 1–5). IEEE. https://doi.org/DOI: 10.1109/AUPEC.2018.8757993

Makolo, P., Zamora, R., & Lie, T.-T. (2021). The role of inertia for grid flexibility under high penetration of variable renewables-A review of challenges and solutions. *Renewable & Sustainable Energy Reviews*, *147*, 111223. DOI: 10.1016/j.rser.2021.111223

Makolo, P., Zamora, R., Perera, U., & Lie, T. T. (2024). Flexible synthetic inertia optimization in modern power systems. *Inventions (Basel, Switzerland)*, *9*(1), 18. DOI: 10.3390/inventions9010018

Maksimović, M., Jokić, S., & Bošković, M. Č. (2025). Innovative horizons for sustainable smart energy: Exploring the synergy of 5G and digital twin technologies. *Process Integration and Optimization for Sustainability*, *9*(2), 431–470. DOI: 10.1007/s41660-024-00478-4

Mansour, D. E. A., Numair, M., Zalhaf, A. S., Ramadan, R., Darwish, M. M., Huang, Q., Hussien, M. G., & Abdel-Rahim, O. (2023). Applications of IoT and digital twin in electrical power systems: A comprehensive survey. *IET Generation, Transmission & Distribution*, *17*(20), 4457–4479. DOI: 10.1049/gtd2.12940

Mantar Gundogdu, B., Gladwin, D. T., Nejad, S., & Stone, D. A. (2019). Scheduling of grid-tied battery energy storage system participating in frequency response services and energy arbitrage. *IET Generation, Transmission & Distribution*, *13*(14), 2930–2941. DOI: 10.1049/iet-gtd.2018.6690

Manwell, J. F., McGowan, J. G., & Rogers, A. L. (2009). *Wind energy explained : theory, design and application*. 689. https://www.wiley.com/en-us/Wind+Energy+Explained%3A+Theory%2C+Design+and+Application%2C+2nd+Edition-p-9781119994367

Marzbnai, F., Osman, A., & Hassan, M. S. (2025). Advances in virtual power plant operations: A review of optimization models. *IEEE Access : Practical Innovations, Open Solutions*, *13*, 131525–131548. Advance online publication. DOI: 10.1109/ACCESS.2025.3591697

Mazare, M., & Ramezani, H. (2024). Enhancing cybersecurity in wind turbines: A resilient reinforcement learning-based optimal control for mitigating FDI attacks. *Applied Energy*, *373*, 123939. https://doi.org/10.1016/j.apenergy.2024.123939

Mazare, M., Ramezani, H., & Smith, A. (2024). Cyber-physical threat mitigation in wind energy systems. *Energy Informatics*, *7*(1), 49. https://doi.org/10.1186/s42162-024-00449-6

Mbasso, W. F., Harrison, A., Dagal, I., Jangir, P., Khishe, M., Kotb, H., Shaikh, M. S., Smerat, A., Fendzi Donfack, E., & Kumar, R. (2025). Digital twins in renewable energy systems: A comprehensive review of concepts, applications, and future directions. *Energy Strategy Reviews*, *61*, 101814. Advance online publication. DOI: 10.1016/j.esr.2025.101814

Mengelkamp, E., Gärttner, J., Rock, K., Kessler, S., Orsini, L., & Weinhardt, C. (2018). Designing microgrid energy markets: A case study: The Brooklyn Microgrid. *Applied Energy*, *210*, 870–880. DOI: 10.1016/j.apenergy.2017.06.054

Mercom Clean Energy Insights. (n.d). https://www.mercomindia.com/wind-installations-india-grew-129-with-878-mw-added-in-q3

Mercom Clean Energy Insights. (n.d.) https://www.mercomindia.com/india-added-wind-capacity-yoy-increase

Mercom Clean Energy Insights. (n.d.). https://www.mercomindia.com/global-offshore-wind-capacity-rises

Mercom Clean Energy Insights. (n.d.). https://www.mercomindia.com/global-wind-installations-2023

Mercom Clean Energy Insights. (n.d.). https://www.mercomindia.com/india-growth-wind-installations-2023

Mikac, R. (2023). Protection of the eu's critical infrastructures: Results and challenges. *Applied Cybersecurity and Internet Governance*, *2*(1), 1–5. DOI: 10.60097/ACIG/162868

Millan, J., Godignon, P., Perpina, X., Perez-Tomas, A., & Rebollo, J. (2014). A survey of wide bandgap power semiconductor devices. *IEEE Transactions on Power Electronics*, *29*(5), 2155–2163. DOI: 10.1109/TPEL.2013.2268900

Mocanu, E., Mocanu, D. C., Nguyen, P. H., Liotta, A., Webber, M. E., Gibescu, M., & Slootweg, J. G. (2018). On-line building energy optimization using deep reinforcement learning. *IEEE Transactions on Smart Grid*, *10*(4), 3698–3708. DOI: 10.1109/TSG.2018.2834219

Mohamed, E., Seresht, N. G., & Rizk, S. A. (2023). Context-driven ontology-based risk identification for onshore wind farm projects: A domain-specific approach. *Advanced Engineering Informatics*, *56*, 101962. DOI: 10.1016/j.aei.2023.101962

Mohammadi, F., Danapour, N., & Rouzbehi, K. (2024). Integrated Wind Turbines and HVDC Power Transmission Lines. *IEEE Transactions on Sustainable Energy*, *15*(3), 2134–2137. DOI: 10.1109/TSTE.2024.3402010

Montoya, M. C., Rubio-Medrano, C. E., & Kareem, A. (2024). On the cybersecurity of smart structures under wind. *Journal of Wind Engineering and Industrial Aerodynamics*, *251*, 105777. https://doi.org/10.1016/j.jweia.2024.105777

Mordor Intelligence, https://www.mordorintelligence.com/industry-reports/north-america-wind-power-market

Morrison, R., Liu, X., & Lin, Z. (2022). Anomaly detection in wind turbine SCADA data for power-curve cleaning. *Renewable Energy*, *184*, 473–486. https://doi.org/10.1016/j.renene.2021.11.118

Murdan, A., Jahmeerbacus, I., & Hassen, S. Z. S. (2023). Challenges of existing grid codes and the call for enhanced standards. *Clean Technologies and Recycling*, *3*(4), 241–256. DOI: 10.3934/ctr.2023015

Musial, W., Spitsen, P., Duffy, P., Beiter, P., Shields, M., Mulas Hernando, D., & Sathish, S. (2023). *Offshore Wind Market Report: 2023 Edition (No. NREL/TP-5000-87232). National Renewable Energy Laboratory*. NREL.

Muyeen, S. M. (2012). Wind energy conversion systems: Technology and trends. *Green Energy and Technology*, *78*. Advance online publication. DOI: 10.1007/978-1-4471-2201-2

Nagababu, G., Puppala, H., Pritam, K., & Kantipudi, M. V. V. P. (2022). Two-stage GIS-MCDM based algorithm to identify plausible regions at micro level to install wind farms: A case study of India. *Energy*, *248*, 123594. DOI: 10.1016/j.energy.2022.123594

Najafzadeh, M., Ahmadiahangar, R., Husev, O., Roasto, I., Jalakas, T., & Blinov, A. (2021). Recent contributions, future prospects and limitations of interlinking converter control in hybrid AC/DC microgrids. *IEEE Access : Practical Innovations, Open Solutions*, *9*, 7960–7984. DOI: 10.1109/ACCESS.2020.3049023

Narayanan, V. L. (2024). Reinforcement learning in wind energy—A review. *International Journal of Green Energy*, *21*(9), 1945–1968. DOI: 10.1080/15435075.2023.2281329

Natarajan, A. (2014). An overview of the state of the art technologies for multi-MW scale offshore wind turbines and beyond. *Wiley Interdisciplinary Reviews. Energy and Environment*, *3*(2), 111–121. DOI: 10.1002/wene.80

National Renewable Energy Laboratory (NREL). (2019). *Key cyber security requirements for wind farms (Systems Engineering Workshop).* https://www.nrel.gov/wind/systems-engineering-workshop-2019

National Renewable Energy Laboratory (NREL). (2024). *Renewable energy and storage cybersecurity research (RESCue) pilot final report* (NREL/TP-5T00-89921). https://www.nrel.gov/docs/fy24osti/89921.pdf

Nghitevelekwa, K., & Bansal, R. C. (2018). A review of generation dispatch with large-scale photovoltaic systems. *Renewable & Sustainable Energy Reviews, 81*, 615–624. DOI: 10.1016/j.rser.2017.08.035

Ngoenmeesri, R., Chidaruksa, S., Wangkeeree, R., & Sirisamphanwong, C. (2023). Power quality enhancement for Thailand's wind farm using 5 MWh Li-ion battery energy storage system. *Heliyon*, *9*(11), e22029. DOI: 10.1016/j.heliyon.2023.e22029 PMID: 38034671

Nguyen, T. T., Vu, T., Paudyal, S., Blaabjerg, F., & Vu, T. L. (2022). Grid-forming inverter-based wind turbine generators: Comprehensive review, comparative analysis, and recommendations. *arXiv preprint arXiv:2203.02105*. https://doi.org//arXiv.2203.02105DOI: 10.48550

Nguyen, T. K., Chang, G. W., Liu, Y. J., & Li, G. Y. (2023). Improving Voltage Ride-Through Capability of Grid-Tied Microgrid With Harmonics Mitigation. *IEEE Transactions on Power Delivery*, *38*(1), 738–747. DOI: 10.1109/TPWRD.2022.3199081

Nguyen, V. H., Tran, Q. T., Besanger, Y., Jung, M., & Nguyen, T. L. (2022). Digital twin integrated power-hardware-in-the-loop for the assessment of distributed renewable energy resources. *Electrical Engineering*, *104*(2), 377–388. DOI: 10.1007/s00202-021-01246-0

Nycander, E., & Söder, L. (2018). Review on European grid codes for wind farms and their implications for wind curtailments. *17th int'l wind integration Workshop.* URN: urn:nbn:se:kth:diva-236565

Odeh, R. P., & Watts, D. (2019). Impacts of wind and solar spatial diversification on its market value: A case study of the Chilean electricity market. *Renewable & Sustainable Energy Reviews*, *111*, 442–461. DOI: 10.1016/j.rser.2019.01.015

Ogunlana, O., & Ezeoha, M. (2021). The future of RE in Nigeria: charting the path for RE in Nigeria. *Available at SSRN* 3960313.

Ogunmoye, K. A., Agupugo, C. P., Ejichukwu, E., Barros, P., & Hayden, M. D. (2025). Protecting the Grid: Cybersecurity strategies for renewable energy integration. https://doi.org/DOI: 10.30574/wjarr.2025.26.3.2234

Ohalete, N. C., Aderibigbe, A. O., Ani, E. C., Ohenhen, P. E., Daraojimba, D. O., & Odulaja, B. A.Nzubechukwu Chukwudum OhaleteAdebayo Olusegun Aderibig-beEmmanuel Chigozie AniPeter Efosa OhenhenDonald Obinna DaraojimbaBukola A. Odulaja. (2023). AI-driven solutions in renewable energy: A review of data science applications in solar and wind energy optimization. *World Journal of Advanced Research and Reviews*, *20*(3), 401–417. DOI: 10.30574/wjarr.2023.20.3.2433

Okedu, K. E., & Muyeen, S. M. (2022). Comparative Performance of DFIG and PMSG Wind Turbines during Transient State in Weak and Strong Grid Conditions Considering Series Dynamic Braking Resistor. *Energies*, *15*(23), 9228. Advance online publication. DOI: 10.3390/en15239228

Okedu, K. E., & Muyeen, S. M. (2022). Comparative performance of DFIG and PMSG wind turbines during transient state in weak and strong grid conditions considering series dynamic braking resistor. *Energies*, *15*(23). Advance online publication. DOI: 10.3390/EN15239228

Oleka, E. U., Ndubisi, S. N., & Ijemaru, G. K. (2016). Electric power transmission enhancement: A case of Nigerian electric power grid. *Amer. J. Elect. Electron*, *4*(1), 33–39. DOI: 10.12691/ajeee-4-1-5

Operator, A. E. M. (2024). *Australian Energy Market Operator. 2023*. January.

Paidi, E. S. N. R., Marzooghi, H., Yu, J., & Terzija, V. (2020). Development and Validation of Artificial Neural Network-Based Tools for Forecasting of Power System Inertia With Wind Farms Penetration. *IEEE Systems Journal*, *14*(4), 4978–4989. DOI: 10.1109/JSYST.2020.3017640

Pang, H., & Wei, X. (2018). Research on key technology and equipment for Zhangbei 500kV DC grid. *2018 International Power Electronics Conference (IPEC-Niigata 2018-ECCE Asia)*, 2343–2351. DOI: 10.23919/IPEC.2018.8507575

Pessanha, J. F. M., Almeida, V. A., Melo, A. C. G., Maceira, M. E. P., Mendonca, M. J. C., & Medrano, L. A. (2024). A Bayesian approach to estimate the capacity factor of offshore wind farms on the Brazilian coast. *Electric Power Systems Research*, *234*, 110772. DOI: 10.1016/j.epsr.2024.110772

Pfeifenberger, J. P., Bai, L., Levitt, A., Plet, C. A., Sonnathi, C. M., Alliance, C. G., United, G., & Energy, P. (2023). *The operational and market benefits of HVDC to system operators*. The Brattle Group and DNV.

Polinder, H., Van Der Pijl, F. F. A., De Vilder, G. J., & Tavner, P. J. (2006). Comparison of direct-drive and geared generator concepts for wind turbines. *IEEE Transactions on Energy Conversion*, *21*(3), 725–733. DOI: 10.1109/TEC.2006.875476

Pontes, L., Costa, T., Souza, A., Dantas, N., Vasconcelos, A., Rissi, G., Dias, R., Mohamed, M. A., Siano, P., & Marinho, M. (2023). Operational data analysis of a battery energy storage system to support wind energy generation. *Energies*, *16*(3), 1468. DOI: 10.3390/en16031468

Pragati, A., Mishra, M., Rout, P. K., Gadanayak, D. A., Hasan, S., & Prusty, B. R. (2023). A comprehensive survey of HVDC protection system: Fault analysis, methodology, issues, challenges, and future perspective. *Energies*, *16*(11), 4413. DOI: 10.3390/en16114413

Prasad, K. M. V. V., Nagababu, G., & Jani, H. K. (2023). Enhancing offshore wind resource assessment with LIDAR-validated reanalysis datasets: A case study in Gujarat, India. *International Journal of Thermofluids*, *18*, 100320. DOI: 10.1016/j.ijft.2023.100320

Press Information Bureau. (n.d.) https://pib.gov.in/PressReleaseIframePage.aspx?PRID=1944759

Press Information Bureau. (n.d.)https://pib.gov.in/PressReleaseIframePage.aspx?PRID=1947133

Price, T. J. (2005). James Blyth - Britain's first modern wind power pioneer. *Wind Engineering*, *29*(3), 191–200. DOI: 10.1260/030952405774354921

Prime, J., Abdulkadir Ahmed, I., Akande, D., Elhassan, N., Escamilla, G., Jamdade, A., Melnikov, Y., & Whiteman, A. (2024). Renewable Energy Statistics 2024, International Renewable Energy Agency. *International Renewable Energy Agency*, 1–1099. https://www.irena.org/Publications/2024/Jul/Renewable-energy-statistics-2024

Prime, J., Ahmed, I. A., Akande, D., Elhassan, N., Melnikov, Y., & Whiteman, A. (2025). *RENEWABLE CAPACITY STATISTICS 2025 STATISTIQUES DE CAPACITÉ RENOUVELABLE 2025 ESTADÍSTICAS DE CAPACIDAD RENOVABLE 2025 About IRENA*. https://www.irena.org/-/media/Files/IRENA/Agency/Publication/2025/Mar/IRENA_DAT_RE_Capacity_Statistics_2025.pdf

Proceedings of the Government of Karnataka. (n.d.) JSW Renewable Energy ltd.pdf

Proceedings of the Government of Karnataka. (n.d.) JSWRenewEnergyFourLtd.pdf

Proceedings of the Government of Karnataka. (n.d.) JSWRenewEnergyFourLtd_2.pdf

Proceedings of the Government of Karnataka. (n.d.). 0_0_51113123412151GO600MWJSWRenewableEnergy(Vijayanagar)Ltd.pdf

Pujana, A., Esteras, M., Perea, E., Maqueda, E., & Calvez, P. (2023). Hybrid-model-based digital twin of the drivetrain of a wind turbine and its application for failure synthetic data generation. *Energies*, *16*(2), 861. DOI: 10.3390/en16020861

Qamsane, Y., Chen, C.-Y., Balta, E. C., Kao, B.-C., Mohan, S., Moyne, J., Tilbury, D., & Barton, K. (2019). A unified digital twin framework for real-time monitoring and evaluation of smart manufacturing systems. *2019 IEEE 15th International Conference on Automation Science and Engineering (CASE)*, 1394–1401, DOI: 10.1109/COASE.2019.8842880

Qays, M. O., Ahmed, M. M., Parvez Mahmud, M. A., Abu-Siada, A., Muyeen, S. M., Hossain, M. L., Yasmin, F., & Rahman, M. M. (2022). Monitoring of renewable energy systems by IoT-aided SCADA system. *Energy Science & Engineering*, *10*(6), 1874–1885. DOI: 10.1002/ese3.1130

Rababaah, A. R. (2023). Comparative study of deep learning models versus machine learning models for wind turbine intelligent health diagnosis systems. *Arabian Journal for Science and Engineering*, *48*(8), 10875–10899. DOI: 10.1007/s13369-023-07810-z

Rahmani-Sane, G., Azad, S., Ameli, M. T., & Haghani, S. (2025). The applications of artificial intelligence and digital twin in power systems: An in-depth review. *IEEE Access : Practical Innovations, Open Solutions*, *13*, 108573–108608. DOI: 10.1109/ACCESS.2025.3580340

Raju, M. S. S., Mohapatra, P., Dutta, S., Mallipeddi, R., & Das, K. N. (2024). Optimal placement of fixed hub height wind turbines in a wind farm using twin archive guided decomposition based multi-objective evolutionary algorithm. *Engineering Applications of Artificial Intelligence*, *130*, 107735. DOI: 10.1016/j.engappai.2023.107735

Ranawaka, A., Alahakoon, D., Sun, Y., & Hewapathirana, K. (2024). Leveraging the synergy of digital twins and artificial intelligence for sustainable power grids: A scoping review. *Energies*, *17*(21), 5342. DOI: 10.3390/en17215342

Ratnam, K. S., Palanisamy, K., & Yang, G. (2020). Future low-inertia power systems: Requirements, issues, and solutions-A review. *Renewable & Sustainable Energy Reviews*, *124*, 109773. DOI: 10.1016/j.rser.2020.109773

Reitz, S., Goshen, L., & Ohlhorst, D. (n.d.). *Trade-offs in German wind energy expansion: building bridges between different interests, values and priorities*. DOI: 10.1186/s13705-022-00365-1

Rekeraho, A., Cotfas, D. T., Cotfas, P. A., Bălan, T. C., Tuyishime, E., & Acheampong, R. (2024). Cybersecurity challenges in IoT-based smart renewable energy. *International Journal of Information Security*, *23*(1), 101–117. DOI: 10.1007/s10207-023-00732-9

Renewable Energy Agency. I. (2019). *Renewable Energy Statistics 2019*. www.irena.org

Renewable Energy Agency. I. (2020). *Wind energy: A gender perspective*. www.irena.org

Renewable Energy World, https://www.renewableenergyworld.com/wind-power/2023-was-a-record-year-for-wind-installations-as-world-ramps-up-clean-energy-report-says/

Renewables 2025 Global Status Report. (2025). https://www.ren21.net/gsr-2025/figures/figure_W_2/

Ren, K., Liu, S., Kong, S., Ma, S., & Feng, R. (2026). Thermal recycling of wind turbine blade composites: Optimizing pyrolysis and oxidation for fiber recovery. *Composites. Part A, Applied Science and Manufacturing*, *202*, 109460. DOI: 10.1016/j.compositesa.2025.109460

Righter, R. W. (1996). *Wind energy in America : a history*. 361. https://wind-works.org/wind-energy-in-america-a-history-by-robert-righter/

Robles, E., Haro-Larrode, M., Santos-Mugica, M., Etxegarai, A., & Tedeschi, E. (2019). Comparative analysis of European grid codes relevant to offshore renewable energy installations. *Renewable & Sustainable Energy Reviews*, *102*, 171–185. DOI: 10.1016/j.rser.2018.12.002

Rodriguez, E. (2025). Digital twin dynamics in smart energy grids: Advanced integration and practical implementations. *Digital Twin Dynamics*, *1*(1), 33–46.

Rodríguez, J., Bernet, S., Wu, B., Pontt, J. O., & Kouro, S. (2007). Multilevel voltage-source-converter topologies for industrial medium-voltage drives. *IEEE Transactions on Industrial Electronics*, *54*(6), 2930–2945. DOI: 10.1109/TIE.2007.907044

Roga, S., Dahiwale, H., Bardhan, S., & Sinha, S. (2024). Wind energy potential assessment: a case study in Central India. *Proceedings of the Institution of Civil Engineers – Energy*, Volume 177, Issue 3, Pages 130-148. DOI: 10.1680/jener.22.00016

Rosca, C. M., & Stancu, A. (2025). A comprehensive review of machine learning models for optimizing wind power processes. *Applied Sciences (Basel, Switzerland)*, *15*(7), 3758. DOI: 10.3390/app15073758

Ryu, D., Lee, S., Yang, S., Jeong, J., Lee, Y., & Shin, D. (2024). Enhancing cybersecurity in energy IT infrastructure through a layered defense approach to major malware threats. *Applied Sciences (Basel, Switzerland)*, *14*(22), 10342. https://doi.org/10.3390/app142210342

Sadiq, M., Ali, S. W., Terriche, Y., Mutarraf, M. U., Hassan, M. A., Hamid, K., Ali, Z., Sze, J. Y., Su, C.-L., & Guerrero, J. M. (2021). Future greener seaports: A review of new infrastructure, challenges, and energy efficiency measures. *IEEE Access : Practical Innovations, Open Solutions*, *9*, 75568–75587. DOI: 10.1109/ACCESS.2021.3081430

Safari, A., Daneshvar, M., & Anvari-Moghaddam, A. (2024). Energy intelligence: A systematic review of artificial intelligence for energy management. *Applied Sciences (Basel, Switzerland)*, *14*(23), 11112. DOI: 10.3390/app142311112

Sakuru, S. K. V. S., & Ramana, M. V. (2023). Wind power potential over India using the ERA5 reanalysis. *Sustainable Energy Technologies and Assessments*, *56*, 103038. DOI: 10.1016/j.seta.2023.103038

Salameh, J. P., Cauet, S., Etien, E., Sakout, A., & Rambault, L. (2018). Gearbox condition monitoring in wind turbines: A review. *Mechanical Systems and Signal Processing*, *111*, 251–281. DOI: 10.1016/j.ymssp.2018.03.052

Saleem, D., ... (2024). *Renewable energy and storage cybersecurity research (RESCue) pilot final report* (NREL/TP-5T00-89921). National Renewable Energy Laboratory. https://www.nrel.gov/docs/fy24osti/89921.pdf

Sang, W., Guo, W., Dai, S., Tian, C., Yu, S., & Teng, Y. (2022). Virtual synchronous generator, a comprehensive overview. *Energies*, *15*(17), 6148. DOI: 10.3390/en15176148

Saraswat, S. K., Digalwar, A. K., Yadav, S. S., & Kumar, G. (2021). MCDM and GIS based modelling technique for assessment of solar and wind farm locations in India. *Renewable Energy*, *169*, 865–884. DOI: 10.1016/j.renene.2021.01.056

Sarkar, M. N. I., Meegahapola, L. G., & Datta, M. (2018). *Reactive power management in renewable rich power grids: A review of grid-codes, renewable generators, support devices, control strategies and optimization algorithms*. IEEE., DOI: 10.1109/ACCESS.2018.2838563

Shafik, W. (2024). Introduction to renewable energy technologies. In *Soft computing in renewable energy technologies* (pp. 1–31). CRC Press. DOI: 10.1201/9781003462460-1

Sharifabadi, K., Harnefors, L., Nee, H.-P., Norrga, S., & Teodorescu, R. (2016). *Design, control, and application of modular multilevel converters for HVDC transmission systems*. John Wiley & Sons., DOI: 10.1002/9781118851555

Sheng, S., & O'Connor, R. (2023). Reliability of wind turbines. In *Wind energy engineering* (pp. 195–211). Academic Press. DOI: 10.1016/B978-0-323-99353-1.00016-5

Sheng, Y., Wang, H., Yan, J., Liu, Y., & Han, S. (2023). Short-term wind power prediction method based on deep clustering-improved Temporal Convolutional Network. *Energy Reports*, *9*, 2118–2129. DOI: 10.1016/j.egyr.2023.01.015

Shittu, H., & Shittu, M. (2024). AI powered digital twins for predictive maintenance and operational optimization of renewable energy systems. *International Journal of Science, Architecture. Technology and Environment*, *1*, 151–163.

Shukla, S., & Singh, B. (2019). MPPT control technique for solar powered direct torque control of induction motor drive with a robust speed and parameters adaptation scheme for water pumping. *IET Renewable Power Generation*, *13*(2), 273–284. DOI: 10.1049/iet-rpg.2018.5390

Siano, P. (2014). Demand response and smart grids—A survey. *Renewable & Sustainable Energy Reviews*, *30*, 461–478. DOI: 10.1016/j.rser.2013.10.022

Singh, M. B., Jain, P., & Singh, P. (2024). Wind Energy. *Wind Energy Storage and Conversion*, 1–16. DOI: 10.1002/9781394204564.ch1

Singh, A. (2023). The Current Scenario of Sustainable Planning and Development of the Indian Built Environment. In Cobîrzan, N., Muntean, R., & Felseghi, R. (Eds.), *Circular Economy Implementation for Sustainability in the Built Environment* (pp. 106–138). IGI Global., DOI: 10.4018/978-1-6684-8238-4.ch005

Singh, A. (2023). The Current Trends and Need of Infrastructure Projects in the Country India. In Popescu, C., Yu, P., & Wei, Y. (Eds.), *Achieving the Sustainable Development Goals Through Infrastructure Development* (pp. 235–261). IGI Global., DOI: 10.4018/979-8-3693-0794-6.ch010

Singh, A. (2023). The Development and Significance of Bengaluru Suburban Rail Project. In Yilmaz, Z., Golem, S., & Costescu, D. (Eds.), *Handbook of Research on Promoting Sustainable Public Transportation Strategies in Urban Environments* (pp. 267–292). IGI Global., DOI: 10.4018/978-1-6684-5996-6.ch013

Singh, A. (2023). The Need and Development of Dams in India as Well as in the World for a Better Future. In Lousada, S. (Ed.), *Geoinformatics in Support of Urban Politics and the Development of Civil Engineering* (pp. 156–180). IGI Global., DOI: 10.4018/978-1-6684-6449-6.ch007

Singh, A. (2023). The Present and Future Role of Artificial Intelligence in the Promotion of Sustainable Practices in the Construction Sector. In Kahraman, C., Sari, I. U., Oztaysi, B., Cebi, S., Cevik Onar, S., & Tolga, A. Ç. (Eds.), *Intelligent and Fuzzy Systems. INFUS 2023. Lecture Notes in Networks and Systems* (Vol. 759). Springer., DOI: 10.1007/978-3-031-39777-6_2

Siram, O., Sahoo, N., & Saha, U. K. (2022). Changing landscape of India's renewable energy and the contribution of wind energy. *Cleaner Engineering and Technology, 8*, 100506. DOI: 10.1016/j.clet.2022.100506

Smallwood, K. S., & Bell, D. A. (2020). Effects of Wind Turbine Curtailment on Bird and Bat Fatalities. *The Journal of Wildlife Management, 84*(4), 685–696. DOI: 10.1002/jwmg.21844

Smart, E. E., Olanrewaju, L. O., Usman, J., Otaru, K., & Umar, D. (2025). Artificial intelligence (AI) in renewable energy forecasting and optimization. *Renewable Energy, 10*, 11.

SolarQuarter. (n.d.). https://solarquarter.com/2024/07/11/indias-solar-power-revolution-leading-with-85474-31-mw-and-57-72-of-renewable-energy-by-june-2024/#:~:text=Wind%20power%20has%20also%20seen%20significant%20growth,embrace%20the%20vast%20potential%20of%20solar%20energy

Somi, S., Seresht, N. G., & Fayek, A. R. (2021). Developing a risk breakdown matrix for onshore wind farm projects using fuzzy case-based reasoning. *Journal of Cleaner Production, 311*, 127572. DOI: 10.1016/j.jclepro.2021.127572

Song, D., Fan, T., Li, Q., & Joo, Y. H. (2024). Advances in Offshore Wind. *Journal of Marine Science and Engineering 2024, Vol. 12, Page 359, 12*(2), 359. DOI: 10.3390/jmse12020359

Song, J., Zhou, X., Zhou, Z., Wang, Y., Wang, Y., & Wang, X. (2023). Review of Low Inertia in Power Systems Caused by High Proportion of Renewable Energy Grid Integration. *Energies, 16*(16), 6042. DOI: 10.3390/en16166042

Song, Z., Hackl, C. M., Anand, A., Thommessen, A., Petzschmann, J., Kamel, O., Braunbehrens, R., Kaifel, A., Roos, C., & Hauptmann, S. (2023). Digital twins for the future power system: An overview and a future perspective. *Sustainability (Basel), 15*(6), 5259. DOI: 10.3390/su15065259

Sourkounis, B., & Tourou, P. (2013). Grid code requirements for wind power integration in Europe. *Energy, 2013*, 1–9. DOI: 10.1155/2013/437674

Souza, M. E. T.Junior, & Freitas, L. C. G. (2022). Power electronics for modern sustainable power systems: Distributed generation, microgrids and smart grids—A review. *Sustainability (Basel), 14*(6), 3597. DOI: 10.3390/su14063597

Stadtmann, F., Rasheed, A., Kvamsdal, T., Johannessen, K. A., San, O., Kölle, K., Tande, J. O., Barstad, I., Benhamou, A., Brathaug, T., Christiansen, T., Firle, A.-L., Fjeldly, A., Frøyd, L., Gleim, A., Høiberget, A., Meissner, C., Nygård, G., Olsen, J., & Skogås, J. O. (2023). Digital twins in wind energy: Emerging technologies and industry-informed future directions. *IEEE Access : Practical Innovations, Open Solutions, 11*, 110762–110795. DOI: 10.1109/ACCESS.2023.3321320

Statista. (n.d.). https://www.statista.com/statistics/1368205/global-wind-capacity-additions-forecast-by-type/

Statista. (n.d.). https://www.statista.com/statistics/264257/number-of-offshore-wind-farms-worldwide-by-country/

Statista. (n.d.). https://www.statista.com/statistics/865745/india-wind-energy-capacity/#:~:text=The%20wind%20energy%20capacity%20in%20the%20south,India%20from%202009%20to%202023%20(in%20megawatts)&text=Figures%20are%20compiled%20from%20several%20editions%20of%20this%20report

Statista. (n.d.). https://www.statista.com/statistics/963139/china-capacity-of-onshore-wind-energy/

Stecca, M., Elizondo, L. R., Soeiro, T. B., Bauer, P., & Palensky, P. (2020). A comprehensive review of the integration of battery energy storage systems into distribution networks. *IEEE Open Journal of the Industrial Electronics Society, 1*, 46–65. DOI: 10.1109/OJIES.2020.2981832

Stetco, A., Dinmohammadi, F., Zhao, X., Robu, V., Flynn, D., Barnes, M., Keane, J., & Nenadic, G. (2019). Machine learning methods for wind turbine condition monitoring: A review. *Renewable Energy, 133*, 620–635. DOI: 10.1016/j.renene.2018.10.047

Strasser, T. I., Widl, E., Kuchenbuch, R. A., Lázaro-Elorriaga, L., Laraudogoitia, B. T., Ginocchi, M., . . . Algaba, E. R. (2024, November). Towards interoperability testing of smart energy systems–An overview and discussion of possibilities. In *IET Conference Proceedings CP904* (Vol. 2024, No. 29, pp. 263–268). The Institution of Engineering and Technology.

Subedi, S., Gui, Y., & Xue, Y. (2025). Applications of Data-Driven Dynamic Modeling of Power Converters in Power Systems: An Overview. *IEEE Transactions on Industry Applications*, *61*(2), 2434–2456. DOI: 10.1109/TIA.2025.3529797

Sun, P., Li, J., Wang, C., & Lei, X. (2016). A generalized model for wind turbine anomaly identification based on SCADA data. *Applied Energy*, *168*, 550–567. DOI: 10.1016/j.apenergy.2016.01.133

Swiss Re Institute, https://www.swissre.com/institute/research/topics-and-risk-dialogues/climate-and-natural-catastrophe-risk/climate-change-wind-power.html

Tagesschau. (2022, March 2). *Cyberattack affects wind turbines in Germany*. https://www.tagesschau.de

Taher, F., & Liu, J., AbdulAzeem, M., & Zhan, Z. (2022). Cybersecurity challenges in offshore renewable energy systems: A survey. *Energies*, *15*(14), 5012. https://doi.org/10.3390/en15145012

Tamilselvan, P., & Wang, P. (2013). Failure diagnosis using deep belief learning based health state classification. *Reliability Engineering & System Safety*, *115*, 124–135. DOI: 10.1016/j.ress.2013.02.022

Tapoglou, E. (2024). *Wind energy circularity challenges*. European Commission.

Tautz-Weinert, J., & Watson, S. J. (2017). Using SCADA data for wind turbine condition monitoring–a review. *IET Renewable Power Generation*, *11*(4), 382–394. DOI: 10.1049/iet-rpg.2016.0248

Tchakoua, P., Wamkeue, R., Ouhrouche, M., Slaoui-Hasnaoui, F., Tameghe, T. A., & Ekemb, G. (2014). Wind turbine condition monitoring: State-of-the-art review, new trends, and future challenges. *Energies*, *7*(4), 2595–2630. DOI: 10.3390/en7042595

Tchakoua, P., Wamkeue, R., Tameghe, T. A., & Ekemb, G. (2013). A review of concepts and methods for wind turbines condition monitoring. *2013 World Congress on Computer and Information Technology (WCCIT)*, 1–9, DOI: 10.1109/WCCIT.2013.6618706

Teodorescu, R., Liserre, M., & Rodriguez, P. (2011). *Grid converters for photovoltaic and wind power systems* (1st ed.). Wiley.

Toke, D. (2005). Will the government catch the wind? *The Political Quarterly*, *76*(1), 48–56. DOI: 10.1111/j.1467-923X.2005.00655.x

Tovstukha, I. (2017). Modern approaches to defining alternative sources of energy. In *International Scientific and Practical Conference* (p. 124).

Tuyen, N., Quan, N., Linh, V., Vu, T., & Fujita, G. (2022). A comprehensive review of cybersecurity in inverter based smart power system amid the boom of renewable energy. *IEEE Access : Practical Innovations, Open Solutions*, *10*, 35846–35875. DOI: 10.1109/ACCESS.2022.3163551

Udo, W. S., Kwakye, J. M., Ekechukwu, D. E., & Ogundipe, O. B. (2024). Optimizing wind energy systems using machine learning for predictive maintenance and efficiency enhancement. *Journal of Renewable Energy Technology*, *28*(3), 312–330.

Udo, W., & Muhammad, Y. (2021). Data-driven predictive maintenance of wind turbine based on SCADA data. *IEEE Access : Practical Innovations, Open Solutions*, *9*, 162370–162388. DOI: 10.1109/ACCESS.2021.3132684

Ukoba, K., Olatunji, K. O., Adeoye, E., Jen, T. C., & Madyira, D. M. (2024). Optimizing renewable energy systems through artificial intelligence: Review and future prospects. *Energy & Environment*, *35*(7), 3833–3879. DOI: 10.1177/0958305X241256293

Ullah, F., Zhang, X., Khan, M., Mastoi, M. S., Munir, H. M., Flah, A., & Said, Y. (2024). A comprehensive review of wind power integration and energy storage technologies for modern grid frequency regulation. *Heliyon*, *10*(9), e30466. Advance online publication. DOI: 10.1016/j.heliyon.2024.e30466 PMID: 38756608

Ullah, M., Guan, Y., Yu, Y., Chaudhary, S. K., Vasquez, J. C., & Guerrero, J. M. (2025). Dynamic Performance and Power Quality of Large-Scale Wind Power Plants: A Review on Challenges, Evolving Grid Code, and Proposed Solutions. *IEEE Open Journal of Power Electronics*, *6*, 1148–1173. DOI: 10.1109/OJPEL.2025.3582012

Ummels, B. C., Pelgrum, E., & Kling, W. L. (2008). Integration of large-scale wind power and use of energy storage in the Netherlands' electricity supply. *IET Renewable Power Generation*, *2*(1), 34–46. DOI: 10.1049/iet-rpg:20070056

van Est, R. (2022). The Success of Danish Wind Energy Innovation Policy: Combining Visionary Politics and Pragmatic Policymaking. *Successful Public Policy in the Nordic Countries: Cases, Lessons. Challenges*, •••, 45–66. DOI: 10.1093/oso/9780192856296.003.0003

Van Hertem, D., Gomis-Bellmunt, O., & Liang, J. (2016). *HVDC grids: for offshore and supergrid of the future*. John Wiley & Sons. DOI: 10.1002/9781119115243

Van Schendel, S., & Varenhorst, I. A. M. (2025). A time-aggregation approach for reducing identifiability in household energy data. *Utilities Policy*, *94*, 101900. https://doi.org/10.1016/j.jup.2025.101900

Vázquez-Canteli, J. R., & Nagy, Z. (2019). Reinforcement learning for demand response: A review of algorithms and modeling techniques. *Applied Energy*, *235*, 1072–1089. DOI: 10.1016/j.apenergy.2018.11.002

Velandia-Cardenas, C., Vidal, Y., & Pozo, F. (2021). Wind Turbine Fault Detection Using Highly Imbalanced Real SCADA Data. *Energies 2021, Vol. 14, Page 1728*, *14*(6), 1728. DOI: 10.3390/en14061728

Vianny, M., Kathirvel, A., Rithik, G., Naren, K., & Anthony Raj, A. (2025). Private AI—An overview. In *AI-driven breakthroughs in antimicrobial resistance*. IGI Global., https://doi.org/10.4018/979-8-3693-7550-1

Vijay, S., Swaminathan, G. V., & Periasamy, S. (2024). Analysis and mitigation of subsynchronous resonance in VSC-HVDC-based offshore wind farms. In Chenniappan, S., Padmanaban, S., & Palanisamy, S. (Eds.), *Power Systems Operation with 100% Renewable Energy Sources* (pp. 149–164)., DOI: 10.1016/B978-0-443-15578-9.00010-8

Voigt, C. C., Bernard, E., Huang, J. C. C., Frick, W. F., Kerbiriou, C., MacEwan, K., Mathews, F., Rodríguez-Durán, A., Scholz, C., Webala, P. W., Welbergen, J., & Whitby, M. (2024). Toward solving the global green–green dilemma between wind energy production and bat conservation. *Bioscience*, *74*(4), 240–252. DOI: 10.1093/biosci/biae023 PMID: 38720909

von Krauland, A.-K., & Jacobson, M. Z. (2024). India onshore wind energy atlas accounting for altitude and land use restrictions and co-located solar. *Cell Reports Sustainability*, *1*(5), 100083. DOI: 10.1016/j.crsus.2024.100083

Vrana, T. K., Flynn, D., Gómez-Lázaro, E., Kiviluoma, J., Marcel, D., Cutululis, N., & Smith, J. C. (2018). Wind power within European grid codes: Evolution, status and outlook. *Wiley Interdisciplinary Reviews. Energy and Environment*, *7*(3), 1–21. DOI: 10.1002/wene.285

Vrana, T., Attya, A., & Trilla, L. (2021). Future-oriented generic grid code regarding wind power plants in Europe. *International Journal of Electrical Power & Energy Systems*, *125*, 106490. DOI: 10.1016/j.ijepes.2020.106490

Vukosavić, S. N. (2018). *Grid-side converters control and design: Interfacing between the AC grid and renewable power sources*. Springer.

Wang, H., Lei, Z., Zhang, X., Zhou, B., & Peng, J. (2019). A review of deep learning for renewable energy forecasting. *Energy Conversion and Management*, *198*, 111799. DOI: 10.1016/j.enconman.2019.111799

Wang, N., Chen, Q., Zhu, L., & Sun, H. (2022). Integration of data-driven and physics-based modeling of wind waves in a shallow estuary. *Ocean Modelling (Oxford)*, *172*, 101978. DOI: 10.1016/j.ocemod.2022.101978

Wang, Z., & Bu, S. (2024). Probabilistic frequency stability analysis considering dynamics of wind power generation with different control strategies. *IEEE Transactions on Power Systems*, *39*(5), 6412–6425. DOI: 10.1109/TPWRS.2024.3362056

Watson, S. C. L., Somerfield, P. J., Lemasson, A. J., Knights, A. M., Edwards-Jones, A., Nunes, J., Pascoe, C., McNeill, C. L., Schratzberger, M., Thompson, M. S. A., Couce, E., Szostek, C. L., Baxter, H., & Beaumont, N. J. (2024). The global impact of offshore wind farms on ecosystem services. *Ocean and Coastal Management*, *249*, 107023. DOI: 10.1016/j.ocecoaman.2024.107023

Westholm, A. (2025). *Ocean Development & International Law The Role of Planning in Offshore Wind Power Deployment*. DOI: 10.1080/00908320.2025.2558538

Wikimedia (n.d.). https://commons.wikimedia.org/wiki/File:DK-84_-_K%C3%B8benhavn_-_Energy_-_Copenhagen_-_Denmark_-_Wind_Farm_(4890897324).jpg

Wikimedia, (n.d.). https://commons.wikimedia.org/wiki/File:Middelgrunden_wind_farm_2009-07-01_edit_filtered.jpg

Wikimedia. (n.d). https://commons.wikimedia.org/wiki/File:Wind_Turbines_and_Power_Lines,_East_Sussex,_England_-_April_2009.jpg

Wikimedia. (n.d.). https://commons.wikimedia.org/wiki/File:A_Wind_Energy_farm_and_the_Cenotaphs,_the_ancient_and_the_modern,_Jaisalmer_Rajasthan_India.jpg

Wikimedia. (n.d.). https://commons.wikimedia.org/wiki/File:Skidby_Mill_in_Winter_-_geograph.org.uk_-_233142.jpg

Wikimedia. (n.d.). https://commons.wikimedia.org/wiki/File:Wind_farm_(1828952868).jpg

Wikimedia. (n.d.).https://commons.wikimedia.org/wiki/File:Biscay_Spain_wind_farm_05_2012_1756.jpg

Wikimedia. (n.d.).https://commons.wikimedia.org/wiki/File:Kachch.png

Wikimedia. (n.d.).https://commons.wikimedia.org/wiki/File:Monte_de_A_Ru%C3%B1a._Foto_-_Alexandre_Varela.jpg

Wikimedia. (n.d.).https://commons.wikimedia.org/wiki/File:Skidby_Mill_1950_(archive_ref_DDX272-11)_(27968260583).jpg

Wikimedia. (n.d.).https://commons.wikimedia.org/wiki/File:Wind_energy_turbines_in_M%C3%B6lsheim,_Germany.jpg

Wikimedia. (n.d.).https://commons.wikimedia.org/wiki/File:Wind_power_in_Gothenburg_3.JPG

Wikimedia/ (n.d.). https://commons.wikimedia.org/wiki/File:Wind_power_in_south_india_-_panoramio.jpg

Wimhurst, J. J., Greene, J. S., & Koch, J. (2023). Predicting commercial wind farm site suitability in the conterminous United States using a logistic regression model. *Applied Energy*, *352*, 121880. DOI: 10.1016/j.apenergy.2023.121880

Wimhurst, J. J., Nsude, C. C., & Greene, J. S. (2023). Standardizing the factors used in wind farm site suitability models: A review. *Heliyon*, *9*(5), e15903. DOI: 10.1016/j.heliyon.2023.e15903 PMID: 37168883

Wind River. (n.d.). *Wind River—Secure development lifecycle/IEC 62443 guidance and resources.* https://www.windriver.com/resource/wind-river-security-IEC62443

Wind, G., & Council, E. (2024). *GWEC | GLOBAL WIND REPORT 2024.*

WindEurope. (2023). *Wind energy cyber resilience report 2023.* WindEurope Association. https://eepublicdownloads.entsoe.eu/clean-documents/Network%20codes%20documents/NC%20CS/WindEurope-position-paper-cybersecurity-framework-fit-for-wind-energy.pdf

Wiser, R. H., Bolinger, M., Hoen, B., Millstein, D., Rand, J., Barbose, G. L., Darghouth, N. R., Gorman, W., Jeong, S., & Mills, A. D. (2020). *Wind energy technology data update: 2020 edition*, https://emp.lbl.gov/publications/wind-energy-technology-data-update

Wiser, R., Millstein, D., Hoen, B., Gorman, W., Rand, J., Seel, J., Barbose, G., Cheyette, A., Darghouth, N., Jeong, S., Kemp, J., O'shaughnessy, E., & Paulos, B. (2025). *Land-Based Wind Energy Technology Data Update Figure File: 2025 Edition.*

Wiser, R., Lantz, E., Mai, T., Zayas, J., DeMeo, E., Eugeni, E., Lin-Powers, J., & Tusing, R. (2015). Wind Vision: A New Era for Wind Power in the United States. *The Electricity Journal*, *28*(9), 120–132. DOI: 10.1016/j.tej.2015.09.016

Wojtaszek, H., Borowski, P. F., Handschke, M., Miciuła, I., Stecyk, A., Bielawa, A., Ozdyk, S., Kowalczyk, A., & Czepło, F. (2025a). Wind Energy in Transition: Development, Socio-Economic Impacts, and Policy Challenges in Europe. *Energies 2025, Vol. 18, Page 2811, 18*(11), 2811. DOI: 10.3390/en18112811

Wood Mackenzie, https://www.woodmac.com/press-releases/2024-press-releases/global-wind-oem-marketshare/#:~:text=China%20accounted%20for%2065%%20of%20global%20wind,global%20rankings%2C%20a%20first%20for%20the%20sector.&text=The%20country%27s%20record%20installations%2C%20backed%20by%20more,not%20shielded%20local%20OEMs%20from%20reducing%20profitability.
World Wind Energy Association, https://wwindea.org/AnnualReport2023

World Economic Forum. (2024, November 15). *Cybersecurity guidance for the global offshore wind industry.* https://www.weforum.org/stories/2024/11/offshore-wind-industry-cybersecurity/

Wozabal, D., Graf, C., & Hirschmann, D. (2016). The effect of intermittent renewables on the electricity price variance. *OR-Spektrum, 38*(3), 687–709. DOI: 10.1007/s00291-015-0395-x

Wretling, V., Balfors, B., & Mörtberg, U. (n.d.). *Balancing wind power deployment and sustainability objectives in Swedish planning and permitting.* DOI: 10.1186/s13705-022-00376-y

Wu, D., Seo, G.-S., Xu, L., Su, C., Kocewiak, L., Sun, Y., & Qin, Z. (2024). Grid integration of offshore wind power: Standards, control, power quality and transmission. *IEEE Open Journal of Power Electronics, 5*, 583–604. DOI: 10.1109/OJPEL.2024.3390417

Xiao, C., Liu, Z., Zhang, T., & Zhang, X. (2021). Deep learning method for fault detection of wind turbine converter. *Applied Sciences (Basel, Switzerland), 11*(3), 1–22. DOI: 10.3390/app11031280

Xu, Y., Zhou, B., Sun, X., Tian, Y., & Jiang, X. (2025). Dynamic modeling and online updating of full-power converter wind turbines based on physics-informed neural networks and Bayesian neural networks. *Electronics (Basel), 14*(15). Advance online publication. DOI: 10.3390/ELECTRONICS14152985

Yadav, B. K., Yadav, R., Jahangiri, M., Priya, S. S., Bajracharya, T. R., & Sudhakar, K. (2024). Decarbonizing airport using solar and wind farm: A case of Biratnagar, Nepal. *e-Prime - Advances in Electrical Engineering. Electronics and Energy, 8*, 100583. DOI: 10.1016/j.prime.2024.100583

Yang, Q., Liu, Y., Chen, T., & Tong, Y. (2019). Federated machine learning: Concept and applications. *ACM Transactions on Intelligent Systems and Technology*, *10*(2), 1–19. Advance online publication. DOI: 10.1145/3298981

Yan, J., Nuertayi, A., Yan, Y., Liu, S., & Liu, Y. (2023). Hybrid physical and data-driven modeling for dynamic operation characteristic simulation of wind turbine. *Renewable Energy*, *215*, 118958. DOI: 10.1016/j.renene.2023.118958

Yan, L., Yongning, C., Haiyan, T., Xinshou, T., Zhankui, Z., & Jianqing, J. (2019, November). Common focus and new requirement on technical standards of RE grid integration. In *2019 Chinese Automation Congress (CAC)*, 3719-3723. DOI: 10.1109/CAC48633.2019.8996943

Yildiz, S. S. (2024). Spatial multi-criteria decision making approach for wind farm site selection: A case study in Balıkesir, Turkey. *Renewable & Sustainable Energy Reviews*, *192*, 114158. DOI: 10.1016/j.rser.2023.114158

Yin, X., Zhu, Y., & Hu, J. (2022). A Comprehensive Survey of Privacy-preserving Federated Learning: A Taxonomy, Review, and Future Directions. *ACM Computing Surveys*, *54*(6), 1–36. Advance online publication. DOI: 10.1145/3460427

Yousefi, H., Moradi, S., Zahedi, R., & Ranjbar, Z. (2024). Developed analytic hierarchy process and multi criteria decision support system for wind farm site selection using GIS: A regional-scale application with environmental responsibility. *Energy Conversion and Management: X*, Volume 22, 100594. DOI: 10.1016/j.ecmx.2024.100594

Yuan, X., Wang, S., & Zhao, X. (2025, March). Dynamic quality management in wind power construction: Integrating digital twins with BIM and IoT. In *2025 IEEE International Conference on Electronics, Energy Systems and Power Engineering (EESPE)* (pp. 916–919). IEEE. DOI: 10.1109/EESPE63401.2025.10986901

Yu, Z., Song, Q., Yuan, Z., Zou, C., Qiao, X., & Liu, W. (2024). Transmission topology and control for ultra-large offshore wind bases integrating multiple offshore low-frequency AC links and onshore HVDC link. *IEEE Transactions on Power Delivery*, *39*(5), 2797–2809. DOI: 10.1109/TPWRD.2024.3432897

Zarei, M. E., Ramirez, D., Veganzones, C., & Rodriguez, J. (2020). Predictive direct control of SPMS generators applied to the machine side converter of an OWC power plant. *IEEE Transactions on Power Electronics*, *35*(7), 6719–6731. DOI: 10.1109/TPEL.2019.2956738

Zeng, M., Li, C., & Zhou, L. (2013). Progress and prospective on the police system of RE in China. *Renewable & Sustainable Energy Reviews*, *20*, 36–44. DOI: 10.1016/j.rser.2012.11.048

Zhang, L., Sun, L., Xu, W., & Gong, R. (2026). Voltage droop control optimization in microgrid integration with VSCHVDC systems and wind power plants for enhanced dynamic stability. *Electric Power Systems Research*, *252*, 112363. DOI: 10.1016/j.epsr.2025.112363

Zhang, X., Tao, J., & Noshadravan, A. (2026). Probabilistic digital twin for reliability-based maintenance optimization of offshore wind turbines. *Renewable Energy*, *256*, 123777. DOI: 10.1016/j.renene.2025.123777

Zhang, Z., Zhao, Y., Qiao, W., & Qu, L. (2014). A space-vector-modulated sensorless direct-torque control for direct-drive PMSG wind turbines. *IEEE Transactions on Industry Applications*, *50*(4), 2331–2341. DOI: 10.1109/TIA.2013.2296618

Zhang, Z., Zhao, Y., Qiao, W., & Qu, L. (2015). A discrete-time direct torque control for direct-drive PMSG-based wind energy conversion systems. *IEEE Transactions on Industry Applications*, *51*(4), 3504–3514. DOI: 10.1109/TIA.2015.2413760

Zhao, J., Lyu, X., Fu, Y., Hu, X., & Li, F. (2016). Coordinated microgrid frequency regulation based on DFIG variable coefficient using virtual inertia and primary frequency control. *IEEE Transactions on Energy Conversion*, *31*(3), 833–845. DOI: 10.1109/TEC.2016.2537539

Zhao, Y., Song, Z., Li, D., Qian, R., & Lin, S. (2024). Wind Turbine Gearbox Fault Diagnosis Based on Multi-Sensor Signals Fusion. *Protection and Control of Modern Power Systems*, *9*(4), 96–109. DOI: 10.23919/PCMP.2023.000241

Zheng, H., & Tang, Y. (2012, December). *Comparison of interconnection standards of RE generation. In 2012 10th International Power & Energy Conference*. IPEC., DOI: 10.1109/ASSCC.2012.6523254

Zhirong, Z., Hao, Y., Feng, W., Fang, Z., & Zhenxiong, W. (2016). A novel control strategy of photovoltaic-battery system for restraining the photovoltaic power fluctuations and suppressing the low frequency oscillations of power system. *2016 IEEE 8th International Power Electronics & Motion Control Conference (IPEMC-ECCE Asia)*, 2978, .DOI: 10.1109/IPEMC.2016.7512770

Zhong, D., Xia, Z., Zhu, Y., & Duan, J. (2023). Overview of predictive maintenance based on digital twin technology. *Heliyon*, *9*(4), e14534. DOI: 10.1016/j.heliyon.2023.e14534

Ziemba, P. (2022). Uncertain Multi-Criteria analysis of offshore wind farms projects investments – Case study of the Polish Economic Zone of the Baltic Sea. *Applied Energy*, *309*, 118232. DOI: 10.1016/j.apenergy.2021.118232

Zin, A. A. B. M. HA, M. P., Khairuddin, A. B., Jahanshaloo, L., & Shariati, O. (2013). An overview on doubly fed induction generators′ controls and contributions to wind based electricity generation. *Renewable and Sustainable Energy Reviews, 27*, 692–708, DOI: 10.1016/j.rser.2013.07.010

Zupok, S., Chomać-Pierzecka, E., Dmowski, A., Dyrka, S., & Hordyj, A. (2025). A Review of Key Factors Shaping the Development of the U.S. Wind Energy Market in the Context of Contemporary Challenges. *Energies 2025, Vol. 18, Page 4224, 18*(16), 4224. DOI: 10.3390/en18164224

About the Contributors

Jackson J. Justo (PhD): Currently working as a Senior Researcher at ITMO University in Russia and a Senior Lecturer with the Department of Electrical Engineering, University of Dar es Salaam, Tanzania. Received his PhD Degree from Dongguk University, Seoul, South Korea (2015). He was with the University of Pretoria, Pretoria, South Africa (2016 – 2018) as Postdoctoral Research Fellow. His current research interests on: High power converter topology, control systems, condition monitoring and fault detection, grid Integration of renewable power generation, AC and DC microgrid, virtual power plants (VPPs) configurations, Intelligent Control Systems, Electrical Machine Drives, Wireless Power Transfer for Electric Vehicle (EVs) Autonomous Charging Systems. Virtual Energy Storage Systems, and Power System Control, Operation and Dynamics.

Galina Demidova is currently with Hangzhou Dianzi University, China. Previously worked as an Associate Professor with the Faculty of Control Systems and Robotics, ITMO University, and was a Post-Doctoral with the Tallinn University of Technology, Tallinn, Estonia. Her current research interests include control systems, wind turbines, adaptive control, fuzzy logic control, motion control, machine learning, hybrid intelligent systems, genetic algorithms, and electromechanical systems.

Francis A. Mwasilu is currently working with Department of Electrical Engineering, University of Dar es Salaam, Tanzania. From 2008 to 2009 he was an Instrumentation Engineer with Illovo Sugar Ltd, Tanzania. From 2009 to 2011, he was a Utilities Engineer with the Japan Tobacco International, Tanzania. Since July 2016, he has been a Lecturer with the Department of Electrical. His current research interests include digital signal processor-based electric machine drives, electric vehicles, smart grid enabling technologies, and renewable energy sources integration into modern power systems.

Dmitry V. Lukichev is leading lector and scientific researcher, who works with industrial projects in various areas of industry who works at ITMO University in St. Petersburg, Russia. He is also visiting Professor of Changchun University of Technology and Hangzhou Dianzi University, China. His current research interests encompass control systems for both DC and AC drives, adaptive control modelling and application and the implementation of neuro-fuzzy algorithms in the control systems of complex dynamic objects. He is author of about 70 scientific articles. Additionally, Professor Dmitry is actively involved in projects focused on the control of wind turbines and generators, further expanding his contributions to innovative control systems. In this proposed book, Prof. Dmitry will play a crucial role in documenting the chapters, ensuring that every detail is accurately captured before they are forwarded to the quality assurance team for final review.

Ramesh C. Bansal has more than 25 years of diversified experience of scholarship of teaching and learning, accreditation, research, industrial, and academic leadership in several countries. Currently he is a Professor in the Department of Electrical Engineering at University of Sharjah and extraordinary Professor at the University of Pretoria. Previously he was Professor and Group Head (Power) in the ECE Department at University of Pretoria (UP), South Africa. Prior to his appointment at UP, he was employed by the University of Queensland, Australia; University of the South Pacific, Fiji; and BITS Pilani, India. Prof. Bansal has significant experience of collaborating with industry and Government organizations. He has made significant contribution to the development and delivery of BS and ME programmes for Utilities. He has extensive experience in the design and delivery of CPD programmes for professional engineers. He has carried out research and consultancy and attracted significant funding from Industry and Government Organisations. Prof. Bansal has published over 450 journal articles, presented papers at conferences, books, and chapters in books. He has Google citations of over 20,000 with h-index of 67. He has supervised 25 PhD, 5 Post Docs and currently supervising several PhD students. His diversified research interests are in the areas of Renewable Energy (Wind, PV, DG, Micro Grid) and Smart Grid. Professor Bansal is an editor of several highly regarded journals including IEEE Systems Journal, ECPS, SGSE. He is a Fellow and Chartered Engineer IET-UK, Fellow Institution of Engineers (India), Fellow SAIEE, and Senior Member of IEEE-USA. In this proposed book, Prof. Bansal will take the advisory role which include providing necessary inputs and steps to take to successfully accomplish this book project.

Sutanu Chatterjee received his B.tech degree in Electronics and Communications Engineering in 2017 from Guru Nanak Institute of Technology, Kolkata, India, and his M.S. degree in Electronics and Nanoelectronics in 2021 from Saint Petersburg

Electrotechnical University, Saint Petersburg, Russia. He is currently a PhD student of ITMO University, Saint Petersburg, Russia and Assistant Researcher at Harbin Engineering University, China. His research interests include the design and simulation of power electronic systems, wireless power transfer systems, and the design and programming of embedded systems.

Aleksandr Mamatov received the B.Sc. degree in Electrical Engineering, Electromechanics, and Electrical Technologies in 2013, and the M.Sc. degree in Electrical Engineering and Power Engineering in 2015 from ITMO University, Saint Petersburg, Russia. He received the Ph.D. degree (Candidate of Technical Sciences) in 2019 in the field of Systems Analysis and Control of Technical Systems. His research focuses on control systems for precision electric drives and inverter-based power converters for various applications. His research interests include advanced control of electric drives, power electronic converters, inverter systems, and high-performance motion control.

Rampelli Manojkumar received his Ph.D. in Power Engineering from the Indian Institute of Technology Guwahati in 2023. He obtained his Master of Technology in Power and Energy Systems from the National Institute of Technology Karnataka, Surathkal, India, in 2015. He is currently an Assistant Professor in the Department of Electrical and Electronics Engineering at BVRIT Hyderabad College of Engineering for Women. He is currently the Editorial Board Member of Scientific Reports, Early Career Editor for the International Journal of Ambient Energy and the Guest Editor for PLOS ONE. He has been a speaker at the IEEE Smart Grid webinar series and has contributed to multiple funded research projects on smart transformer-based microgrid systems and distributed generation management. He has published 9 SCI journal papers, including articles in IEEE Transactions on Power Systems, IEEE Transactions on Industrial Informatics, and Sustainable Cities and Society. His research expertise spans energy management, hybrid AC/DC distribution systems, energy storage, and power electronics applications in modern distribution systems. Additionally, he is an active reviewer for around 20 SCI journals, including IEEE Transactions on Industrial Informatics, IEEE Systems Journal, Journal of Energy Storage, Applied Energy, etc. He has also been actively involved in faculty training programs focused on teaching and learning enhancement, including instructional design, formative assessment, and teaching analytics, thereby strengthening his interdisciplinary expertise in education and technology integration.

Samson W. Mlay completed an M.Sc. in Power Electronics and Electrical Drives in 2025 at the University of Dar es Salaam, Tanzania. His academic work focuses on efficient and reliable power conversion technologies for modern energy applications. His research interests include the analysis and control of DC-DC converters and multilevel inverters, emphasizing advanced topologies and performance optimization. He is also engaged in electric vehicle technologies, particularly inverter control and fast-charging DC station systems. Additionally, his research explores optimization and control of grid-connected renewable energy systems, including power-to-gas integration for enhanced energy sustainability.

Aviti Thadei Mushi is a Tanzanian electrical engineer and academic. He is a Senior Lecturer in the Department of Electrical Engineering at the University of Dar es Salaam (UDSM). He earned his PhD in Electrical & Computer Engineering from Yokohama National University in 2017, where his doctoral thesis focused on nonlinear deadbeat control for boost converters. He also holds an MSc in Power Systems and High Voltage Engineering and a BSc in Electrical Engineering, both from UDSM. During the MSc studies, Dr Mushi researched outdoor polymeric insulators in tropical climates and shared insights to the electrical power utility that these insulators had a potential to last longer and make the transmission lines cheaper compared to the traditional insulators. During the PhD studies, Dr Mushi focused on applications of innovative control algorithms for DC-DC converters that are used in rocket propulsion systems as well as EVs. His research showed remarkable improvement over the existing algorithms in both reference tracking response and disturbance rejection. Dr. Mushi's research covers renewable energy, rural electrification, smart grids, high-voltage engineering, power converters, and telemedicine. He uses these research results to inform his teaching at the university. He has co-authored works on several themes such as - microgrids in the Democratic Republic of Congo; design of a hybrid renewable energy system for International Airport; improving power quality in industrial networks using dynamic voltage restorers; and on-line partial discharge detection in high-voltage cable joints, comparing different diagnostic techniques. In addition, he has co-authored several textbooks and reference books that are used by the primary and secondary schools of Tanzania. He served as Chairperson of the University of Dar es Salaam Academic Staff Assembly during the 2020–2022 term. During his tenure, he managed to keep the UDASA Newsletter alive with timely publications. He spearheaded transformations that brought UDASA to its members and made them avid participants in its activities. He groomed several leaders as well. During this time, he served as the member to the University Council and was responsible for the governing of the UDSM Management and Senate. He also served as a committee member of Estates Department of UDSM advising on equitable housing for staff and students, as well as overall estates management of UDSM properties. As a lecturer, he teaches courses in power electronics, high-voltage technology, renewable energies, and smart grid systems. He has supervised and guided several students in their B.Sc. and M.Sc. Engineering studies. Further, he is guiding students in their B.Sc., M.Sc. and PhD in Engineering studies. All students assigned to him for supervision receive a dedicated support that enables them to undertake research with determination. Apart from academic duties, Dr. Mushi works on several research projects on SUNRISE, telemedicine, renewable energies, microbial fuel cells, high voltage technologies, power electronics and others. He consults on affordable rural electrification, high voltage outdoor insulation systems, and all kinds of electrifications for buildings, industries, and systems. These services are offered to both public and private sectors.

Dieudonne Musongya Bisimwa received the B.Sc. degree in electrical engineering from the Université Libre des Pays des Grands Lacs, DR Congo, in 2014, and the M.Sc. degree in Power Systems Engineering from the University of Rwanda, Rwanda, in 2023, he worked as a Teaching Assistant, since November 2015 at the Université Libre des Pays des Grands Lacs, DR Congo. He is currently a PhD student in Electrical Engineering at the University of Dar Es Salaam, since March 2025.

Ibrahim Mwammenywa is a researcher, consultant and lecturer in ICT and Electrical Engineering, currently working with the University of Dar es Salaam. He holds a Doctor of Engineering (Dr.-Ing.) in Electrical Engineering from the University of Paderborn (UPB), Germany. Ibrahim received a Bachelor of Science in Telecommunication Engineering degree from the University of Dar es Salaam (UDSM), Tanzania, in 2014, and a Master's in Information and Communications Science and Engineering from Nelson Mandela – African Institution of Science and Technology, Tanzania, in 2019. He is a fully licensed professional engineer by the Engineers Registration Board (ERB) of Tanzania. He actively works as an ICT and Electrical service engineer (consultant) through the Bureau for Industrial Cooperation (BICO) of UDSM. Ibrahim is also an active academic and researcher in the fields of electronics, electrical engineering, and ICT. His area of interest is in sensor systems using the Internet of Things (IoT) as well as the integration of ICT solutions into the electric grid, i.e., the smart grid.

Nikolai Poliakov received the B.S. and M.S. degrees in electrical engineering from ITMO University, Saint Petersburg, Russia, in 2009 and 2011, respectively. He defended the Ph.D. thesis at ITMO University, in 2016. He is currently an Associate Professor with the Faculty of Control Engineering and Robotics, ITMO University. His current research interests include power converters design, power electronics, power efficiency, battery management systems, wireless power transfer systems, and control theory and its applications.

Kübra Nur Şahin is currently pursuing her Ph.D. degree in the Department of Industrial Engineering at Abdullah Gül University. Her research focuses on data-driven optimization and reinforcement learning applications in sustainable energy systems. Her work reflects a strong commitment to innovation and a passion for translating advanced computational techniques into practical energy solutions. Her research interests include smart grid optimization, renewable energy integration, and intelligent control strategies for energy-efficient systems.

Aditya Singh ISDS, is an Independent Researcher, Editorial Review Board Member in 6 American International Journals (IJSSMET, IJSDS, IJPAE, IJESGT, IJDSST, IJRLEDM), Reviewer in 10 American, 6 Singaporean, 1 Indonesian, 1 Swiss International Journals (IJUPSC, IJCVIP, IRMJ, JTA, IJAIBM, JITR, IJSSSP, IJAIEF, IJEOE, IJITN, JDSIS, AAES, JCCE, SWT, JOPR, JCWR, IJEECS, Buildings), Alumni Mentor registered by Amrita University Alumni (Amrita Vishwa Vidyapeetham, India), Ex-Research Mentor at Academy Innova World, India & Ex-Alumni Mentor (LPU) at present. He had completed his regular B. Tech degree in Civil Engineering from Lovely Professional University, India, in 2020, completed regular M.Tech (Structural & Construction Engineering) in 2025 from Amrita Vishwa Vidyapeetham, while simultaneously maintaining an uninterrupted global professional career from 2020 onwards, which was built on his research journey as a scholar from late 2018 to mid 2020 during his B.Tech days. He is currently affiliated with Amrita Vishwa Vidyapeetham, India. He has 35 publications under his name, in which he is a single author in 24 of them. Most of his publications are Indexed in Scopus. He had also worked as a Reviewer/Editorial Review Board Member for 218 times in the evaluation of manuscripts so far in international conferences, international books and international journals. His global presence as a professional could also be seen with 36 Conference Presentations in 30 Conferences in 14 countries (Thailand, Sri Lanka, Egypt, India, Cambodia, Bangladesh, Turkey, UK, Germany, Austria, Italy, Czech Republic, UAE, Jordan) as Paper Presenter & Invited Speaker/Eminent Speaker; as well as his position as Program Committee Member, Scientific Committee Member, Session Chair in 5 International Conferences (India, UK, Canada, Czech Republic, Jordan), and Reviewer in total 11 International Conferences (including countries Thailand, Cambodia, USA, India, China, Turkey, Denmark), and Keynote Speaker in Faculty Development Programme, Editorial Advisory Board(Scopus indexed International Research Book), etc. He had also won 2 International Awards((Thailand) & (Cambodia)) as a Reviewer in 2023 for his excellent performance as a Reviewer, which are lifetime achievements in this particular category. He had worked on 70 positions so far in his professional career, including Journals, Conferences, Research Books of International stature. He is also a lifetime full member of ISDS (Japan) & International Association of Engineers

(IAENG), and 14 IAENG Societies (ISAI, ISB, ISCS, ISDM, ISEE, ISIME, ISINDE, ISISE, ISICWS, ISME, ISOR, ISSC, ISSE, ISWN), AMIE, M.ASCE, SEI, ISDTDE (Founding Member).

Stampinus Mberwa Stephano received the B.Sc. degree in Electrical Engineering in 2018, and the M.Sc. degree in Power Systems and High Voltages in 2021, both from the University of Dar es Salaam, Dar es Salaam, Tanzania. He is currently pursuing the Ph.D. degree in Electrical Engineering. His research focuses on finite control set model predictive control (FCS-MPC) for grid-connected and grid-forming inverters, adaptive neural-network-based control, and neural-network-based state estimation with reduced sensor configurations for inverter-based power systems. His research interests include model predictive control, intelligent control of power electronic converters, renewable energy systems, inverter-based power systems, and real-time digital implementation of advanced control algorithms.

Cherala Vikram received his Ph.D. in Power Systems from the Indian Institute of Technology Hyderabad in 2024. He completed his M.Tech in Energy Technology from the National Institute of Technology Hamirpur in 2018 and his B.Tech in Electrical and Electronics Engineering from Kakatiya Institute of Technology and Science, Warangal in 2012. He is currently working as a Data Scientist at NDS Infoserve, where he develops and deploys Large Language Models (LLMs) and AI-based solutions for the healthcare sector. His work focuses on automating medical code detection, optimizing data pipelines, and deploying scalable AI systems on Microsoft Azure Cloud. Previously, he served as a Research Assistant at IIT Hyderabad in the AI Center of Excellence (AICOE) and Smart Grids and Energy Systems group, where he worked on AI-driven peer-to-peer (P2P) energy sharing systems, 5G smart meter data analytics, and renewable energy integration. He has also published several journal and IEEE conference papers in the areas of energy management, electric vehicles, and sustainable systems. Dr. Vikram's research interests include AI for energy systems, smart grids, and renewable energy optimization. He is a recipient of the TCS Research Scholarship (2019–2023) and is actively involved in social and rural development initiatives promoting education and community engagement.

Yue Zhao received the bachelor's degree in Materials Science and Engineering from Jiangsu University, Jiangsu, China (2022). Currently, she is pursuing the master's degree in Electronic Information at Harbin Engineering University, Qingdao, China. She also has a joint training experience at ITMO University, St. Petersburg, Russia (2025–2026). Her current research interests include: Wireless

Power Transfer, Inverter Control, and Unmanned Aerial Vehicle (UAV) Charging Systems.

Qingyi Zhou received a Bachelor of Engineering degree in Communication Engineering from Northeast Forestry University in 2024, and currently a postgraduate student majoring in Electronic and Information Engineering at Harbin Engineering University. The research interests focus on: topology and control of power electronic converters, optimization of wireless power transfer systems.

Index

A

B

C

D

E

G

H

I

J

M

N

O

P

R

S

V

W

www.ingramcontent.com/pod-product-compliance
Lightning Source LLC
LaVergne TN
LVHW060005031026
849161LV00025B/1488
* 9 7 9 8 2 6 0 0 0 1 0 6 6 *